Self-Organization, Emerging Properties, and Learning

NATO ASI Series

Advanced Science Institutes Series

A series presenting the results of activities sponsored by the NATO Science Committee, which aims at the dissemination of advanced scientific and technological knowledge, with a view to strengthening links between scientific communities.

The series is published by an international board of publishers in conjunction with the NATO Scientific Affairs Division

A	**Life Sciences**	Plenum Publishing Corporation
B	**Physics**	New York and London
C	**Mathematical and Physical Sciences**	Kluwer Academic Publishers
D	**Behavioral and Social Sciences**	Dordrecht, Boston, and London
E	**Applied Sciences**	
F	**Computer and Systems Sciences**	Springer-Verlag
G	**Ecological Sciences**	Berlin, Heidelberg, New York, London,
H	**Cell Biology**	Paris, Tokyo, Hong Kong, and Barcelona
I	**Global Environmental Change**	

Recent Volumes in this Series

Volume 255—Vacuum Structure in Intense Fields
edited by H. M. Fried and Berndt Müller

Volume 256—Information Dynamics
edited by Harald Atmanspacher and Herbert Scheingraber

Volume 257—Excitations in Two-Dimensional and Three-Dimensional Quantum Fluids
edited by A. F. G. Wyatt and H. J. Lauter

Volume 258—Large-Scale Molecular Systems: Quantum and Stochastic Aspects—Beyond the Simple Molecular Picture
edited by Werner Gans, Alexander Blumen, and Anton Amann

Volume 259—Science and Technology of Nanostructured Magnetic Materials
edited by George C. Hadjipanayis and Gary A. Prinz

Volume 260—Self-Organization, Emerging Properties, and Learning
edited by Agnessa Babloyantz

Volume 261—Z° Physics: *Cargèse 1990*
edited by Maurice Lévy, Jean-Louis Basdevant, Maurice Jacob, David Speiser, Jacques Weyers, and Raymond Gastmans

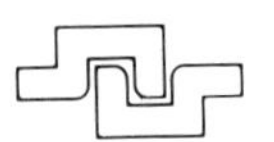

Series B: Physics

Self-Organization, Emerging Properties, and Learning

Edited by
Agnessa Babloyantz
Free University of Brussels
Brussels, Belgium

Plenum Press
New York and London
Published in cooperation with NATO Scientific Affairs Division

Proceedings of a NATO Advanced Research Workshop on
Self-Organization, Emerging Properties, and Learning,
held March 12-14, 1990,
in Austin, Texas

Library of Congress Cataloging-in-Publication Data

NATO Advanced Study Research Workshop on Self-Organization, Emerging
Properties, and Learning (1990 : Austin, Tex.)
Self-organization, emerging properties, and learning / edited by
Agnessa Babloyantz.
p. cm. -- (NATO ASI series. Series B, Physics ; v. 260)
"Proceedings of a NATO Advanced Research Workshop on Self
-Organization, Emerging Properties, and Learning, held March 12-14,
1990, in Austin, Texas"--T.p. verso.
"Published in cooperation with NATO Scientific Affairs Division."
Includes bibliographical references and index.
ISBN 0-306-43930-1
1. Self-organizing systems--Congresses. 2. Learning--Congresses.
I. Babloyantz, A. (Agnessa) II. North Atlantic Treaty Organization.
Scientific Affairs Division. III. Title. IV. Series.
Q325.N37 1990
006.3--dc20
91-21415
CIP

ISBN 0-306-43930-1

Printed in the United States of America

SPECIAL PROGRAM ON CHAOS, ORDER, AND PATTERNS

This book contains the proceedings of a NATO Advanced Research Workshop held within the program of activities of the NATO Special Program on Chaos, Order, and Patterns.

Volume 208—MEASURES OF COMPLEXITY AND CHAOS
edited by Neal B. Abraham, Alfonso M. Albano, Anthony Passamante, and Paul E. Rapp

Volume 225—NONLINEAR EVOLUTION OF SPATIO-TEMPORAL STRUCTURES IN DISSIPATIVE CONTINUOUS SYSTEMS
edited by F. H. Busse and L. Kramer

Volume 235—DISORDER AND FRACTURE
edited by J. C. Charmet, S. Roux, and E. Guyon

Volume 236—MICROSCOPIC SIMULATIONS OF COMPLEX FLOWS
edited by Michel Mareschal

Volume 240—GLOBAL CLIMATE AND ECOSYSTEM CHANGE
edited by Gordon J. MacDonald and Luigi Sertorio

Volume 243—DAVYDOV'S SOLITON REVISITED: Self-Trapping of Vibrational Energy in Protein
edited by Peter L. Christiansen and Alwyn C. Scott

Volume 244—NONLINEAR WAVE PROCESSES IN EXCITABLE MEDIA
edited by Arun V. Holden, Mario Markus, and Hans G. Othmer

Volume 245—DIFFERENTIAL GEOMETRIC METHODS IN THEORETICAL PHYSICS: Physics and Geometry
edited by Ling-Lie Chau and Werner Nahm

Volume 256—INFORMATION DYNAMICS
edited by Harald Atmanspacher and Herbert Scheingraber

Volume 260—SELF-ORGANIZATION, EMERGING PROPERTIES, AND LEARNING
edited by Agnessa Babloyantz

PREFACE

This volume contains the proceedings of the workshop held in March 1990 at Austin, Texas on Self-Organization, Emerging Properties and Learning. The workshop was co-sponsored by NATO Scientific Affairs Division, Solvay Institutes of Physics and Chemistry, the University of Texas at Austin and IC^2 Institute at Austin. It gathered representatives from a large spectrum of scientific endeavour.

The subject matter of self-organization extends over several fields such as hydrodynamics, chemistry, biology, neural networks and social sciences. Several key concepts are common to all these different disciplines. In general the self-organization processes in these fields are described in the framework of the nonlinear dynamics, which also governs the mechanisms underlying the learning processes. Because of this common language, it is expected that any progress in one area could benefit other fields, thus a beneficial cross fertilization may result.

In last two decades many workshops and conferences had been organized in various specific fields dealing with self-organization and emerging properties of systems. The aim of the workshop in Austin was to bring together researchers from seemingly unrelated areas and interested in self-organization, emerging properties and learning capabilities of interconnected multi-unit systems. The hope was to initiate interesting exchange and lively discussions.

The expectations of the organiziers are materialized in this unusual collection of papers, which brings together in a single volume representative research from many related fields. Thus this volume gives to the reader a wider perspective over the generality and ramifications of the key concepts of self-organization.

The meeting could not have been so succesfull without the great effort of all involved in organizing the gathering. Special thanks go to L. Reichel, acting director, Ilya Prigogine Center for Studies in Statistical Mechanics and Complex Systems, the University of Texas in Austin and Amy Kimmons from the same institution. The beautifull and comfortable setting of the IC^2 Institut and the delicious lunches served during the meeting were the finishing touches to a very stimulating meeting.

The preparation of this volume was also a collective effort which required substantial help from A. Destexhe, D. Gallez, P. Maurer and J.A. Sepulchre.

Finally I would like to acknowledge the valuable help of N. Galland from Solvay Institutes for solving the various "after workshop" problems.

A. Babloyantz

Brussels, July 1991

INTRODUCTION

In the last decades the concept of self-organization has been of great value for the understanding of a wealth of macroscopic phenomena in many fields such as physics, chemistry, biology and also the social sciences. Broadly speaking, self-organization describes the ability of systems comprising many units and subject to constraints, to organize themselves in various spatial, temporal or spatiotemporal activities. These emerging properties are pertinent to the system as a whole and cannot be seen in units which comprise the system.

In the field of hydrodynamics, phenomena such as convective patterns, turbulence, and climatic variability could be tackled with this concept. Oscillations and waves in unusual chemical reactions were shown to be emerging properties which resulted from spontaneous molecular cooperation. In biological systems it was suggested that self-organization is the basic principle governing the emergence of genetic material from simple molecules. The same principle is shown to be at work at cellular, embryonic and cortical levels, regulating the key processes of life. Recently, it was also found that the self-organizational ability of a network of simple processors, mimicking in a primitive manner cortical tissue, may be used in pattern recognition and other cognitive and learning tasks.

The purpose of the workshop was to bring together researchers interested in self-organizing systems from different disciplines such as physics, physiology and neural networks. The discussion evolved around three distinct domains however, all built around similar concepts. The areas of investigation comprised self-organization, dynamics of networks of interacting elements, experimental and theoretical modeling of networks of neuronal activity, and the investigation of the role of dynamical attractors in cognition and learning.

Although the emerging properties in fluid dynamics or in reaction-diffusion systems are well documented, the investigation of networks of coupled units is more recent and still in its infancy. Several speakers dealt with this subject and showed some new and unexpected emerging properties in these networks. The usual collective properties such as target waves and counter-rotating spiral waves are also seen in these discrete networks. However the most intriguing question was how the propagating waves would behave in the presence of obstacles.

In the present volume, J.A. Sepulchre and A. Babloyantz introduce the concept of self-organization and the ensuing emerging properties. They consider the more specific case of a network of oscillatory units. The units may be either in a quiescent but unstable state or in an oscillating regime. For both cases, they show new and unexpected properties. However, the most intriguing part of their investigations considers the properties of target waves in the presence of obstacles. They show how in a finite network, target waves of a given wavelength could be triggered that cannot travel across inter-obstacle distances of a critical size. They discuss the relevance of these emerging properties for such problems as path finding and robotics.

H. Haken and A. Wunderlin reconsider the subject matter of spontaneous self-organization of systems on the macroscopic scale. They introduce the necessary mathematical formulations in terms of the theory of differential equations. They show the conditions which lead to highly ordered spatial, temporal and spatio-temporal structures which emerge as global properties on scales which are much larger than the corresponding characteristic length and time scales of the subsystems. Finally, they show how these emerging properties could be used in problems involving pattern recognition and associative memory.

R. Kapral considers the emerging properties of a class of discrete models, namely coupled map lattices. He investigates some of the rich bifurcation structures generic to spatially distributed systems. He probes the underlying general features and scaling behaviour which give rise to the phenomena. He discusses more specifically the onset of period doubling, intermittency, oscillations and bistable states in these coupled map lattices which may also generate counter-rotating spiral wave pairs and target patterns.

The paper by Kelso *et al.* shows the role of emerging spatio-temporal patterns in the laws of coordinated action. The authors aim to identify the principles that govern how the individual components of a complex, biological system are put together to form recognizable functions. They discuss a discrete form of collective variable dynamics which is known as phase attractive circle map. The experimental and theoretical results they obtained led them to propose that the phase attraction is a crucial feature of cooperation and coordination in biological systems.

This last paper relates self-organization to biological function. The relationship becomes more apparent in the next few contributions where the role of emerging properties of biological networks in perception is considered.

One of the most exciting discoveries of recent years in the field of physiology is the experimental evidence that synchronized field potentials appear as a result of input into the visual cortex. This stimulus specific-activity of high frequency was first seen in the cat visual cortex by Eckhorn and coworkers and by Singer and his group, both working in Germany. Similar activity was also measured in the rabbit olfactory bulb by Freeman and his colleagues in Berkeley. These findings gave rise to many imaginative models which try to account for such synchronization processes. Several papers in this volume deal with experimental as well as theoretical aspects of cortical synchronization.

Eckhorn and Schanze describe in detail how, in cat visual cortex, stimulus-specific collective modes of oscillatory activity of 35-80 Hz are seen among distributed cell assemblies of the same and of different areas of the cortex. They suggest that such synchronization supports pre-attentive feature grouping. An object is labeled as a coherent entity when the cells it activates fire in synchrony. To account for their numerous experiments, the authors construct neural network models and discuss the relevance of their finding to natural vision.

Kurrer *et al.* reconsider the same phenomena and investigate more specifically the figure ground separation problem of image processing in the brain. Their approach is theoretical and starts from single neurons considered as excitable elements with stochastic activity. They show that a set of weakly coupled neurons can show synchronous activity when subjected to coherent excitation. They also suggest that synchronous firing can be used in visual cortex to segment images.

Cortical synchronization is also the focus of attention of Traub and Miles who discuss the structure of guinea-pig hippocampus, which anatomically is the simplest cortex. They show how this cortex may exhibit a repertoire of emerging collective modes of population activity. They relate these collective modes to cellular parameters such as the conductances of the inhibitory synapses. They also speculate on the behavioural significance of these collective modes.

The mechanism of visual information processing has been reconsidered by D. S. Tang in the context of information theory. He assumes that an information processing neural network can be viewed as an ensemble of distributed decision-making units residing in different functional blocks. These subunits are linked through a dense network of communication channels which operate in parallel. The model addresses early visual information processing in the retina and demonstrates that there is an emergence of statistically independent operational modes for feature detection.

The experimental data on synchronization seen in the previous papers are qualitative and probably do not belong to the family of oscillatory dynamics called limit cycle behaviour. In order to describe pseudo-periodic behaviour, the concept of deterministic chaos was introduced in the field of nonlinear dynamics several years ago. In such chaotic systems, although there is a sensitivity to initial conditions, the system visits and remains in a portion of phase space called the attractor. If the attractor of a system becomes available, many of its dynamical characteristics such as attractor dimension, Lyapunov exponents and metric entropies can be evaluated. In recent years several algorithms were developed whereby a systems attractor could be assessed from the measurement in time of a single parameter. Such an analysis was applied to electroencephalographic recordings of human EEG during several key behavioural states. The same type of analysis was also performed on animal recordings. In all cases it was shown that the EEG follows deterministic chaotic dynamics. Moreover the evaluation of attractor dimensions showed that a high cognitive state corresponds to a high value of correlation dimension of underlying attractor. In a lower cognitive state this value decreases. These findings led to several models which show the emergence of chaotic attractors in networks of model cortices. Of crucial importance also is the understanding of mechanisms by which chaotic dynamics process information. The next few papers describe the existence of chaotic attractors in the global activity of human cortex and discuss their role in cognition.

A. Destexhe and A. Babloyantz introduce a simple model of the cortical tissue, subject to periodic inputs which mimic thalamic activity. They show that in the absence of input, the dynamics of the system are turbulent and desynchronized. The onset of the pacemaker input organizes the system into a more coherent oscillatory behaviour. From this spatio-temporal chaotic activity they construct a global network variable which is reminiscent of the EEG activity. They assess the temporal coherence of this mean activity with the assistance of techniques from nonlinear time series analysis and compare them with previous data computed from electroencephalograms.

J.S. Nicolis is also concerned with the role of deterministic chaos. He investigates the mechanisms by which multifractal attractors mediate perception and cognition. These concepts are illustrated for the case of highly nonlinear linguistic filtering processes.

X. J. Wang discusses the possibility of chaotic dynamics in intrinsic global oscillatory dynamics of a larger network of neuron like analog units. However he revisits the classical connectionist models in a different light. Here the aim is to account for some aspects of ground state activity of mammalian brain during wakefulness. The activity of the network is probed by using an entropic quantity called ε-entropy. He shows that the activity of the network is intermediate between low dimensional chaos and random noise.

Throughout history, mankind has desired to construct artificial devices which can perform some of the tasks attributed to the human mind. The finding that the brain tissue is a dense network of highly interconnected cells led to the hypothesis that human intelligence results from network properties. This led in the 1950s to the concept of artificial neural networks, a field which experienced several ups and downs.

In 1982 Hopfield revived the interest of the scientific community in the investigation of neural networks. The field under the general label of «connectionism» witnessed an explosion in the number of researchers, conferences and scientific journals. The investigation of neural networks involves the fields of statistical physics, artificial intelligence, neurophysiology and psychology. The goal is to mimic in artificial systems several natural abilities of the human brain such as construction of associative memories, pattern recognition, optimization problems, etc. Such abilities are related to the emerging properties of a network of many units working in parallel, thus mimicking in some way the structure of human cortex. These networks are endowed with various learning rules which give them the ability to perform the desired tasks. Several papers in this volume address the subject matter of learning algorithms in neural networks and their applications in robotics.

The paper by C. Torras surveys in a clear and concise manner the learning capabilities of networks. A distinction is made between learning tasks, learning rules and network learning models. Learning tasks are discussed in terms of pattern reproduction, pattern association, feature discovery and reward maximization. Various neuronal learning algorithms, such as correlational rules, error minimization rules, and reinforcement rules are considered. Finally, an application in the field of robotics is presented.

P. Gaudiano and S. Grossberg also are interested in arm trajectory control. They present a biological model of unsupervised, real time, error-based learning which is a recent addition to their previous models. This so-called «Vector

Associative Map model» is designed to overcome some of limitations of previous error-based learning models such as perceptrons and back propagation algorithms which have been supervised. The paper describes as an example a self-organizing neural network that controls arm movement trajectories during visually guided reaching. Possible generalizations relevant to other tasks are also discussed.

N. Reeke does not believe that the usual algorithms defining events and object classification in neural networks are relevant in animal cognition processes. He offers an alternative approach, namely the theory of neuronal group selection. The model is related to the concepts of Darwinian selection and does not require *a priori* assumptions about the stimulus world. One of his models addresses again the problem of coherent oscillations in the visual cortex, considered in several papers of this volume. The second model is a neurally based automaton, capable of autonomous behaviour involving categorization and motor acts with an eye and an arm.

Berthommier *et al.* investigate the asymptotic properties of neural networks in various operating conditions. The networks are used for object detection, (such as for example tumour tracking) and offer interesting medical applications.

C. Van den Broeck offers an opposite view to all the preceding papers and introduces an equilibrium theory of learning. He suggests that learning may be the outcome of the law of maximum entropy subject to constraints imposed by the teaching from examples. In a Boolean network of building blocks, if a critical threshold of teaching is exceeded a phase transition towards « understanding » occurs.

P. Shuster shows that selection and adaptation to environmental conditions are already observed in populations of RNA molecules replicating *in vitro*. This complex optimization behaviour is explored as the simplest case of a relation between genotype and phenotype. Here again optimization process takes place on a fitness landscape built on the free energy of two dimensional folding patterns of RNA molecules.

Mandell and Selz introduce a nonthermodynamic formalism for biological information processing. They suggest hierarchical lacunarity as a new coding principle. The formalism is applied to the evolutionary divergence of homological polypeptide chains in hemoglobin, myoglobin, and cytochrome *c*. This approach brings together many concepts from different fields of physics and mathematics.

The concept of self-organization and the role of emerging properties as governing modes of behaviour in a multi-unit and multiply connected ensemble of independent and autonomous entities is also of great importance in human as well as insect societies. Learning and communication are eminently nonlinear processes. So are the economic exchanges which operate at a planetary level. Therefore it is not surprising that self-organization and the emerging properties are seen in social systems and play a crucial role. In theories dealing with social sciences and economics there is a shift from linear thinking towards a global nonlinear approach.

The capacity of insect societies to build structures and solve problems is truly impressive. Such capacity raises the old crucial question about what is learned and what is acquired by such simple creatures. It seems that emerging properties due to interactions in a colony of insects may bring an element of answer to the learned versus acquired dilemma.

J.L. Deneubourg *et al.* address precisely the question of functionality of collective modes of organization in insect societies. They show that problem solving in insect societies does not necessarily require a «prior knowledge». In systems consisting of large number of units, the interaction among units and the environment generates collective behaviour which is the solution to the given problem. They also discuss the extensions and applications of their findings to robotics and transportation problems.

Ping Chen addresses the problems of imitation, learning and communication in social systems. He offers a stochastic model of human collective behaviour. In the framework of the Ising model he demonstrates the interplay between independent decision making and group pressure. This may lead to central or polarized distribution functions which describe the middling or polarization in social phenomena.

Research in all the areas considered in the workshop is currently in progress. We believe that many promising alleys will be opened in the future in each domain, and that these will converge into and benefit other fields of research.

Agnessa Babloyantz

Brussels July 1991

CONTENTS

Spatio-temporal Patterns and Network Computation

J.A. Sepulchre, A. Babloyantz

Free University of Brussel
Service de Chimie Physique CP 231
Bvd. du Triomphe, 1050 Brussel
Belgium

1. Introduction

Self-organization of matter has been the subject matter of intense research in last decades [1,2,3]. This concept expresses the fact that an ensemble of units in interaction may exhibit collective behavior not seen in separate entities. These emerging states are the property of the system as a whole. Self-organizational ability and the ensuing emerging properties were shown to play a crucial role in hydrodynamics, in chemistry and above all in biology. Indeed it has been shown the possibility that self-organization is the basic principle governing the emergence of genetic material from simple molecules [4]. The same principle is shown to be at work at cellular, embryonic and cortical level, regulating the key processes of life [3].

The realization that the unique ability of human brain to perform a wide variety of perceptual and cognitive tasks, reside in the self-organizing ability of nerve cells led to the extremely active field of artificial neural networks. In this line of research, one tries to construct ever more complex networks of neuronal like elements, thus mimicking to some degree the architecture of human cortex. The emerging properties of the network are the substrata for such «intelligent» tasks as pattern recognition, associative memory and various optimization problems.

Self-Organization, Emerging Properties, and Learning
Edited by A. Babloyantz, Plenum Press, New York, 1991

In the classical field of neural networks, in most cases, global properties which perform intelligent tasks are time independent fixed points. This fact is the basis of non applicability of the theory to many time-dependent problems. However in recent years few exceptions to this general trend appeared which do not rely on a hamiltonian like function but rather make use of the dynamical states of the network [5]. The research reported in this paper belongs to this category.

We consider the time dependent emerging properties of a network of nonlinear elements connected via a linear connectivity function to their first neighbors. These nonlinear elements could be considered as simple processors, therefore the network may be seen as a parallel processor which can perform fault tolerant computation.

The elementary processors considered here are oscillating elements which for a critical value of their parameters may be found in an unstable quiescent or in a stable oscillatory state. We show that the unstable network may sustain front propagation associated to and followed by propagating waves which either propagate in the same or in opposite direction to the front.

The dynamics of networks with interconnected oscillatory units are also investigated. We show that a local frequency shift generates target waves in the network. These waves show unexpected and interesting properties in presence of obstacles build inside the network. We show that a critical inter-obstacle distance may inhibit wave propagation. We discuss the possibility of using these properties in artificial networks for path finding and robotics.

2. Network of oscillators

Let us consider N oscillating elements each described by two variables X and Y. The oscillators form a two dimensional square lattice, where each element is only connected to its first neighbors. The equations describing the evolution of the oscillator at site (k,j) reads :

$$\frac{dX_{(k,j)}}{dt} = F_1(X_{(k,j)}, Y_{(k,j)}) + C\,L(X)$$

$$\frac{dY_{(k,j)}}{dt} = F_2(X_{(k,j)}, Y_{(k,j)}) + C\,L(Y) \qquad (1)$$

here L is a linear operator defined as :

$$L(X)(k,j) = X(k+1,j) + X(k,j+1) + X(k-1,j) + X(k,j-1) - 4X(k,j)$$

$$k,j = 2,\ldots,N-1$$

F_1 and F_2 describe the law of interaction between the two variables of the oscillator. C is a constant which accounts for the strength of the interaction. The boundary cells are subject to « zero flux » boundary conditions.

Let us note that the linear connectivity operator L can be thought as a discrete representation of the continuous operator ∇^2 describing diffusion processes.

In eqs.(1) we assume that the dynamics of each oscillator is described by the following functions F_1 and F_2 :

$$F_1(X, Y) = A - (B+1)X + X^2 Y$$
$$F_2(X, Y) = BX - X^2 Y \tag{2}$$

Here A and B are constant quantities. This model has been studied extensively in the literature as a paradigm for homogeneous oscillating chemical media [1,6].

The set of eqs.(1) and (2) have a steady state solution given by $X_0 = (X_0,Y_0) = (A;B/A)$. If we consider B as a bifurcation parameter, a straightforward linear stability analysis shows that $B_c=1+A^2$ is a critical point which separates the domain of stable homogeneous state from a domain where this state is unstable and small fluctuations drive the system into an oscillatory regime. The latter appears for values $B > 1+A^2$ following a Hopf bifurcation.

3. Wave propagation in unstable networks

Let us imagine an experiment where all the oscillators of the network have a value of $B > B_c$ and are held in an unstable homogeneous state. Now we perturb slightly one of the variables, say $X=X_0+\delta X$ in a small number of oscillators, in the centre of the domain. Because of the unstable nature of the medium, the disturbance gives rise to a front which propagates outwardly into the network.

This propagating front is the boundary between the elements which are already in a stable oscillatory regime and the elements which are still in the unstable non oscillatory steady state. This front is associated with concentric circular waves which may propagate inwards behind the front in opposite direction to the front (fig. 1). This unusual direction of propagation may be seen more easily by considering a radial cross section of the two dimensional system as shown in fig. 2.

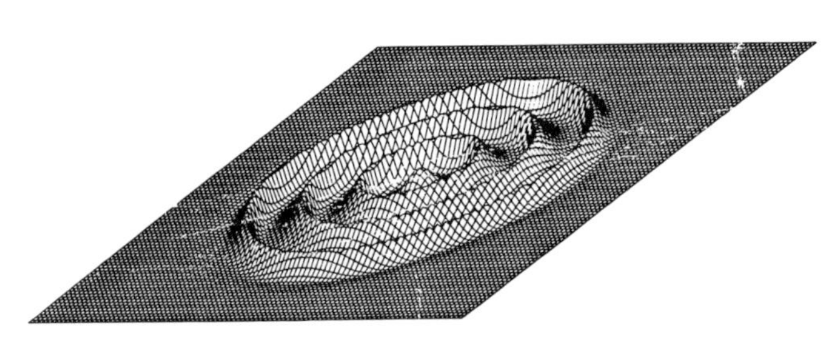

Fig.1 - Spatial distribution of the X variable of the Brusselator (A=2; B=5.3 ; C=0.08). Behind the front which propagates in the unstable network (X=2 ; Y=2.65), phase waves appear, and travel inward towards the centre.

Near the Hopf instability point, where $B=B_c(1+\varepsilon)$ $(0 < \varepsilon \ll 1)$, the dynamics may be described by a generic equation known as normal form. If Tr and Δ denotes respectively the trace and the determinant of the jacobian matrix of the function $\mathbf{F} = (F_1,F_2)$, then $Tr= \alpha\varepsilon$ and the set of eqs.(1) and (2) may be described by the following amplitude equation in terms of the slowly varying envelope W(k,j) [7]:

$$\tau_o \frac{dW}{dt} = \varepsilon W - (g_r + i g_i) |W|^2 W + \xi_o^2 L(W) \quad (3)$$

$$\text{where : } \mathbf{X} - \mathbf{X}_0 = \boldsymbol{\eta} W e^{i\omega_o t} + \boldsymbol{\eta}^+ W^+ e^{-i\omega_o t} \quad (4)$$

Here ε is a measure of the degree of supercriticality and

$$\tau_o^{-1} = \frac{\alpha}{2} \;;\; \xi_o^2 = C\,\tau_o \;;\; \omega_o = \sqrt{\Delta(B_c)} \;;\; \beta = g_i/g_r$$

The vector η is the critical mode, i.e. the eigenvector associated with the eigenvalue $i\omega_o$ of the jacobian matrix of function **F**. The constants g_r and g_i can be related to the parameters of the model. In the case of a supercritical bifurcation we have $g_r > 0$. When the nonlinear dispersion g_i is normal, i.e., when the period of the oscillations increases with the amplitude, then $g_i > 0$. This is the case for the Brusselator model, whereas there are models where one finds $g_i < 0$.

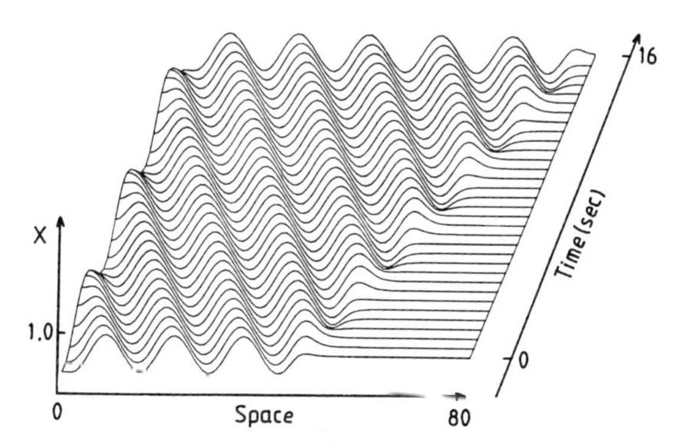

Fig.2 - Spatio-temporal evolution of a radial cross section of a two-dimensional system described by Eqs.(1) and (2) The nonlinear dispersion is of the same sign as in the Brusselator model. Therefore the front and the wave velocity are of the same magnitude but of the opposite direction.

One can show that for sufficiently localised perturbation, the velocity of the front approaches the marginal stability value v*, i.e. the slowest stable propagation velocity [8]. Moreover, the velocity c of propagation of the waves behind the front is given by [9] :

$$c = v^* \left[1 - \frac{\beta\omega_o\tau_o}{2\varepsilon(\sqrt{\beta^2+1} - 1)}\right]$$

In this expression, as $0 < \varepsilon << 1$, the value of the fraction is greater than 1. On the other hand, as $\omega_o > 0$ and $\tau_o > 0$, the velocities c and v* have different signs only if $\beta = g_i / g_r$ is positive. Consequently, the velocity of the front v* and the velocity c of the crests behind the front will be opposite if g_i is positive, as in the case of the Brusselator. There are other models, for which g_i is negative, and the reverse situation prevails [10].

4. Wave propagation in an oscillatory network

In the preceding section we considered a network with oscillatory elements which were held initially at unstable nonoscillatory state. We showed under what condition fronts propagate in this system.

In this section we consider again the same network, however presently all units oscillate and moreover they are synchronized into a homogeneous stable bulk oscillation. Again the system is subject to zero-flux boundary conditions.

Target waves can be generated in this network around centres where local and permanent inhomogeneities are introduced in the network [11]. This may be achieved for example by imposing a local frequency or amplitude increase in a very small number of oscillators that we call a « pacemaker centre » [12,13]. These centre initiates target waves in the network which again propagate in concentric rings starting from the pacemaker centre and gradually taking over the entire network.

The frequency of these waves can be controlled by changing the characteristics of the pacemaker. Therefore it is possible to control the wave propagation in the network. As we shall see in the sequel this important property may be used for computational purposes.

Here again in the vicinity of the Hopf bifurcation the network can be described with the help of the amplitude equation (3). After introducing the terms relative to the pacemaker region into the eq. (3) and rescaling the variables, we get :

$$\frac{\partial Z}{\partial \tau} = (1 + i\Delta\omega(k,j))\, Z - (1 + i\beta)\, |Z|^2\, Z + \kappa\, L(Z) \qquad (5)$$

with $W = (\varepsilon/g_r)^{1/2}\, Z$; $\tau = \varepsilon t/\tau_o$; $\kappa = \xi_o{}^2 / \varepsilon$

Here $\Delta\omega(k,j)$ takes a nonzero constant value $\Delta\omega$ only in a small cluster of cells defining the pacemaker. Several shapes of pacemaker were considered and the same qualitative results were obtained for all. In the following the pacemaker is chosen cross-shaped, build on five adjacent cells.

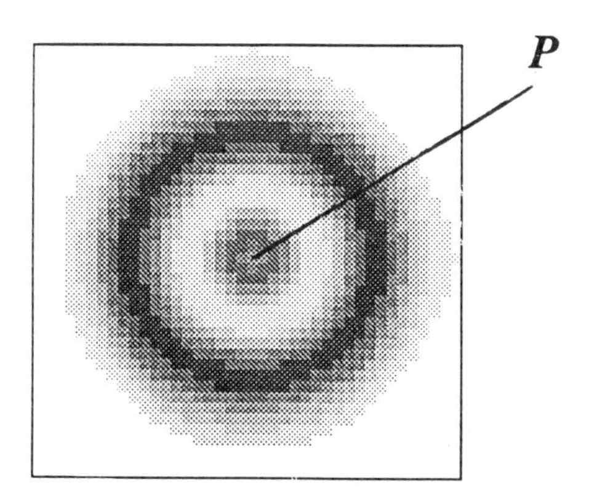

Fig. 3. - Target waves in an oscillatory media. A pacemaker region at *P* with a frequency shift $\Delta\omega = 1$, generates target waves with frequency Ω. The parameters of eq. (5) are $\beta=1$, $\kappa = 0.8$. The size of the network is 40*40.

Numerical simulations of eq. (5) show that when $\Delta\omega = 0$, whatever initial conditions, the asymptotic solution for Z is the homogeneous oscillation at frequency $\Omega_o = -\beta$. However, when $\Delta\omega \neq 0$, target waves are generated continuously under the influence of the pacemaker and a frequency locked solution of eq. (5) appears (fig. 3). Moreover one sees also a radial phase distribution around the pacemaker centre, giving rise to concentric waves.

Let us study the relation between the frequency Ω of these waves and the frequency shift $\Delta\omega$ of the pacemaker. Our simulations reported in fig. 4 show that Ω is a highly nonlinear function of $\Delta\omega$. The frequency Ω is also a function of the coupling coefficient κ, the nonlinear dispersion β and the size of the pacemaker region. It is seen from this relationship that Ω exhibits a maximum for a given value of $\Delta\omega_r$ and decreases for $\Delta\omega > \Delta\omega_r$. The existence of a maximum in the curve of fig. 4 is of importance when there are several pacemaker centres in the medium, as seen in the next paragraph.

We now consider the case when *n* simultaneous pacemaker regions are active with different local frequency shifts $\Delta\omega_1 < \Delta\omega_2 < ... < \Delta\omega_n$. These values correspond respectively to wave frequencies $\Omega(\Delta\omega_i)$ (i=1,...,n). Target waves start to propagate from all centres (fig. 5(a)). However, after a while, one sees that one of the centres takes over the entire network (fig. 5(b)). As it was shown previously, the pacemaker

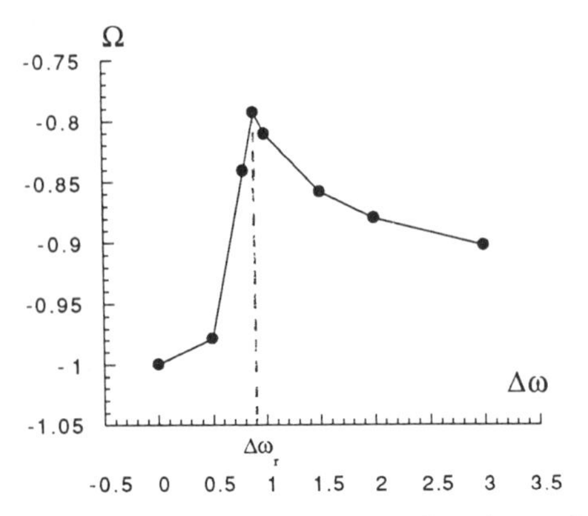

Fig. 4 - Variation of the frequency Ω as a function of local frequency shift Δω . Same parameters as in fig. 3 .

centre which wins the competition is the one whose frequency Ω is the largest [12]. One would expect that $\Omega(\Delta\omega_n)$, with the largest $\Delta\omega_i$,should be the winner if $\Omega(\Delta\omega)$ was a monotonic increasing function of the frequency shift. However, because of the decreasing part of function depicted in fig. 4, the largest frequency $\Omega(\Delta\omega_i)$, whose value is the closest to $\Omega(\Delta\omega_r)$, is the winner.

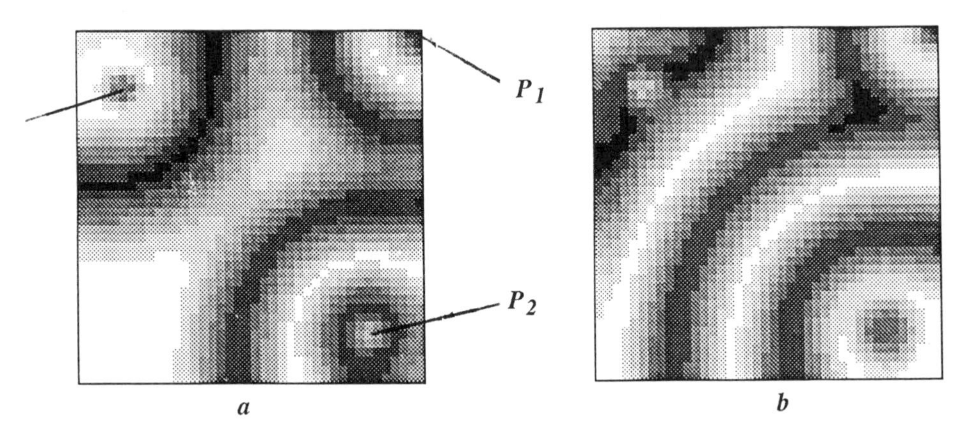

Fig. 5 - (a) Interaction of three cross shaped pacemakers in a network of 40*40 units ($\Delta\omega_1 = 0.8$, $\Delta\omega_2 = 1.0$, $\Delta\omega_3 = 3.0$). (b) After a transient regime, the pacemaker with the closest frequency to $\Omega(\Delta\omega_r) = \Omega(0.9)$ inhibits the other pacemakers. Same parameters as in fig. 3 .

5. Wave propagation in the presence of obstacles

In the last section target waves propagated freely in the network till they reached the boundaries of the system. In this section we investigate their behavior whenever they encounter obstacles inside the network. The latter are introduced in the system by deleting oscillators in the network. Zero-flux boundary conditions are restored after deletion.

5.1 Propagation from one window

Let us first introduce a partition in the network which divides the system into two separate parts related only by an array of oscillators which we shall call « window ». The number of units m connecting the two compartments will be called « window length » ($0 \leq m \leq N$). In our numerical experiments, the length of the window is taken as a control parameter and as we shall see, the phenomena observed are critically dependent not only on the window length but also on the position of the window relative to the pacemaker position.

Let us consider the case where the window position is in the centre of the wall and the source is aligned perpendicular to the window. In the same manner as in the preceding section, we produce target waves with frequency Ω in the first compartment. We want to know wether the oscillators of the second compartment will lock onto the frequency of the oscillators in the first compartment. The simulations show that the frequency locking appears only if the window length m is larger than a critical value m_c. Below this value a new solution is seen in the system(fig. 6(a-b)).

Let us measure the phase difference ϕ between two oscillators of compartment I and II. For a window length of $m > m_c$, this phase difference tends to a stationary state, therefore in this case, the frequency locking between the two compartments occurs (fig. 7(a)). This behavior is not surprising since, in the case of a large window, all happens as if we were in the presence of a square network with somewhat modified configuration. However, as the size of the window m decreases, the frequency locked solution disappears at $m = m_c = 5$ (fig. 7(b)).

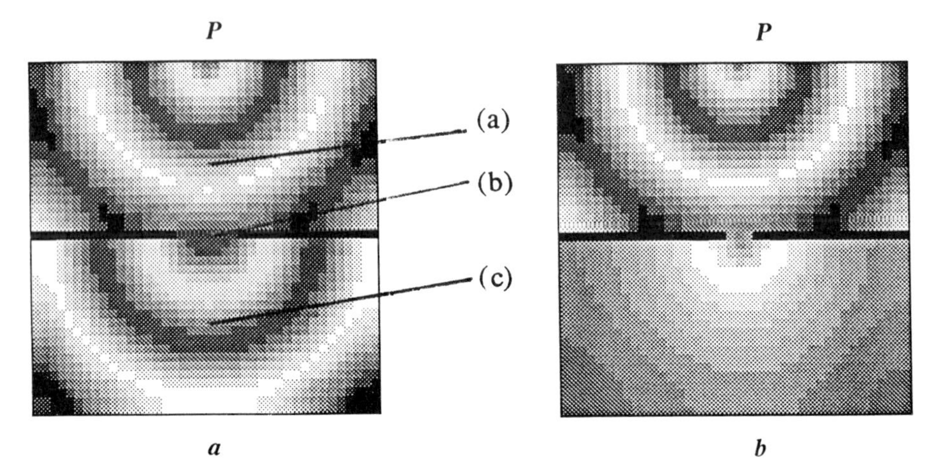

Fig. 6 - A square network of 40*40 oscillators is partitioned by a wall and communicates through a window of size m. The parameters of eq. (5) are $\beta = 1$, $\kappa = 0.8$. A pacemaker region at P generates target waves in compartment I, with frequency $\Omega = -0.8103$. (a) $m = 1.4\, m_c$: new targets emerge from the window in the second compartment with a frequency Ω. (b) $m = 0.6\, m_c$, : in compartment II a fraction of a wavelength is seen.

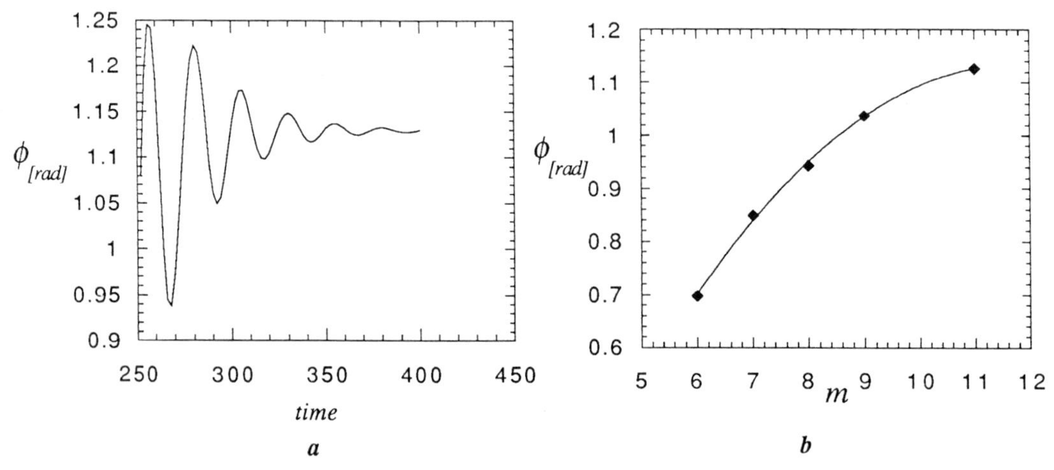

Fig. 7 - Phase difference ϕ between cells (20,10) and (20,20) in the 40*40 network depicted in fig. 6 (a). For $m > m_c$, ϕ converges in time to a steady state (here $m = 8$ and $m_c = 5$). (b) Steady states values of ϕ as a function of m, for frequency locked solutions.

If in the network of fig. 6, $m \leq m_c$, there is no more frequency locked solution. For this case, the time evolution of three oscillating cells is monitored. The first cell (a) is in the compartment I and far away from the partition. The second cell (b) is in the opening and a third cell is taken far away from the window in compartment II. Figure 8(a) shows the time behavior of the real part *Re* Z(t) for these cells. As

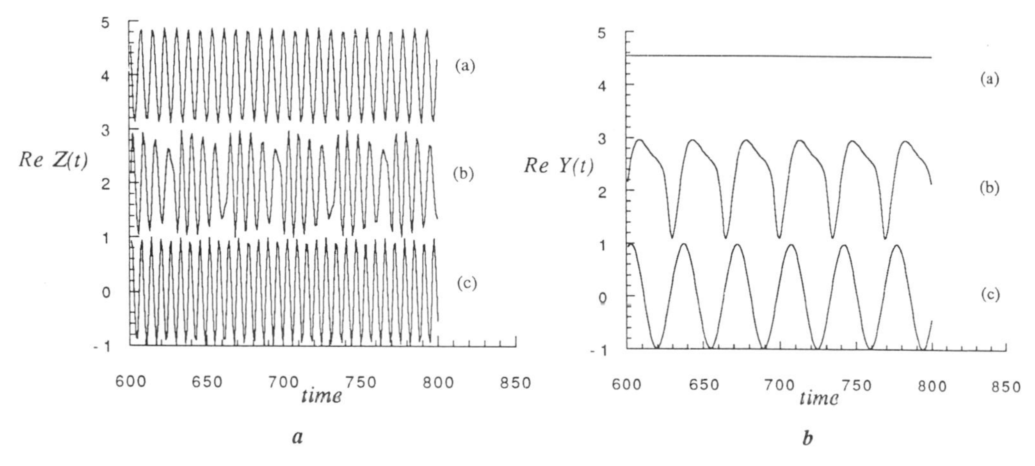

Fig. 8 - Time variation of *Re* Z(t) and *Re* Y(t) at various points in the two compartments of fig. 6(b).

expected, cell (a) has periodic dynamics with frequency Ω. The cell at (b) shows some irregular dynamics, apparently quasi-peridodic. At cell (c) again an oscillatory behavior is seen with a frequency which is higher than in (a).

In order to better understand this behavior in the vicinity of the opening and in the second compartment, it is convenient to perform the change of variable $Z = Y \exp(i\Omega t)$. In terms of variable Y, the frequency locked solution which exists for $m > m_c$, appears as an inhomogeneous steady state $Y = Y_S$ of eq. (5). However, for smaller window, $m \leq m_c$, the steady state Y_S disappears in the second compartment and the dynamics of Y becomes oscillating with a frequency γ (fig. 8(b)). Therefore, at point (b) and in the vicinity of the window, the dynamics of Z is quasi-periodic, with both frequencies γ and Ω. At point (c), i.e. in compartment II and far from the opening, Z follows periodic dynamics with the frequency $\Omega' = \Omega + \gamma$.

For a fixed value of Ω numerical simulations reported in fig. 9 show the functional relationship between γ and m. From the points of fig. 9, after a curve fitting procedure, we see that the function $\Omega'-\Omega_0 = \Omega+\beta+\gamma$ is proportional to m^4. As m decreases, the wavelength of target waves increases. Thus if the size of the network is much smaller than the wavelength of the target waves, one sees only a weak concentration gradient which is only a fraction of the wavelength (fig. 6(b)). Therefore, in a finite system, target patterns are not observed in compartment II when the size m of the window becomes too small. This results from the fact that the wavelength is much larger than the size N of the network.

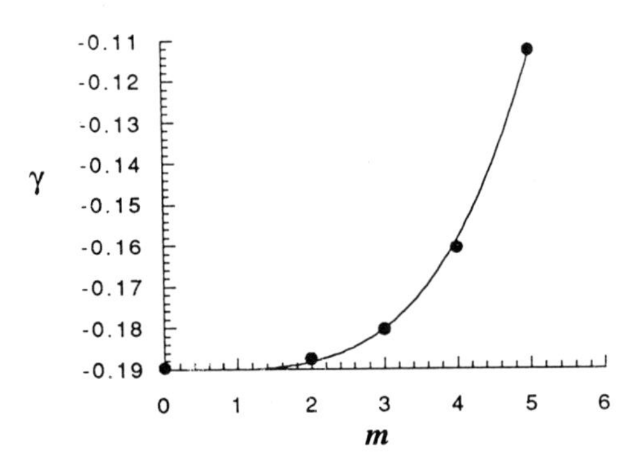

Fig. 9 - Frequency γ of Y(t) in compartment II as a function of the window size. The parameters are the same as in fig. 6.

Our simulations show that the relative position of the pacemaker and of the window may be of importance to the type of propagation seen in compartment II. To see this we simulate the reaction in a network where the pacemaker is not aligned perpendicular to the partition. Figure 10 shows the variation of γ as a function of m in this case. Contrary to fig. 9 in the present situation we observe a plateau region for moderate values of m. For values of $m<m_a$ and $m>m_b$ the behavior of the system is similar to the case discussed in fig. 6. However for window size $m_a \leq m \leq m_b$ one observes the onset of spiral wave activity in compartment II (see fig. 12).

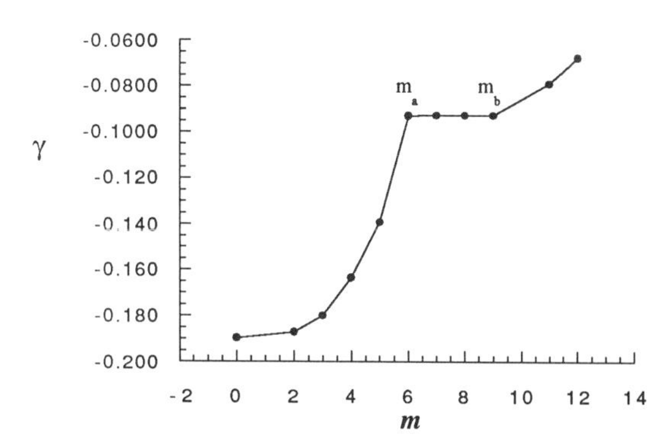

Fig. 10 - Frequency γ in the compartment II as a function of the window size when the pacemaker is not aligned with the opening. The parameters are as in fig. 6. Spiral waves appear if the window size $m_a \leq m \leq m_b$.

5.2 Propagation from two windows

Let us now consider the same network of oscillators as described above. Presently, we introduce two openings m_1 and m_2 in the system. Again a single local frequency shift $\Delta\omega$ is created at P. Target waves start to propagate in compartment I and reach successively the two openings m_1 and m_2. For κ=0.8 and $\Delta\omega$= 3, the two windows act as new pacemaker regions and generate in turn target waves in the second compartment with frequencies $\Omega'(m_1) \approx \Omega'(m_2) \approx \Omega$. After a while, the target waves propagating from the two centres collide and cusp-like structures are formed (fig. 11(a)).

In the next experiment, we increase slightly the extent of the pacemaker region, increase the connectivity coefficient κ and decrease $\Delta\omega$. A higher value of Ω, as compared with the preceding case, is obtained in the first compartment for the propagating target waves. In this experiment, when the wavefront reaches the windows m_1 and m_2, new fronts are again generated. However, very soon the waves emerging from the window m_2 take over the entire system and inhibit all propagation from m_1 (fig. 11(b)).

A tentative explanation to this phenomenon could be given as follows. In this case one observes that $\Omega'(m_1) < \Omega'(m_2) < \Omega$. We saw in a preceding paragraph that if two pacemakers emit simultaneously, then the fastest waves inhibit the slower propagation and take over the entire system. As $\Omega'(m_1) < \Omega'(m_2)$ it is reasonable to think that a similar

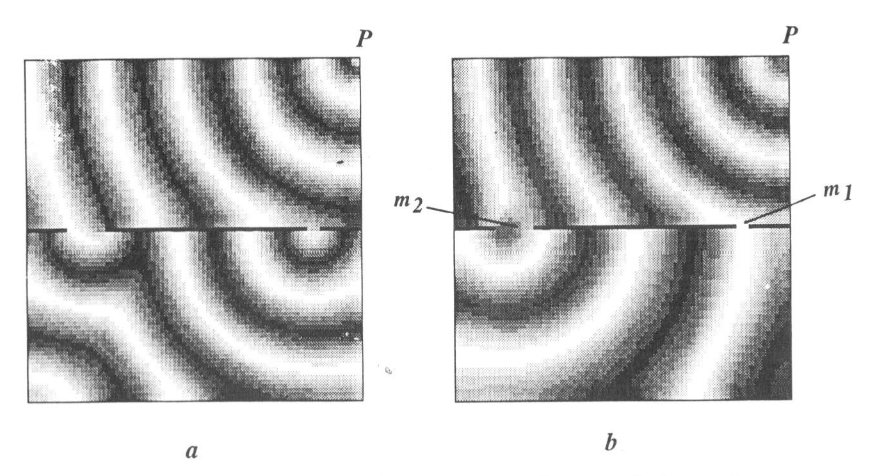

Fig. 11 - Wave propagation in the presence of two windows. Parameters are : β=1, window lengths m_1= 3, m_2= 9 and the network size is 80*80.

(a) Waves propagate into compartment II from both windows. A cusp like structure is formed. (κ = 0.8, Δω=3).

(b) Waves propagating from the largest window inhibit the propagation from the small window. (κ = 1.42, Δω=1)

explanation prevails here. The frequency of the waves emerging from m_2 is fastest and, therefore, they take over the network and inhibit the slowly evolving target waves which emerge from m_1.

In section 5.1 we showed that for a specific configuration of source *P* and a single window length *m*, spiral waves may appear in the second compartment. Let us again consider two windows m_1 and m_2. When the fronts starting from compartment I reach m_1, spiral type activity is generated in the second compartment. However when the front reaches m_2 cusp like fronts are seen in compartment II (fig. 12(a)). In time the tip of the spiral moves toward the opposite boundary and finally disappears, giving rise to the ordinary target waves as shown in fig. 12(b).

6. Computation with target waves

The properties of target waves, specially their unexpected behavior in the presence of obstacles may be useful for performing computational tasks. Here we sketch very rapidly the possible applications of target waves in path finding and robotics.

Let us consider again our network of fig. 11 where a partition is introduced in the system. To this network, we can associate a physical space [14]. For example a room partitioned by walls and

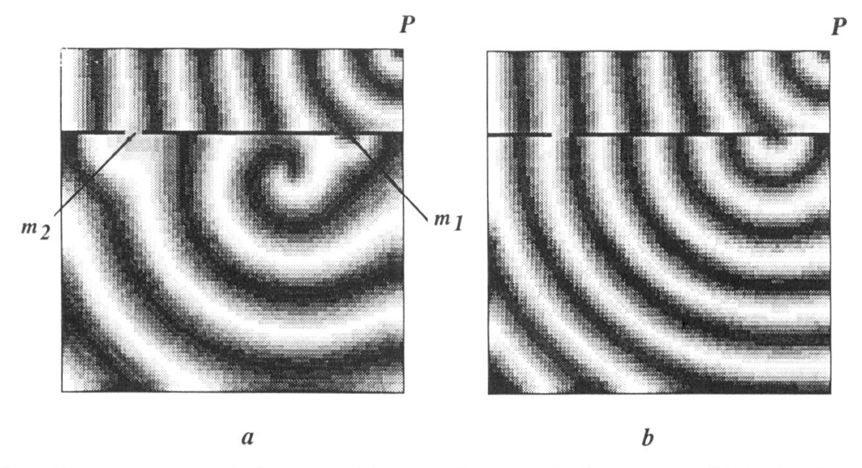

Fig. 12 - A two window partition produces spiral waves which in time give way to target waves. All happens as if again the propagation from the small window is inhibited. Parameters are as in fig. 11.

communicating via two doors. The floor of the room is also divided into small units. To each unit in the floor of the room there corresponds an oscillating element in the network. The walls of the room are represented in the network by the deleted oscillators.

In fig. 13 a mobile *R* located in the second compartment must find the shortest path to the point *G* in the first compartment.

To the point *G* there corresponds a point *P* in the network space. If this point is made to be a pacemaker centre, target waves will start to propagate in the network, in the manner described in the preceding sections. The crest of these waves may become information carriers indicating a direction of propagation.

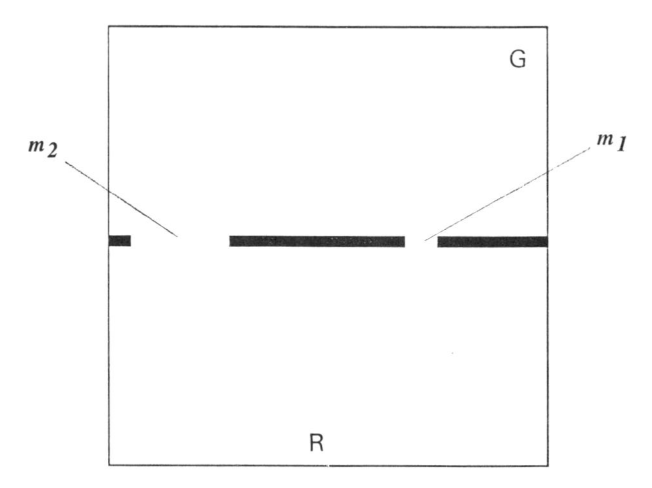

Fig. 13 - Physical space associated with the network.

To see this we consider the discrete phase gradient of the waves. The phase of each oscillator is defined as $\Phi(k,j) = \mathrm{Arctg}(Y(k,j)/X(k,j))$ and finite difference approximation of $\nabla\Phi$ leads to :

$$V^1(k,j) = \frac{X(k,j)(Y(k+1,j)-Y(k-1,j)) - Y(k,j)(X(k+1,j)-X(k-1,j))}{2(\ X(k,j)^2 + Y(k,j)^2\ )}$$

$$V^2(k,j) = \frac{X(k,j)(Y(k,j+1)-Y(k,j-1)) - Y(k,j)(X(k,j+1)-X(k,j-1))}{2(\ X(k,j)^2 + Y(k,j)^2\ )} \tag{6}$$

Relations (6) provide a vector field which is perpendicular to the crests and points to the direction of the pacemaker centre P. Thus if the vector field is computed in each cell as they are reached by the wave, a trajectory is constructed, which joins every point in the compartment II to the point P. Due to the mapping of the network space into the physical space, a family of trajectories are also created in the physical space relating every point in compartment II to the goal-point G.

Such a mapping could be used for example for finding the shortest path between G and a point S where the trajectory starts. The frequency shifts $\Delta\omega$ in P, m_1 and m_2 are chosen such that the waves can propagate from the two openings. The vector field corresponding to this problem is depicted in fig. 11 and it is seen that the shortest path is C_1 which passes through the door m_1.

However a more interesting case is the one in which for technical reason such as for example spatial extension of mobile object R the door m_1 is too small and a suitable path must be found between point S and P.

Such a situation is depicted in fig. 14. In this case the $\Delta\omega$ at P is chosen such that the waves only propagate through larger door m_2. Again the vector field indicates that the path is the curve C_2 for every location S of mobile object R in the second compartment. However if the mobile is in a very small restricted area in compartment II and close to the door m_1, the network indicates a path C_2' which is shorter than C_2 but useless in the present situation as m_1 is too small to let pass the object. We feel that this ambiguity may be resolved by choosing a more appropriate $\Delta\omega$. Work in this direction is in progress.

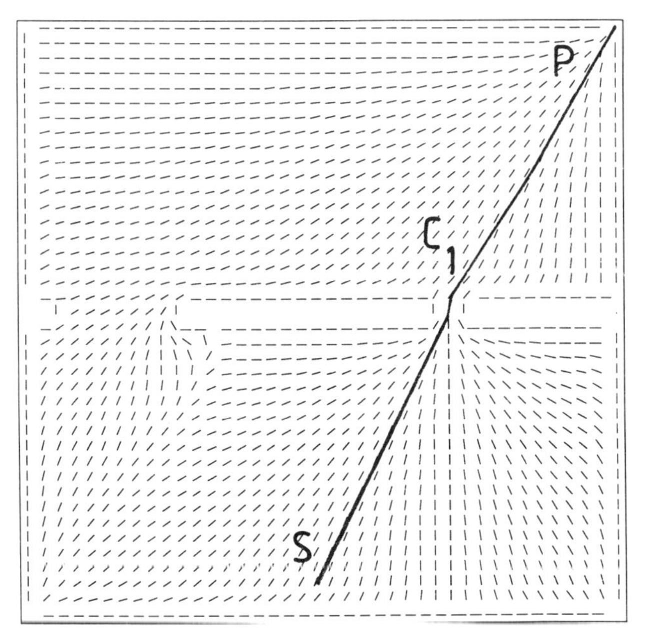

Fig. 14 - The vector field computed from eqs. (6) defines the trajectory from compartment II to point P in compartment I.

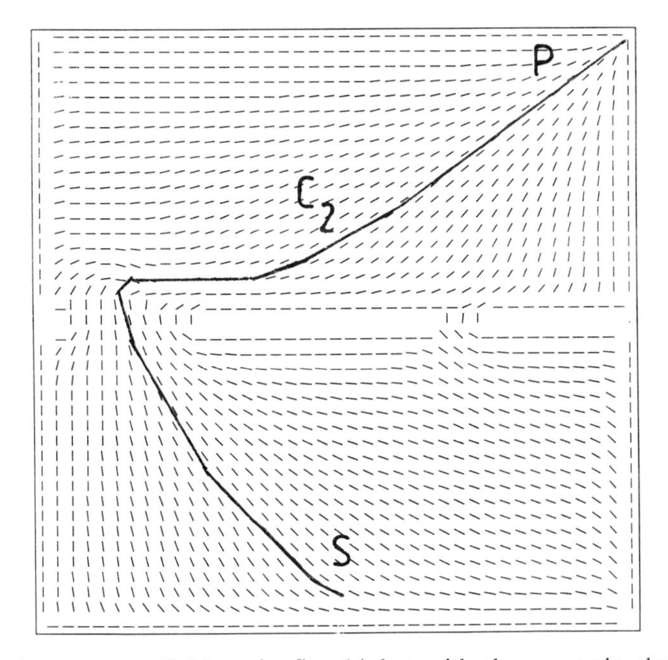

Fig. 15 - Same vector field as in fig. 14 but with the constrain that the door m_1 should be avoided.

Conclusions

In this paper we have investigated the onset of several collective modes in networks of interconnected oscillators. We could show the propagation of target waves, fronts and spiral waves, similar to the collective behavior seen in continuous media, where local oscillatory activity and diffusion are present.

These waves, in the presence of obstacles, show interesting behavior which are of value in applications pertaining to such problems as path finding or robotics. We believe that the network of oscillators endowed with unusual dynamical properties described in this paper, alone, or in multiple layers are an important addition to the existing neural networks which use only fixed points of the network dynamics. Networks of oscillators introduced as an extra layer in the existing networks which generalize Hopfield type approach, may enhance and broaden the scope of the computational ability of the latter.

When the distance d between the oscillators is decreased, in the limit $d \longrightarrow 0$, the network is isomorph to a continuous media subject to oscillating chemical reactions and diffusion processes. The investigation of nonlinear dynamical systems in continuous media, has been the focus of research for many years. Many analytical as well as computer results are reported in the literature [15]. We suggest that some of these analytical results could be extended to the discontinuous system studied in this paper. However, discontinuous networks have their own specific properties which are absent in continuous systems. Computer work in this direction is on the way. Analytical results pertaining to the solutions of a set of coupled nonlinear ordinary differential equations are desirable in probing these networks.

References

[1] G. Nicolis & I. Prigogine, Self-organization in Nonequilibrium Systems, (Wiley, New York, 1977)

[2] G. Nicolis & I. Prigogine, Exploring Complexity. (Freeman, New York, 1989)

[3] A. Babloyantz, Molecules, Dynamics and Life, (Wiley, New York, 1986).

[4] Prigogine, G. Nicolis and A. Babloyantz, Physics Today 25 (1972) 11

[5] B. Baird , Physica D 22 (1986) 150

[6] I. Prigogine & R. Lefever, J. Chem. Phys. 48 (1968) 1695

[7] D. Walgraef, G. Dewel and P. Borckmans, J Chem Phys. 78 Part I (1983) 3043

[8] C. Dee and J.S. Langer, Phys. Rev. Lett. 50 (1983) 383 ; W. van Saarloos, a) Phys. Rev A 37 (1988) 211 ; b) Phys. Rev. A 39 (1989) 6367

[9] J.A. Sepulchre, G. Dewel & A. Babloyantz, Phys. Lett. A 147 (1990) 178

[10] K. Maginu, J. Diff. Eqs., 31 (1979) 130

[11] A.T. Winfree, Theoret. Chem. 4 (1978) 1

[12] P.S. Hagan, Advances in applied Mathematics, 2 (1981) 400

[13] Y. Kuramoto, Chemical Oscillations, Waves, and Turbulence, (Springer-Verlag, Berlin Heidelberg New York Tokyo1984)

[14] L. Steels in: Proceedings of the European Conference in Artificial Intelligence 88, Y. Kadratoff Ed., Pitman Publishing, London (1989)

[15] M. Kubicek and M. Marek, Computational Methods in Bifurcation Theory and Dissipative Structures, (Springer-Verlag, New York, 1983)

APPLICATION OF SYNERGETICS TO PATTERN FORMATION AND PATTERN RECOGNITION

Hermann Haken and Arne Wunderlin

Institut für Theoretische Physik und Synergetik
Universität Stuttgart
Federal Republic of Germany

INTRODUCTION

Synergetics is a systematic study and analysis of complex systems that are composed of many subsystems [1,2]. These systems show the ability to self-organize spontaneously on macroscopic scales, i.e. we may observe highly ordered spatial, temporal, or spatio-temporal structures on scales which are much larger than the corresponding characteristic length and time scales of the subsystems. As a result, we conclude that there have to be very many subsystems coherently involved in order to produce these highly regular states on characteristic scales of the composed system. Synergetics offers a unifying viewpoint to the emergence of such patterns and the underlying processes. Furthermore, powerful mathematical methods have been developed to understand and predict the spontaneous occurrence of order, qualitatively as well as quantitatively.

Aspects central to the theory are the notions of instability, fluctuations, and in particular, order parameters and slaving. Processes of self-organization in an open system always arise via an instability. Similarly, switching between self-organized states of qualitatively different macroscopic behavior is again associated with instabilities. A complex system "notices" the presence of an instability through the fluctuations. Fluctuations may be considered as continually testing the stability of its present state under externally given conditions. Eventually, slaving guarantees that a macroscopically evolving ordered state can be characterized by only a few collective variables, the so-called order parameters, that are generated by the system itself. It is understood that these order parameters enslave the subsystems in a way that the latter act in a coherent fashion and thus maintain the macroscopically ordered state. In addition they may be considered as a quantitative measure of the evolving order.

The slaving principle has been successfully applied to various systems treated in the natural sciences as well as to problems of human sciences [1,2]. Laser action, for example, may be considered as a fundamental paradigm of slaving. The non-linear interaction between the laser-active atoms, which play the part of the subsystems here, and the light field

Self-Organization, Emerging Properties, and Learning
Edited by A. Babloyantz, Plenum Press, New York, 1991

generate a highly ordered state of the atoms that is connected with coherent laser action. Laser action is a particularly important example because its complete theory may be derived rigorously from first principles of physics, i.e. quantum mechanics and quantum electrodynamics. In addition, we note that non-linear optical systems can be understood along similar lines. Furthermore, the mathematical formulation of this fundamental principle has been applied to hydrodynamic systems, e.g. the Bénard and Taylor problem, solidification processes, flame instabilities, atmospheric movements, plasma instabilities etc. There are applications in magnetohydrodynamics where the discussion of the generation of macroscopic stellar and planetary magnetic fields proves especially interesting.

Applications, however, are not only restricted to physical systems. Indeed, interdisciplinary aims are one of the most appealing aspects of synergetics. In chemistry, the Belousov-Zhabotinski reaction, the Briggs-Rauscher reaction and model reactions like the Brusselator and Oregonator have been analyzed using the slaving principle [1,2]. It has also been applied to biological problems, especially to questions posed in morphogenesis and on the coordination of human hands movements. Furthermore it is expected, as we shall discuss, to become an important tool in the analysis of brain functioning [3].

Finally, we mention applications in the social sciences: for example the problem of formation of public opinion in sociology [4], and its use in the discussions of quantitative and qualitative models in economics [5].

This article is organized as follows. First we shall give a mathematical formulation of the central principle of synergetics, the slaving principle. This principle allows us especially to apply methods of the qualitative theory of differential equations to the description of spontaneously emerging macroscopically ordered structures observed in a large variety of systems. We then mention the application of these results to wave-like phenomena in synergetic systems which exhibit spatial symmetries, and to structures in largely extended systems. Finally, we present some ideas of pattern recognition combined with a realization of an associative memory by a synergetic computer, concepts which have been developed quite recently [6].

ON THE SLAVING PRINCIPLE

In the following we present an overview of mathematical methods of synergetics applied to systems where the details of the subsystems as well as their interactions are completely known. Here complete knowledge means that we know all the variables of the subsystems that we denote by U_i (i= 1, 2 ...), and the evolution of the single subsystems through their interactions. If we put together all the variables into a state vector $\underset{\sim}{U}$ that is an element of a state space Γ,

$$\underset{\sim}{U} = (U_1, U_2, \ldots) \quad . \tag{2.1}$$

Its time evolution is typically governed by an equation of motion of the following general type

$$\dot{\underset{\sim}{U}} = \underset{\sim}{N}(\underset{\sim}{U},\nabla,\{\sigma\}) + \underset{\sim}{F} \quad . \tag{2.2}$$

Here $\underset{\sim}{N}$ denotes a nonlinear vector field depending on $\underset{\sim}{U}$, which may additionally depend on spatial derivatives symbolized through the ∇-symbol as well as a set of external (contol) parameters which have been abbreviated by $\{\sigma\}$. Symbolically, $\tilde{F}$ summarizes the presence of fluctuations which are considered as a result of the residual action of the already eliminated microscopic degrees of freedom and/or of not completely controlable external conditions. We note that this typical form of the evolution equations (2.2) can be modified in various ways: We may include time-delay effects and non-local interactions in space, or we can formulate equations for the description of time-discrete processes, etc.

We shall analyze the equations (2.2) to some extent. Generally, we are unable - at least by analytical methods - to describe the complete global behavior of the solutions of (2.2) even if we confine ourselves to the purely deterministic part represented by $\tilde{N}$. We therefore concentrate to more simple subregions of the state space Γ which become of fundamental importance for an understanding of the system's long term behavior: We consider stationary states. The most simple stationary state which we may expect as a solution of (2.2) describes a situation homogeneous in time and space. This state is usually referred to as the thermodynamic solution branch of the system [14]. However, we may even consider stationary states which are not homogeneous in space or in time (periodic and quasi-periodic states) as a basic solution [2]. For the moment being, however, we confine ourselves to the most simple stationary state which is homogeneous in time and space. We construct this state as a solution of

$$\dot{\tilde{U}} = 0 \quad \text{--->} \quad \tilde{U} = \tilde{U}_o \quad . \tag{2.3}$$

We now consider the stability of the state (2.3). When we change the external conditions by varying the control parameters, the stationary state $\tilde{U}_o$ may lose its stability. Mathematically, this means that we are looking for instability regions in the parameter space spanned by $\{\sigma\}$. A general method of achieving this, substantiated by the Hartman-Grobman theorem, is provided by linear stability analysis: We consider the behavior of small deviations $\tilde{q}$ from the stationary state $\tilde{U}_o$, i.e.

$$\tilde{U} = \tilde{U}_o + \tilde{q} \tag{2.4}$$

and confine ourselves to the linearized deterministic version of (2.2). We then immediately find the equation of motion for $\tilde{q}$:

$$\dot{\tilde{q}} = L\tilde{q} + O(\|\tilde{q}\|^2) \quad . \tag{2.5}$$

The elements of the linear matrix L which still depend on the control parameters $\{\sigma\}$ are given by

$$L = (L_{ik}) \quad \text{and} \quad L_{ik} = \partial N_i / \partial U_k |_{\tilde{U}=\tilde{U}_o} \quad . \tag{2.6}$$

Obviously, the matrix L is a function of the stationary state $\tilde{U}_o$ and may depend on gradients of $\tilde{U}$. In cases of a time dependent reference state $\tilde{U}_o$, L also becomes time dependent. In the situation considered here, there exists a well-known method to solve (2.5). The ansatz

$$\tilde{q} = \tilde{q}_o \exp(\lambda t) \tag{2.7}$$

transforms (2.5) into a linear algebraic eigenvalue problem from which we obtain the set of eigenvalues λ_i and the corresponding eigenvectors, which we denote by $\tilde{v}_i$. We assume that they form a complete set. Instabilities in the space of the external parameters are then indicated by the condition

$$\mathrm{Re}\ [\lambda_i(\{\sigma\})] = 0 \ . \tag{2.8}$$

We summarize the information obtained from a linear stability analysis of the basic solution $\underset{\sim}{U}_o$. Firstly, we find the regions of instability in the space of the external parameters. Secondly, we obtain the eigenvectors $\underset{\sim}{v}_i$ which we interpret as collective modes of the system and finally, we know locally the directions in Γ along which the stationary state becomes unstable.

The next step a systematic treatment needs is a non-linear analysis of (2.2). We note that the solutions of the full non-linear equations can be written in the form

$$\underset{\sim}{U} = \underset{\sim}{U}_o + \underset{\sim}{q} \quad \text{and} \quad \underset{\sim}{q} = \sum_i \xi_i(t)\, \underset{\sim}{v}_i \ . \tag{2.9}$$

We now can transform the original equation (2.2) into equations for the amplitudes $\xi_i(t)$ of the collective modes:

$$\dot{\xi}_i(t) = \Lambda_{ik}\xi_k(t) + H_i(\{\xi_i\}) + \tilde{F}_i \ . \tag{2.10}$$

Here $\Lambda = (\Lambda_{ik})$ is a diagonal matrix, or is at least of the Jordan canonical form, and summation is understood over dummy indices. H_i contains all nonlinear terms, and $\tilde{F}_i$ denotes the correspondingly transformed fluctuating forces. We emphasize that the set of equations (2.10) is still exact.

We now arrive at the central step, which results in the formulation and application of the slaving principle. As mentioned previously, slaving means that in the vicinity of a critical region in the space of external parameters only a few modes, the so-called order parameters, dominate the behavior of the complex system on macroscopic scales. Obviously, the linear stability analysis indicates which amplitudes of collective modes will finally become the order parameters. Indeed, the linear movement yields a separation in the time scales: The modes which become unstable will move on a very slow time scale, τ_u, because of $|\mathrm{Re}\lambda_u| = 1/\tau_u$, where we have identified the index i with u to exhibit unstable behavior. In the vicinity of the critical point, τ_u becomes extremely large. On the other hand, the modes still remaining stable have a comparably short characteristic time scale which we denote by τ_s. Therefore in the vicinity of an instability region we observe the following hierarchy in the time scales

$$\tau_u \gg \tau_s \ . \tag{2.11}$$

Taking into account the interaction of stable and unstable modes, we expect that in the long time behavior of the system the stable variables behave in a way that is completely determined by the unstable modes alone: the order parameters. This observation may be cast into mathematical terms. We split the set of $\{\xi_i\}$ into the set of modes which become unstable and combine it to a vector $\underset{\sim}{u}$ and the remaining set of stable modes which we collect into the vector $\underset{\sim}{s}$. Accordingly, we split the set of equations (2.10) into

$$\dot{\underset{\sim}{u}} = \Lambda_u \underset{\sim}{u} + \underset{\sim}{Q}(\underset{\sim}{u},\underset{\sim}{s}) + \underset{\sim}{F}_u \ , \tag{2.12}$$

$$\dot{\underset{\sim}{s}} = \Lambda_s \underset{\sim}{s} + \underset{\sim}{P}(\underset{\sim}{u},\underset{\sim}{s}) + \underset{\sim}{F}_s \ . \tag{2.13}$$

Slaving of the stable modes through the order parameters now can be expressed through the formula

$$\tilde{s}(t) = \tilde{s}(\tilde{u}(t),t) \ . \tag{2.14}$$

Equation (2.14) describes the fact that the values of the stable modes are completely determined by the instantaneous values of the order parameters u and can be considered as a generalization of the center manifold [2,7,8]. With the result (2.14), the complete set of equations (2.12) and (2.13) can be drastically simplified. In fact we can use this result to eliminate the stable modes from the equations for the order parameters and arrive at the equation

$$\dot{\tilde{u}} = \Lambda_u \tilde{u} + \tilde{Q}(\tilde{u},\tilde{s}(\tilde{u},t)) + \tilde{F}_u \ , \tag{2.15}$$

which is the final order parameter equation. We observe that the idea of slaving generally leads to a drastic reduction in the degrees of freedom of the system and yields a complete description of the system on macroscopic scales. The next step, then, consists in the solution of the order parameter equation (2.15) and in the identification of the macroscopically evolving ordered states which correspond to definite patterns or functioning of the system. The central problem which still remains to be solved resides in the explicit construction of (2.14).

In the construction of (2.14), presupposed by (2.11), our interest is devoted to the long term behavior of the slaved modes. We can formally integrate (2.13) in the following way:

$$\tilde{s} = \int_{-\infty}^{t} \exp[\Lambda_s(t-\tau)]\{\tilde{P}(\tilde{u},\tilde{s}) + \tilde{F}_s\}_\tau d\tau \ . \tag{2.16}$$

By means of (2.16), we define the operator $(d/dt - \Lambda_s)^{-1}$:

$$\tilde{s} = (d/dt - \Lambda_s)^{-1} \ [\tilde{P}(\tilde{u},\tilde{s}) + \tilde{F}_s] \ . \tag{2.17}$$

Here we mainly consider the purely deterministic case where $\tilde{F}_s$ vanishes (more general situations are given in [2,7,8], where the method has been extented especially to the stochastic case). To proceed further we introduce the operators

$$(d/dt) = \tilde{Q}\partial/\partial\tilde{u} \tag{2.18}$$

and

$$(d/dt - \Lambda_s)^{-1}_{(o)}\tilde{P} = \int_{-\infty}^{t} \exp[\Lambda_s(t-\tau)] \ \tilde{P}(\tilde{u},\tilde{s})_t d\tau \ . \tag{2.19}$$

A partial integration in (2.16) can now be performed using the operators defined above and expressed through the operator identity

$$\begin{aligned}(d/dt - \Lambda_s)^{-1} &= (d/dt - \Lambda_s)^{1}_{(o)} \\ &- (d/dt - \Lambda_s)^{-1} \ (d/dt) \ (d/dt - \Lambda_s)^{-1}_{(o)} \ .\end{aligned} \tag{2.20}$$

We now apply (2.20) to develop a systematic iteration scheme to construct (2.14). We use the ansatz

$$\underset{\sim}{s} = \sum_{n=2}^{N} \underset{\sim}{C}^{(n)} \quad , \tag{2.21}$$

where $\underset{\sim}{C}^{(n)}$ is precisely of order n in $\underset{\sim}{u}$. Correspondingly, we define $\underset{\sim}{P}^{(n)}$ and $(d/dt)^{(n)}$ (compare (2.13) and (2.18)). Inserting (2.20) and (2.21) into (2.16) we obtain the solution in the form

$$\underset{\sim}{C}^{(n)} = \sum_{m=0}^{n-2} (d/dt - \Lambda_s)^{-1}_{(m)} \underset{\sim}{P}^{(n-m)} \quad , \tag{2.22}$$

where we have defined

$$(d/dt - \Lambda_s)^{-1}_{(m)} = (d/dt - \Lambda_s)^{-1}_{(o)} \sum \prod_i [(-d/dt)^{(i)} (d/dt - \Lambda_s)^{-1}_{(o)}] \tag{2.23}$$

and the product over i has to be taken in such a way that $i \geq 1$ and $\sum i = m$. The sum indicates summing over all different products. Equation (2.22) provides us with a systematic method for the construction of (2.14) and has, as already mentioned, been generalized in various ways. Especially we can take into account fluctuations [2,8]. Furthermore, we need not restrict ourselves to the stationary time-independent reference states $\underset{\sim}{U}_o$. Finally, we may also take into account, in largely extended systems, slow spatial variations by including the effects of finite bandwidths [9].

Here we have presented, in an elementary way, a mathematical method introduced by the idea of slaving and order parameters that applies when all the details about the subsystems and their interaction are known. We have discovered a drastic reduction in the degrees of freedom by restricting ourselves to the order parameters which are exclusively responsible for the macroscopically evolving structures.

SOME RECENT APPLICATIONS IN PATTERN FORMATION

Recent applications of the ideas of slaving have especially been worked out for systems which exhibit symmetries (for a review see[10]). Clearly, various forms of spatial patterns which arise as a result of spontaneuos self-organization processes are related to the geometry of the system. Symmetries manifest themselves mathematically in the structure of the order parameter equations. It turns out that it becomes possible to derive normal forms of the order parameter equations which classify instabilities in the presence of symmetry. Thus, symmetry considerations play an important role in the investigation of the order parameter equations when they are present.

Examples are provided by the oscillatory instability in systems with SO(2) symmetry. Here the normal form of the order parameter equation is given by [10]

$$d_t \, \xi_u(t) = (\varepsilon + i\omega_0) \, \xi_u(t) - \xi_u(t) \, h(|\xi_u(t)|^2) \quad . \tag{3.1}$$

where $h(0) = 0$ but is otherwise an arbitrary function and may be approximated by $a|\xi_u(t)|^2$ close to instability, and ε is a function of the control parameter. The corresponding state vector $\underset{\sim}{U}$ describes a rotating wave. This form has been applied to the instability of the Taylor vortex flow in the idealized case of infinitely long cylinders and corresponds there to the oscillatory instability which is experimentally observed in the so-called wavy vortex flow.

Another example for the application is provided by the baroclinic instability that seems to be the most important instability responsible for the formation of coherent atmospheric patterns dominating the weather activities in the middle latitudes. On the basis of the two layer model introduced by Phillips [11] the geometry of the middle latitudes is approximated by a rotating rectangular channel. Using appropriate boundary conditions this system also shows SO(2) symmetry. Again normal forms of the order parameter equation can be given which correspond to a simultaneous destabilisation of several modes when control parameters are varied. A competition of modes then can again lead to the rotating wave solution discussed above. States consisting in the superposition of two different rotating waves as well as quasiperiodic behavior and frequency-locked states have been found. In a model of four competing modes a transition from quasi-periodic to chaotic behavior was predicted [12].

Waves have furthermore been discussed in systems with O(2) and O(3) symmetry. The method recently has also been applied to waves with cellular structures in largely extended media [13].

PATTERN RECOGNITION AND THE SYNERGETIC COMPUTER

The demonstration of a successful macroscopic description of complex systems by order parameters and the close relation between the order parameter equations and normal forms have supported the development of phenomenological and macroscopic approaches [4,6]. These were developed for a treatment of complex systems that are built of subsystems that are of a complex structure by themselves. This means that we do not know all the details of their behavior as well as of their interactions. To give an example of these macroscopic approaches, still resting on the order parameter concept, we shall discuss aspects of a synergetic computer and its relation to brain functioning.

In trying to understand the functioning of the human brain one powerful method consists in the aim to construct a machine that shows properties and abilities similar to those of the brain. As it is well known nowadays, sequential data processing of a usual computer cannot be a candidate that can become able to simulate the behavior of the human brain. Indeed, the strategies of the human brain are quite different from the method of recognizing for instance a picture or a melody step by step. The strategy seems to be that patterns are identified as a whole in contrast to the method by sequentially combining elements to the picture.

In this situation the macroscopic description of complex systems and the order parameter concept open quite new perspectives that have been applied to the so-called synergetic computer. To test the abilities and the functioning of a synergetic computer the problem of pattern recognition was used [6]. We shall explain the fundamental ideas of the functioning of a synergetic computer first by using the simple example of the convection instability.

In the most simple (idealized) case of the convection instability in a rectangular geometry we have a pure role pattern and two different realizations of this streaming pattern of the fluid which both are equally valued. This can be seen immediately from the corresponding order parameter equation that reads (cf. [2])

$$\dot{\xi}_u = \varepsilon\xi_u - \xi_u^3 \quad . \tag{4.1}$$

The result of the theory can be understood from the discussion of the overdamped motion of a particle in a symmetric double well potential. The valleys of the curve can be interpreted as the stable positions of a moving ball where it comes to rest. Similarly a maximum of the curve characterizes an unstable situation. When we now attribute the valleys to the two available streaming patterns discussed above, we see that both are of equal rank but only one can actually be realized.

Following the ideas of pattern recognition, we can identify these two patterns of motion with so-called prototype patterns which the fluid can "recognize". How can this system now recognize an offered pattern? The offered pattern is usually disturbed and can be incomplete. We can achieve this situation by stirring the fluid in the layer. Then a number of the collective modes become excited. But they all die out and the final long term motion of the fluid is represented again through our ball in a double well potential. And our ball will move to the minimum which is closest to its initial position. During this movement the pattern is reconstructed i.e. the pattern offered is recognized.

There are several comments in order: We have presented our idea by means of a very simple example. But this proposal can be generalized (see below) and a complete theory of this process has been developed and successfully applied to the problem of recognition of faces and other objects. An important property of this concept is that our system acts as an associative memory: incomplete and disturbed patterns that are offered are completed and corrected and thus brought in accordance with the originally stored patterns ("prototype patterns"). If the prototype pattern consists of a person's face and name, and the offered pattern is only the face, the synergetic computer acting as an associative memory provides us with the name of the person, i.e. the computer has recognized him or her.

It is important to mention that the dynamics of the recognition process is governed by the order parameters which characterize the ordered states of the system. These ordered states are, as we have seen, attributed to the prototype patterns.

We shall now explain the model of a synergetic computer and use the problem of pattern recognition. As a pattern we consider two-dimensional pictures, photographs etc. We digitalize these objects by a grid and enumerate the cells of the grid as components of a high-dimensional vector. The components are taken proportional to the grey values in each cell. The form of this vector is given by

$$\underset{\sim}{v} = (v_1, v_2, \ldots, v_n) \tag{4.2}$$

as the representative of a prototype pattern. In applications we have to consider several prototye patterns and we have to distinguisch these by an additional index, say k

$$\underset{\sim}{v}^{(k)} \quad . \tag{4.3}$$

The basic idea consists in the construction of a complex system which is able to attribute to a given pattern, represented by a vector $\underset{\sim}{q}(0)$, the corresponding prototype pattern. This can be achieved by subjecting the vector $\underset{\sim}{q}$ to the following dynamics

$$dq_j/dt = -\partial V/\partial q_j \quad , \tag{4.3}$$

where j enumerates the components of $\underset{\sim}{q}$. Here the potential V consists of three parts:

$$V = V_1 + V_2 + V_3 \quad . \tag{4.4}$$

The first part is connected to the so-called learning matrix [6]

$$V_1 = 1/2 \sum_k \lambda_k (\underset{\sim}{v}^{(k)} \underset{\sim}{q})^2 \tag{4.5}$$

and supports that the offered pattern $\underset{\sim}{q}(0)$ evolves towards the subspace that is spanned by the prototype patterns. (Here for the case of simplicity we have assumed that the different prototype patterns are orthogonal to each other. This restriction can be removed by using adjoint prototype patterns in the case where they are not orthogonal. So in general one only needs linear independence between the prototype patterns.) The second part V_2

$$V_2 = B/4 \sum_{k \neq k'} (\underset{\sim}{v}^{(k)} \underset{\sim}{q})^2 (\underset{\sim}{v}^{(k')} \underset{\sim}{q})^2 \tag{4.6}$$

(B is a constant) discriminates between the different patterns. Finally V_3 guarantees global stabilization of the system:

$$V_3 = C/4 \left(\sum_k (\underset{\sim}{v}^{(k)} \underset{\sim}{q})^2\right)^2 \quad . \tag{4.7}$$

Equations (4.3) have been implemented on a serial computer and applied to recognize faces [6]. It turned out that the computer is able to single out persons from groups, can recognize deformed and rotated offered patterns, has the ability of learning, and can model the oscillatory behavior of the human brain in the recognition of ambiguous patterns. The chosen dynamics models a massiv parallel working computing device.

CONCLUSIONS

We have demonstrated the usefulness of the order parameter concept in connection with the slaving principle when pattern formation and pattern recognition are treated. Especially the "top down" approach in analyzing brain functioning which is provided by the synergetic computer seems a promising ansatz to simulate strategies of human recognition from a completely new point of view.

REFERENCES

[1] H. Haken: "Synergetics. An Introduction" 3rd. edition, Springer, Berlin, Heidelberg 1983

[2] H. Haken: "Advanced Synergetics", Springer, Berlin, Heidelberg 1983

[3] E. Basar, H. Flohr, H. Haken, A.J. Mandell (Eds.): "Synergetics of the Brain", Springer, Berlin, Heidelberg 1984

[4] A. Wunderlin, H. Haken: in "Synergetics - From Microscopic to Macroscopic Order", Ed. E. Frehland, Springer, Berlin, Heidelberg 1982

[5] W. Weidlich, G. Haag: "Concepts and Models of Quantitative Sociology" Springer, Berlin, Heidelberg 1982

[6] H. Haken: Pattern Formation and Pattern Recognition - An Attempt at a Synthesis, in "Pattern Formation by Dynamical Systems and Pattern Recognition", ed. by H. Haken, Springer, Berlin, Heidelberg (1979)

H. Haken: Computers for Pattern Recognition and Associative Memory, in "Computational Systems - Natural and Artificial", ed. by H. Haken, Springer, Berlin, Heidelberg 1987
A. Fuchs and H. Haken: Biol. Cybern. 60, 17, (1988); Biol. Cybern. 60, 107, (1988)
H. Haken: Synergetics in Pattern Formation and Associative Action, in "Neural and Synergetic Computers", ed. by H. Haken, Springer Berlin, Heidelberg 1988

[7] A. Wunderlin, H. Haken: Z. Phys. B44, 135, (1981)

[8] H. Haken, A. Wunderlin: Z. Phys. B47, 179, (1982)

[9] M. Bestehorn: Dissertation Universität Stuttgart (1988)

[10] M. Bestehorn, R. Friedrich, A. Fuchs, H. Haken, A. Kuhn, A. Wunderlin: "Synergetics Applied to Pattern Formation and Pattern Recognition", in: "Optimal Structures in Heterogeneous Reaction Systems", Springer Series in Synergetics vol. 44 (ed. P.J. Plath) Springer-Verlag Berlin (1989)

[11] N.A. Phillips: J. Meteor. 8, 391, (1951); Tellus 6, 273, (1954)

[12] H. Haken, W. Weimer: Z. Geomorph. N. F., Suppl.-Bd. 67, 103, (1988)

[13] R. Friedrich, H. Haken: Phys. Rev. A34, 2100, (1986);
R. Friedrich, H. Haken: in "The Physics of Structure Formation", edited by W. Güttinger and G. Dangelmayr, Springer Berlin, Heidelberg 1987
V. Steinberg, E. Moses, J. Fineberg: Proceedings of the International Conference on "The Physics of Chaos and Systems Far From Equilibrium", Nucl. Phys. B2, 109, (1987);
P. Kolodner, A. Passner, H. L. Williams, C. M. Surko: Proceedings of the International Conference on "The Physics of Chaos and Systems Far From Equilibrium", Nucl. Phys. B2, 97, (1987)
M. Bestehorn, R. Friedrich, and H. Haken: Z. Phys. B72, 265, (1988)
M. Bestehorn, R. Friedrich, and H. Haken: Z. Phys. B, to appear

[14] P. Glansdorff, I. Prigogine: Thermodynamic Theory of Structure, Stability and Fluctuations" Wiley, New York (1971)
G.Nicolis, I. Prigogine: "Self-Organization in Non-Equilibrium Systems" Wiley, New York (1977)

COUPLED MAP LATTICES: ABSTRACT DYNAMICS AND MODELS FOR PHYSICAL SYSTEMS

Raymond Kapral

Chemical Physics Theory Group
Department of Chemistry
University of Toronto
Toronto, Ontario M5S 1A1, Canada

INTRODUCTION

Complicated spatio-temporal structures can arise when large numbers of simple dynamical elements are coupled. There are many physically interesting systems that fall into this category. Numerous examples can be found in biology where self-organization occurs at the cellular level, the brain where interactions among neurons are responsible for its activity, the heart where patterned excitation leads to normal rhythms and the converse to fibrillation. One can include the equations of continuum hydrodynamics and reaction-diffusion equations in this category since they can be considered to arise from a coupling among local fluid elements. A range of descriptions has been used to study the dynamical behavior of such systems.

In this paper a class of discrete models, coupled map lattices (CML), which possess a rich spatio-temporal bifurcation structure is studied.[1-4] CML models are constructed by supposing that a set of local discrete-time dynamical elements (maps) occupy the sites of some lattice and interact with each other. More specifically, if $\mathbf{x}(\mathbf{i},t)$ is the vector of dynamical variables at site $\mathbf{i}=(i_1,i_2,\cdots,i_d)$ on a d-dimensional lattice at time t, the time and space evolution of the system is described by equations of the form

$$\mathbf{x}(\mathbf{i},t+1)=\mathbf{f}(\mathbf{x}(\mathbf{i},t))+\mathbf{C}(\mathbf{x}(\mathbf{j},t))\,, \tag{1.1}$$

where $\mathbf{f}$ is the (nonlinear) local map specifying the site dynamics in the absence of coupling and $\mathbf{C}$ is a function that specifies the coupling among maps. The argument of $\mathbf{C}$ is $\mathbf{x}(\mathbf{j},t)$ where $\mathbf{j}\in\mathcal{N}$ and $\mathcal{N}$ is some neighborhood of $\mathbf{i}$. The coupling function may be linear or nonlinear.

Models like this are used every time a partial differential equation (PDE) is solved on a computer; for instance, the simple explicit Euler scheme discretizes space and time and the resulting equation can be cast into the form of (1.1). Naturally the aim in such methods is to obtain a faithful solution to the PDE, and space and time steps must be sufficiently

Self-Organization, Emerging Properties, and Learning
Edited by A. Babloyantz, Plenum Press, New York, 1991

small to acomplish this. In contrast to this we view the CML as a dynamical system in its own right and as such it is interesting to study its spatio-temporal bifurcation structure. CMLs have a number of advantages (or at least differences) when compared to simple cellular automata.[5] Obviously for the CML while space and time are discrete the system dynamical variables can take on a continuum of values. In cellular automaton models the dynamical variables are also discrete and time evolution occurs by a simple updating rule. The fact that the **f** and **C** functions that determine the CML dynamics are usually simple analytic functions makes bifurcation analysis possible; in addition, these functions can be tuned continuously giving rise to changes in the dynamical behavior analgous to that produced by rule changes for cellular automata.

It is thus interesting to examine the properties of CMLs and to look for parallels in their behavior that are similar to those of physical systems whose accurate description may be much more complex. However, there is an additional way in which such models can be used to study physical systems whose most natural representation is in terms of PDEs. We shall show that it is possible to construct CMLs which can reproduce gross features of the bifurcation structure without going to the extreme limit of very small space and time steps. Viewed in this way CMLs can provide computationally efficient models that capture the robust aspects of the complicated spatio-temporal behavior of real systems.

There are two main sections to this paper. In Sec. 2 we briefly examine some general aspects of the bifurcation structure of simple CMLs in order to illustrate the range of behavior they display. In Sec. 3 we focus on specific physical systems, in particular bistable systems and oscillatory chemical systems, in order to show how CMLs may be constructed to describe the observed phenomena.

BIFURCATION SCENARIOS

Period doubling

As an example consider the following coupled quadratic map lattice with nearest-neighbor coupling on a simple cubic lattice:

$$x(\mathbf{i},t+1) = \lambda x(\mathbf{i},t)[1-x(\mathbf{i},t)] + \gamma \sum_{\mathbf{j}\in\mathcal{N}} [x(\mathbf{j},t) - x(\mathbf{i},t)] \,. \tag{2.1}$$

Clearly λ governs the bifurcation structure of a local map while γ determines the strength of the coupling among maps. One of the most straightforward aspects of the bifurcation structure that can be studied is the stability of the spatially homogeneous solutions to inhomogeneous perturbations. Obviously the spatially homogeneous fixed points are identical to those of an isolated map since the coupling term vanishes in this case; however, their stability does depend on the coupling strength and this has some interesting features which we now describe.

For $\gamma = 0$ we have an array of isolated quadratic maps which undergo the well-known period-doubling bifurcation route to chaos as λ is tuned between $\lambda_0 = 3$ where a nontrivial period 1 solution is born and λ_∞ where the period-doubling sequence accumulates. Thanks to Feigenbaum[6] we know that the sequence of bifurcation points λ_n corresponding to period 2^n orbits scale as $\lambda_\infty - \lambda_n \sim \delta^{-n}$ where $\delta = 4.669\cdots$ for any map with a locally quadratic extremum. For the CML the stability of these period-doubled orbits depends on the coupling strength and this dependence scales in a universal way. The origin of this scaling behavior is easy to see if (2.1) is Fourier transformed to obtain

$$\xi(\mathbf{k},t+1) = \lambda\xi(\mathbf{k},t) - \lambda\sum_{\mathbf{l}}\xi(\mathbf{k}-\mathbf{l},t)\xi(\mathbf{l},t) + \gamma h(\mathbf{k})\xi(\mathbf{k},t)\,, \tag{2.2}$$

where

$$\xi(\mathbf{k},t) = N^{-d}\sum_{\mathbf{j}} e^{-2\pi\mathbf{k}\cdot\mathbf{j}/N}x(\mathbf{j},t)\,, \tag{2.3}$$

and for a two dimensional system ($d = 2$)

$$h(\mathbf{k}) = 4\{cos[\pi(k_1+k_2)/N]cos[\pi(k_1-k_2)/N] - 1\}\,, \tag{2.4}$$

with similar definitions for other dimensions. It is now straightforward to carry out a linear stability analysis of the period-doubled homogeneous solutions x^*_β, $\beta = 1,\cdots,2^n$ to inhomogeneous perturbations. Rather than seek the bifurcation points, the scaling structure is most clearly revealed in the structure of the superstable curves, the curves in the $\lambda\gamma$-plane where the spatially homogeneous period-doubled orbits are most stable. These curves are given by the condition[1]

$$\prod_{\beta=1}^{2^n}\{\lambda(1-2x^*_\beta) + \gamma h(\mathbf{k})\} = 0\,. \tag{2.5}$$

Perhaps the most interesting feature of these curves is the fact that the orbit scaling of an isolated quadratic map, which is governed by Feigenbaum's α, $\alpha = -2.5029\cdots$, also determines the γ scaling in the $\lambda\gamma$-plane. Thus there is a direct relation between the scaling properties of an isolated quadratic map and the scaling properties of the CML for this period doubling sequence.

This connection can be seen in the following way: For a superstable period 2^n orbit one element of the orbit lies at $x^*_1 = 1/2$ while the orbit element closest to it with index $\beta = 2^{n-1} - 1$ is a distance d_n away. For an isolated map it is known that d_n scales as $d_n \sim \alpha^{-n}$. So letting $x^*_\beta = 1/2 + d_n$ for this β value in (2.5) we have the superstable curve for this fixed point

$$\lambda^s_n 2d_n + \gamma h(\mathbf{k}) = 0\,, \tag{2.6}$$

where λ^s_n is the superstable value of λ for the period 2^n orbit. Given the scaling relation for d_n we have $\gamma_n \sim \alpha^{-n}$. A similar analysis can be carried out for other families of period-doubled orbits.

It should be noted that the scaling relation for the coupling strength depends on the general structure of the coupling term used in the CML. For example, Kuznetsov and Pikovsky[7] considered a CML with forward difference diffusive coupling and carried out a renormalization group analysis which showed that the spatial length must be scaled by a factor of $\sqrt{2}$ for each period doubling bifurcation.

Intermittency

CMLs have also been used to provide insight into the nature of spatio-temporal intermittency.[8–11] This line of investigation was initiated by Kaneko.[8] As in the other applications presented here the main advantage of studying CMLs is their simplicity in comparison with PDEs. This simplicity allows one to isolate important features that are responsible for the intermittency phenomenon. The conditions on the CML for the appearance of spatio-temporal intermittency have been discussed earlier.[10] Roughly speaking, an isolated map should support regular dynamics which is stable to small perturbations but may exhibit chaotic transients. If the laminar state of the system is stable to small perturbations but but unstable with respect to finite amplitude perturbations then the system is likely to display spatio-temporal intermittency.

A very simple 1-d CML that possesses these features was studied by Chaté and Manneville.[10] It takes the form of (1.1) with a forward difference diffusive coupling:

$$x(i,t+1) = f(x(i,t)) + \gamma \sum_{j\in\mathcal{N}} [f(x(j,t)) - f(x(i,t))] \,, \tag{2.7}$$

where $j = i+1, i-1$, the neighbors of i, and

$$f(x) = \begin{cases} rx & \text{if } x \in [0,1/2] \\ r(1-x) & \text{if } x \in [1/2,1] \\ x & \text{if } x > 1 \end{cases} \tag{2.8}$$

Notice that this map possesses a continuum of stable fixed points for $x > 1$, while the system evolves to the stable fixed points through a chaotic dynamics on $[0,1]$. Phase points leak from the chaotic region to the fixed point region through the "hole" centered at $x = 1/2$ of size $1-2/r$. Hence the leak rate and thus the length of the chaotic transient can be controlled by r. The coupling parameter γ scales the strength of disorderd initial conditions which act as a perturbation on the isolated site dynamics. It is found that if γ exceeds a threshold value the system evolves to a state with spatio-temporal intermittency rather than to a laminar phase.

The coupled logistic lattice exhibits spatio-temporal intermittency in the vicinity of the spatially homogeneous period-3 orbit.[8,11] It is well known that for an isolated map the tangent bifurcation from chaos to period 3 is characterized by temporal intermittency.[12] For $\lambda \preceq 1+\sqrt{8}$ period 3 is stable but possesses chaotic transients; thus, the isolated map has the features described above. However, this system possesses only three attracting fixed points with additional local dynamics rather than the continuum of fixed points with no additional local dynamics of the Chaté-Manneville model. This may obscure some features of the spatio-temporal intermittency. Nevertheless, the spatially homogeneous laminar state is stable to small perturbations but unstable to large perturbations which give rise to spatio-temporal intermittency in this system.[8,11] For single-site, finite-amplitude perturbations the magnitude of the critical perturbation amplitude has been estimated and provides information about the critical amplitude for disordered initial states.[11]

Due to the simplicity of the CML connections with probabalistic cellular automata can be established[8,10] and conjectures of Pomeau[13] concerning the connection between directed percolation and spatio-temporal intermittency are more easily investigated than for PDE models.

There is a large body of literature dealing with other aspects of the bifurcation structure of CMLs which will not be described here, but Ref. 4 may be consulted for additional references to work in this area.

MODELS FOR PHYSICAL SYSTEMS

In the preceding section CMLs were shown to display interesting bifurcation structure, some of which has counterparts in physical systems. However, many physical systems are described in terms of the PDEs of continuum mechanics and the CMLs described above bear no direct relation to specific PDEs. We now describe how CMLs may be constructed that are capable of reproducing the gross robust features of some spatially distributed systems.

As noted above, at the most trivial level a simple Euler discretization of a PDE constitutes a CML of the dynamics. Of course if space and time increments are made sufficiently small to reproduce all aspects of the solution structure of the PDE then no gain in either conceptual simplicity or computational efficiency is achieved. On the other hand if space and time increments are too large then the solution structure will differ from that of the corresponding PDE. However, it is possible to construct CMLs that do capture the robust features of a given PDE, avoiding at the same time spurious features in the bifurcation structure. We consider two such examples below.

Bistable states

A generic PDE for the desription of phase separation dynamics for a system where the order parameter χ is not conserved is the time dependent Ginzburg-Landau (TDGL) equation[14]

$$\frac{\partial \chi(\mathbf{r},t)}{\partial t} = -\chi^3(\mathbf{r},t) + \epsilon\chi(\mathbf{r},t) + D\nabla^2\chi(\mathbf{r},t) \,. \tag{3.1}$$

The essential features of this equation are a cubic "force" term corresponding to a symmetric quartic "potential" and diffusive coupling of the elments of the order parameter field. One of the most studied processses is the phase separation that occurs from the unstable state resulting from a critical quench.[14] For $\epsilon < 0$ the potential has a single minimum at $\chi = 0$. If ϵ is suddenly changed to $\epsilon > 0$ the deterministic potential now has two minima at $\chi = \pm\sqrt{\epsilon}$ and a maximum at $\chi = 0$. A random distribution of χ values about $\chi = 0$ will evolve into domains of the two stable homogeneous states. The long time evolution of these domains is governed by domain wall curvature effects and is described by the Allen-Cahn theory.[15]

Since the main features necessary for the existence of this phase separation phenomenon are an underlying local bistability and diffusive coupling among local elements it is not difficult to construct a CML with these features.[16] A cubic CML with diffusive coupling behaves like the TDGL on long distance and time scales. Consider the cubic CML

$$\chi(\mathbf{i},t+1) = -\chi^3(\mathbf{i},t) + (\epsilon - 1)\chi(\mathbf{i},t) + \gamma \sum_{\mathbf{j}\in\mathcal{N}} [\chi(\mathbf{j},t) - \chi(\mathbf{i},t)] \,. \tag{3.2}$$

Through a straightforward linear stability analysis of the homogeneous stable states to inhomogeneous perturbations it is not difficult to determine the region in the $\epsilon\gamma$-plane where such phase separation should occur undisturbed by other bifurcations, namely the bifurcation to a short wavelength alternating or checkerboard state which does not exist for the PDE. This analysis can be found in Ref. 16. Simulations of the cubic CML demonstrate that the domain formation and evolution processes mimic those of the TDGL model if the parameters ϵ and γ are suitably chosen. Furthermore analysis of the structure of the nonequilibrium correlation function and its Fourier transform show that these quantities for the CML possess the scaling properties predicted by the Allen-Cahn theory.[16] This provides evidence that the cubic CML in this parameter range is in the same universality class as the TDGL. As a result, with this model it should be possible to investigate a range of scaling behaviors which would be difficult to explore using the full PDE since such studies entail averages over many realizations of the domain growth process over long times. In fact such CMLs have also been used[17] to study the scaling behavior of the phase separation processes in conserved order parameter systems where the underlying PDE is the Cahn-Hilliard[14] equation and the scaling structure has been the subject of some controversy. It is also interesting to inquire into the structure and dynamics of the discrete interface in the CML. Since the maps that describe the stationary inhomogeneous structure are conservative 2-d maps with a chaotic structure, the planar interfacial profile has a rather different origin from that in the corresponding continuum models.[16,18] The generalizations of these considerations to higher dimensions where curvature effects come into play is also an interesting topic.

Oscillating systems

As a second example we examine the response of an oscillatory medium to inhomogeneous perturbations. Consider a spatially distributed oscillatory chemical system which we imagine to be divided into cells corresponding to a local nonequilibrium description of the concentration fields. The perturbations to the system are applied as follows: suppose initially that all local cells are perfectly synchronized and the medium is undergoing a bulk oscillation. At some phase ϕ of this bulk oscillation a randomly chosen fraction p of the cells are perturbed by altering the local concentrations so that in those cells the system is shifted off the limit cycle. In a chemical system this may be accomplished by injecting chemicals locally or exposing local regions of the medium to heat or light pulses. Perturbations of this type can be characterized by three parameters: the phase ϕ at which the perturbation is applied, the strength of the local perturbation, say Δ, and the average fraction p of cells to which the perturbation is applied.

For certain ranges of these parameters the system will relax back to the spatially homogeneous bulk oscillation state but with a phase shift. In this circumstance it is possible to carry out a spatial version[19] of a phase resetting[20] study. The system is perturbed as described above at a phase ϕ of the oscillation and the final phase ϕ' is determined. More specifically, for each realization of the random seeding process the local phase of each cell may be monitored stroboscopically in units of the period of the limit cycle. These local phases may be averaged over the system and over realizations of the initial perturbation process. In the limit of infinite time we obtain the average asymptotic phase $< \phi' >$. The information about the system's response is then contained in the phase transition curve (PTC) which is a plot of the average asymptotic phase $< \phi' >$ versus the initial phase ϕ for various values of the local perturbation strength Δ and seeding probability p.

This is a computationally demanding procedure since the spatially distributed system must be monitored for long times and averages over many realizations of the intial perturbation process must be carried out. Hence, it is useful to attempt to construct a CML that captures the essential aspects of the spatio-temporal dynamics of such systems. We first present some specific results for the Brusselator reaction-diffusion system and then describe how a CML corresponding to this system can be constructed.

The Brusselator reaction-diffusion equations are[21]:

$$\frac{\partial u(\mathbf{r},t)}{\partial t} = A - (B+1)u(\mathbf{r},t) + u^2(\mathbf{r},t)v(\mathbf{r},t) + D_u\nabla^2 u(\mathbf{r},t)$$

$$\frac{\partial v(\mathbf{r},t)}{\partial t} = Bu(\mathbf{r},t) - u^2(\mathbf{r},t)v(\mathbf{r},t) + D_v\nabla^2 v(\mathbf{r},t)\,, \tag{3.3}$$

For $A = 1$ and $B = 3.5$ the spatially homogeneous Brusselator possesses a globally attracting limit cycle solution of the relaxation oscillation type. We consider (3.3) for $D_u = D_v = 10^{-5}$. Perturbations to the Brusselator were carried out as described above with a local perturbation that shifts the concentration of u: $(u_0, v_0) \to (u_0+\Delta u, v_0)$, where (u_0, v_0) is a point on the limit cycle. The results[19] for such a phase resetting study for $\Delta u = 10$ show that for most values of p and ϕ the system relaxes to a homogeneous state and the final average phase can be defined. However, there is a region in the $p\phi$-plane where most realizations of the evolution process do not relax to a homogeneous state: spatio-temporal structures are formed and the final phase cannot be defined.

As a consequence of the above considerations one may now design perturbations to excite spatio-temporal structures in the system. The prescription is as follows: From a knowledge of the PTC of the *homogeneous* system, which is not difficult to construct, one may estimate the local perturbation strength Δ necessary to obtain a type 0 PTC; i.e., a PTC where the ϕ' versus ϕ plot has an average slope of zero. Decreasing the amplitude of Δ will ultimately lead to a type 1 PTC where the average slope of the ϕ' versus ϕ plot is unity.[20] In between the system must exhibit a "singular" response at some ϕ value where the final phase is undefined.[20] Possessing this knowledge, in the spatially distributed system we may fix the local perturbation amplitude Δ so that the homogeneous system exhibits a type 0 response. Then for initial phases near where the homogeneous system has a "singular" response, as the seeding probability p is tuned between zero and one, the spatially distributed system should show a strong response and form spatio-temporal structures. These features have been confirmed for the Brusselator.[19]

In order to investigate these processes for the Brusselator in two spatial dimensions we have constructed a CML that reproduces the gross aspects of the pattern formation and phase dynamics described above.[22] The Brusselator CML may be written in the form

$$\mathbf{c}(\mathbf{i}, t+\tau) = \mathbf{f}_\tau(\mathbf{c}(\mathbf{i},t)) + \gamma \sum_{\mathbf{j}\in\mathcal{N}} [\mathbf{f}_\tau(\mathbf{c}(\mathbf{j},t)) - \mathbf{f}_\tau(\mathbf{c}(\mathbf{i},t))]\,, \tag{3.4}$$

where $\gamma = \tau D/(\Delta q)^2$ with τ and Δq the time and space increments, respectively, and we have taken the diffusion coefficients to be equal, $D_u = D_v = D$. Here the nonlinear mapping function $\mathbf{f}_\tau$ is constructed by integrating the reaction part of the Brusselator reaction-diffusion equation, i.e. (3.3) with $D = 0$, over the time interval τ. This type of procedure was first used by Oono and Puri[23] in a study of phase separation for the conserved and nonconserved order parameter cases. This integration cannot be performed analytically for the Brusselator so it was carried out numerically on a grid of (u, v) values; the table of generated values constitutes the numerical $\mathbf{f}_\tau$ map.

The results of two sample simulations of the CML (3.4) for a 2-d system are shown in Fig. 1. Panel (a) shows a case where the system relaxes to a homogeneous state, while panel (b) is a case where spatial patterns are excited. In this case they consist of a collection of counter-rotating spiral wave pairs and target patterns. This type of response can be understood on the basis of the relaxation character of the oscillator. For

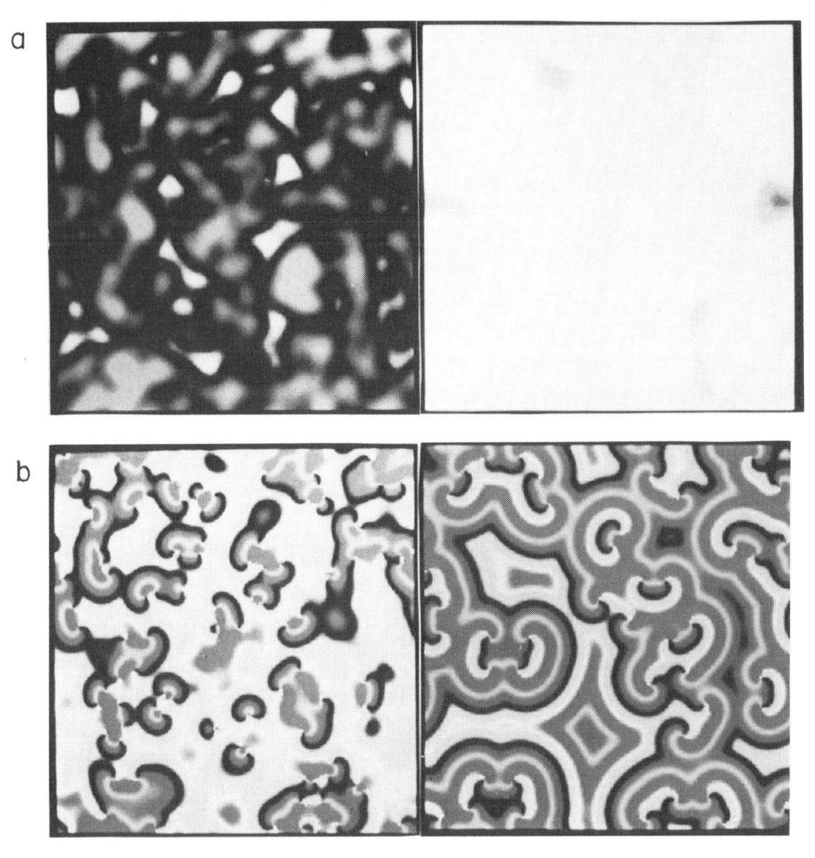

Figure 1. (a) top: the system 5 cycles after application of the perturbation; bottom: the system after 80 cycles. The system evolves to a homogeneous state. Parameter values: $\Delta = 1$, $p = 0.45$ and $\phi = 0.5$. (b) same as (a) but for parameters: $\Delta = 1$, $p = 0.45$ and $\phi = .525$. The system now evolves to a state with spatio-temporal structure. The simulations were carried out on a 200×200 square lattice.[22]

other types of oscillatory systems where the oscillations are harmonic in character, e.g. the complex Ginzburg-Landau model, the spatio-temporal states either fail to form or correspond to phase and amplitude turbulence in the system.[24] Many quantitative aspects of these processes remain to be investigated and CMLs provide a convenient way to carry out such studies.

CONCLUSION

The CMLs described here constitute an interesting class of models for the study of spatio-temporal dynamics. They are much easier to simulate than PDEs and are perhaps more amenable to theoretical analysis. As abstract dynamical models they allow one to investigate some of the rich bifurcation structure generic to spatially distributed systems and to seek underlying general features and scaling behavior giving rise to the phenomena. In addition, CMLs can be constructed that reproduce the gross pattern formation processes described by specific PDEs. In this way properties that are statistical in character, entailing the examination of many realizations of the evolution process, or studies in two or three dimensions may be carried out easily.

ACKNOWLEDGEMENTS

The work described above was carried out in collaboration with I. Waller, G.-L. Oppo, D. Brown, M.-N. Chee and S.G. Whittington, and it is a pleasure for me to thank them for all their contributions. I am especially grateful to M.-N. Chee for providing Fig. 1. This research was supported in part by grants from the Natural Sciences and Engineering Research Council of Canada.

REFERENCES

1. I. Waller and R. Kapral, Phys. Rev. A **30**, 2047 (1984); R. Kapral, Phys. Rev. A **31**, 3868 (1985).
2. K. Kaneko, Prog. Theor. Phys. **72**, 480 (1984); Prog. Theor. Phys. **74**, 1033 (1985).
3. R.J. Deissler, Phys. Lett. A **100**, 451 (1984).
4. For reviews with further references see, J.P. Crutchfield and K. Kaneko, in *Directions in Chaos*, ed. Hao Bai-lin, (World Scientific, Singapore, 1987); K. Kaneko, Physica D **34**, 1 (1989); R. Kapral, M.-N. Chee, S.G. Whittington and G.-L. Oppo, in *Measures of Complexity and Chaos*, ed. N.B. Abraham, A.M. Albano, A. Passamante and P.E. Rapp (Plenum, NY, 1990), pg. 381.
5. See, for instance, *Theory and Applications of Cellular Automata*, ed., S. Wolfram, (World Scientific, Singapore, 1986).
6. M.J. Feigenbaum, J. Stat. Phys. **19**, 25 (1978); **21**, 669 (1979).
7. S.P. Kuznetsov, Radiofizika **29**, 888 (1986); S.P. Kuznetsov and A.S. Pikovsky, Physica D **19**, 384 (1986).
8. K. Kaneko, Prog. Theor. Phys. **74**, 1033 (1985).
9. J.D. Keeler and D. Farmer, Physica D **23**, 413 (1986).
10. H. Chaté and P. Manneville, Physica D **32**, 409 (1988).
11. G.-L. Oppo and R. Kapral, Phys. Rev. A **33**, 4219 (1986).
12. Y. Pomeau and P. Manneville, Comm. Math. Phys. **7**, 189 (1980).
13. Y. Pomeau, Physica D **23**, 3 (1986).
14. J.D. Gunton, M. San Miguel and P. Sahni, in *Phase Transitions and Critical Phenomena*, eds. C. Domb and J.L. Lebowitz, (Academic Press, NY, 1983).
15. S.M. Allen and J.W. Cahn, Acta Metall. **27**, 1085 (1979).
16. R. Kapral and G.-L. Oppo, Physica D **23**, 455 (1986); G.-L. Oppo and R. Kapral, Phys. Rev. A **36**, 5820 (1987).
17. T. M. Rogers, K.R. Elder and R.C. Desai, Phys. Rev. B **37**, 9638 (1988); K.R. Elder, T.M. Rogers and R.C. Desai, Phys. Rev. B **38**, 4725 (1988).
18. S. Puri, R. Kapral and R.C. Desai, *Conserved Order Parameter Dynamics and Chaos in a One-Dimensional Discrete Model*, to be published.
19. M.-N. Chee, R. Kapral and S.G. Whittington, *Inhomogeneous Perturbations and Phase Resetting in Oscillatory Reaction-Diffusion Systems*, J. Chem. Phys., June (1990).
20. A.T. Winfree, *The Geometry of Biological Time*, (Springer, NY, 1980).

21. I. Prigogine and R. Lefever, J. Chem. Phys. **48**, 1695 (1968).

22. M.-N. Chee, R. Kapral and S.G. Whittington, *Phase Resetting Dynamics for a Discrete Reaction-Diffusion Model*, J. Chem. Phys., June (1990).

23. Y. Oono and S. Puri, Phys. Rev. Lett. **58**, 836 (1986).

24. X.-G. Wu, M.-N. Chee and R. Kapral, to be published.

THE SELF-ORGANIZED PHASE ATTRACTIVE DYNAMICS OF COORDINATION

J.A.S. Kelso, G.C. DeGuzman, T. Holroyd

Program in Complex Systems and Brain Sciences
Center for Complex Systems
FAU, Boca Raton, FL 33431

1.0 *Introduction*

One of the most fundamental, but least understood features of living things is the high degree of coordination among the system's many parts. Our aim is to identify the principles that govern how the individual components of a complex, biological system are put together to form recognizable functions. Stable and reproducible functions are observed on many scales of observation in biology and psychology. In the field of neuroscience, they may be exemplified in patterns generated by neural circuitry, in physiology by the beating of the heart, in behavior by actions such as locomotion, speech or reaching for objects. The interacting components that make up these functions are diverse in structure and possess their own dynamics. Our concern here is more in the form that the interactions take than the material composition of the components themselves. In particular, when the elements (e.g. genes, neurons, muscles, joints) interact in such a way as to generate *dynamic patterns*, we speak of *coordination* or *cooperation*.

It is the lawful basis of this coordinated behavior that is our main focus. In Prigogine's terms (this volume) the laws we seek at a chosen level of description deal expressly with *events*, pattern switching (bifurcations), multistability, hysteresis, intermittency and so forth. We may even hope that such events will encapsulate *essential* biological properties like stability (persistence), flexibility, adaptability, anticipation, intentionality, learning and development. Such features render physics, in its present form, incomplete but on the other hand represent a tremendously exciting scientific challenge.

This paper describes some recent experimental and theoretical studies of behavioral patterns and the dynamical laws underlying these patterns. Behavior, largely defined by kinematic measures and relations, is chosen as the window into these dynamics. Our view, however, is that if the appropriate questions are asked, dynamics can be identified on several levels of description -- that the linkage across levels (e.g. between neural and behavioral patterns) is by virtue of shared dynamics, not because any single level has ontological priority over any other. The dynamics that we shall define are for spatiotemporal *patterns*, and hence are abstract and mathematical yet always realized and instantiated by physical structures. Thus, the approach is operational: dynamical laws are formulated for observable variables and predictions are made that can be experimentally tested. On occasion, theoretical models not only cover the empirical facts, but due to their nonlinear character, reveal phenomena that are not easily observed in experiment yet interesting in their own right.

Self-Organization, Emerging Properties, and Learning
Edited by A. Babloyantz, Plenum Press, New York, 1991

In this paper we first (Section 2) briefly mention a) the theoretical approach which draws on the concepts and tools of pattern formation and self-organization in nonequilibrium systems; and b) the experimental model system that motivated the application of these methods. In Sections 3 and 4 we describe new experimental and theoretical results that lead us to propose that *phase attraction* is a crucial feature of cooperative phenomena in biological systems composed of elements whose individual trajectories and intrinsic frequencies are quite variable. The experiments deal with multifrequency processes in which the components synchronize at only certain event times and may not have a well defined frequency relationship. A discrete form of the collective variable dynamics, called a *Phase Attractive Circle Map* accommodates our experimental results and generates new possibilities. In Section 5 we establish the connection between less rigid, more flexible forms of coordination that occur when the individual components are allowed to express their inherent variability, and phase slippage in the map which is due to *intermittency*, a generic feature of dynamical systems near tangent bifurcations. In Section 6 we draw attention to additional theoretical results obtained through numerical experiments that arise due to the presence of phase attraction in the dynamics. We conclude that with phase attraction, complex biological systems can exhibit both stability *and* flexibility whereas without it they cannot. In particular, by living *close* to mode-locked regions, (fixed points of the phase dynamics) but not in them, an optimal form of coordination results which contains the right mix of adaptability and coherence. Also, the system stays in contact with its future by staying near critical points, i.e. it is endowed with a predictive property. The term *anticipatory dynamical system* seems an appropriate characterization of this crucial feature of living things and warrants further study.

2. *Nonequilibrium phase transitions in human behavioral patterns*

Nonequilibrium phase transitions are at the core of pattern formation in physical, chemical and biochemical systems (e.g. Haken, 1977/1983; Nicolis & Prigogine, 1977; 1989). Among the dynamical features observed in such systems are the onset of oscillations, period doubling bifurcations, intermittency and chaos. A central theme of the theory of nonequilibrium phase transitions is the reduction of the number of degrees of freedom that occurs near critical points where patterns form or change spontaneously. Emerging patterns may be characterized by only a few collective variables or order parameters whose dynamics are low dimensional, but nonlinear. Hence, the system may exhibit behavioral complexity, including multiple patterns, multistability even deterministic chaos. The modern theory of nonlinear dissipative dynamical systems therefore embraces both the disorder-order and order-order principles advocated by Schrödinger (1944) in *What is Life* and adds the evolutionary order-to-chaos principle.

In biological systems collective variables defining patterned states are not known *a priori*. Rather they have to be identified and their dynamics studied through a detailed stability analysis. Points of pattern change offer a special entry point for implementing such an approach because they allow: 1) a clear differentiation of one pattern from another; and 2) the study of stability and loss of stability. Around phase transitions or bifurcations phenomenological description turns to prediction; the essential processes governing stability, change and even pattern selection can be found. Nontrivial predictions including enhanced fluctuations and relaxation time of the order parameter may be subjected to experimental test. Also, the control parameter(s) that promote instabilities can be found.

Our work has studied patterns of coordination in humans as a window into principles of biological self-organization. In a quite large number of cases, relative phase, ϕ, was identified as an order parameter or collective variable capturing the ordering relations among the individual components (Jeka & Kelso, 1989; Kelso, 1990; Schöner & Kelso, 1988a/1989 for reviews). Multistability and transitions among phase-locked states were observed at critical values of a continuously changed control parameter, in this case movement frequency. *En route* to these transitions, theoretically predicted features such as enhancement of fluctuations, critical slowing down of the order parameter and the switching time distribution were observed in experiments (for detailed references, see above reviews). Once the pattern dynamics were found by mapping observed patterns onto attractors of the dynamics, it was possible to derive them by cooperative (nonlinear) coupling of the individual oscillatory components thereby effecting a micro- to macro-relation. At bifurcation points, self-organization becomes apparent: different behavioral patterns arise as stable states of the coupled nonlinear dynamics. It is worth emphasizing that the coupling functions are quite unspecific to the patterns of coordination that result.

In our original experiments (Kelso, 1981; 1984) human subjects were instructed to move homologous limbs (e.g. fingers, wrists) rhythmically at a common frequency. Only two forms of temporal patterning were performed stably, in phase (homologous muscles contracting synchronously) and antiphase (homologous muscles contracting in an alternate fashion). A bifurcation was discovered as frequency of movement was increased: the antiphase pattern loses stability and a spontaneous switch to in-phase occurs. As mentioned above, the relative phase, ϕ, between the two actively moving components proves to be a good collective variable. Haken, Kelso & Bunz (1985) determined the dynamics of ϕ from a few basic postulates. The simplest mathematical form is

$$\dot{\phi} = -\frac{\partial V}{\partial \phi} \tag{1}$$

Complying with periodicity and symmetry requirements, the potential

$$V(\phi) = -a\cos(\phi) - b\cos(2\phi) \tag{2}$$

has attractors corresponding to the observed patterns at $\phi=0$ and $\phi=\pi$, and captures the bifurcation or phase transition in that above the critical point (b/a=.25) both patterns are stable (bistability). Below it, only the in-phase mode is stable. After the transition is over, the system stays in the in-phase mode when the direction of the control parameter is reversed (hysteresis). Local measures of the in-phase and anti-phase modes) allow for the easy determination of the a, b parameters in (2), (see Schöner et al. 1986; Scholz et al. 1987).

The collective variable dynamics (1), (2) for ϕ can be derived by nonlinearly coupling the individual components. The latter may be mapped onto limit cycle attractors of the following functional form (whose parameters were again determined by detailed experiments) (Kay, Kelso, Saltzman & Schöner, 1987; Kay, Saltzman & Kelso, in press).

$$\ddot{x}+\alpha\dot{x}+\omega^2 x+\beta\dot{x}^3+\gamma\dot{x}x^2=0 \tag{3}$$

where $\alpha > 0$, $\omega > 0$, $\beta > 0$ and $\gamma > 0$ are model parameters. Using the simplest nonlinear coupling between oscillators of type (4),

$$F(x_1, \dot{x}_1, x_2, \dot{x}_2) = (\dot{x}_1 - \dot{x}_2)\{a' + b'(x_1 - x_2)^2\} \tag{4}$$

Haken et al. (1985) derived a closed form dynamics for the relative phasing patterns and transitions among them. An essential point is that this pattern formation and change occurs through change of a single, non-specific parameter, the oscillation frequency.

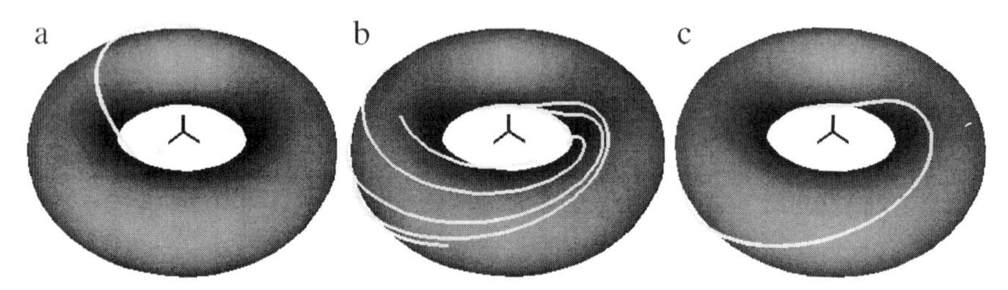

Figure 1 Shows a transition from a stable anti-phase limit cycle (a) to a stable in-phase limit cycle (c) as a is varied from 0.4 to -0.4. Parameter values are: $\omega = 2.0$, $b = 0.0$, (a) $a = 0.4$, (b) & (c) $a = -0.4$.

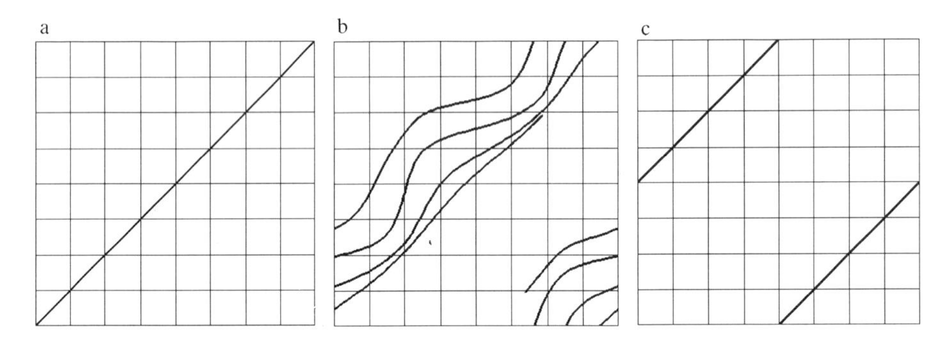

Figure 2 Shows the same transition as in Fig. 1, but only the surface of the torus is shown (see text for details).

A nice way to summarize the entire phenomenon is through the torus plots displayed in figure 1. The state spaces of the two oscillators $(x_1, \dot{x}_1)$ and $(x_2, \dot{x}_2)$ are plotted against each other. For a rational frequency relationship between the two oscillators, in the present case 1:1, the relative phase trajectory is a closed limit cycle and corresponds to a mode-locked state. In fig. 1(a) we show a stable anti-phase limit cycle, transiting (fig. 1b) to a stable in-phase trajectory (fig. 1c). A flat representation of the torus displays the phase of each oscillator in the interval $[0, 2\pi]$ on horizontal and vertical axes. The constant relative phase between the oscillators is reflected by straight lines. In the flat representation shown in fig. 2(a) the phase relation is π, i.e. one oscillator's phase is zero and the other is π. The apparent discontinuity is not real but due to the 2π periodicity of the phase. Thus, when you fall off the top edge of the plane in fig. 2(a) you reappear at the bottom. In figs. 2(b) and 2(c) the transition to the in-phase relationship (zero phase lag between the oscillators) is shown.

The insights gained from detailed theoretical and empirical study of this particular model system have been generalized in a number of ways. On the one hand, a theoretical language and strategy for dynamic patterns has been developed (e.g. Kelso & Schöner, 1987, 1988; Schöner & Kelso, 1988a,b). On the other, numerous examples and analyses of phase transitions and pattern formation have been found in the laboratory, from speech production (Tuller & Kelso, 1990) to action-perception patterns (Kelso, DelColle & Schöner, 1990), to visually coordinated motions between two people (Schmidt, Carello & Turvey, 1990). Extensions of the approach to learning, both theoretical (Schöner & Kelso, 1988b; Schöner, 1989) and empirical (Kelso, 1990; Zanone & Kelso, to appear) have occurred. Posture, discrete and rhythmical behaviors have been accommodated in the same theoretical picture (Kelso & Schöner, 1988; Schöner, in press) and earlier observations of interlimb coordination (Kelso, Southard & Goodman, 1979) modelled. Moreover, remarkable parallels exist between this work on humans and dynamic patterns of neural activity in vertebrate and invertebrate networks. For example, of all the possible neuronal patterns that such networks could generate only a few kinds of spatiotemporal order are actually observed. Such temporal constraints reflect tremendous information compression, sometimes referred to as 'degeneracy in the pattern code' (Kristan, 1980). Interestingly, even *single* neurons have been demonstrated to display many of the features of neuronal and behavioral *patterns*, including multistability, bifurcations and even deterministic chaos (e.g. Matsumoto et al. 1987). Thus, from the point of view of dynamic patterns, there is no ontological priority of one level of observation over another. The language of dynamics and the strategy of identifying order parameters for patterns and their equations of motion apply at several spatiotemporal and temporal scales in a level-independent fashion.

3. *Phase Attractive Dynamics: Relative Coordination*

Phase attraction, a preference for a limited set of phase relations between multiple components, is a strong feature of cooperative phenomena in many biological systems. von Holst (1937/1973) coined the term *absolute coordination* when two or more components oscillate at the same frequency and maintain a fixed-phase relation. The model system described in Sect. 2 rationalizes the phenomenon of absolute coordination in terms of self-organized phase-locked states. When the symmetry of the dynamics (2) is broken, however, it is possible to predict and observe other, perhaps more important, effects as we shall see. In fact, all the interesting phenomena e.g. intentional switching, adjustment to the environment, learning, multicomponent coordination etc. (reviews cited in Section 2) arise when the symmetry of the dynamics is broken or lowered. For example, when human subjects must coordinate hand motion with an external periodic signal (Kelso, DelColle & Schöner, 1990) the symmetry of the relative phase dynamics under the operation $\phi \rightarrow -\phi$ can no longer be assumed. Limiting the expansion of the equation of motion to second order we now find:

$$\dot{\phi} = \Delta\omega - a \sin \phi - 2b \sin 2\phi + \sqrt{Q}\, \xi_t \qquad (5)$$

where the parameters a and b are as in (2) and $\Delta\omega$ (the zero order term) corresponds to the frequency difference between the metronome signal and the limb. For reasons discussed elsewhere (e.g. Schöner, Haken & Kelso, 1986) a Gaussian white noise process, ξ_t, is added as a stochastic force of strength Q. Note that for $\Delta\omega = 0$ the system is identical to the relative phase dynamics for interlimb coordination patterns (eq. 2).

Equation (5) reproduces experimental observations of loss of entrainment (no frequency or phase-locking) and what von Holst (1937/1973) called *relative coordination*. Relative coordination refers to conditions where a tendency for phase attraction persists even though the oscillatory components have different periodicities. A good way to intuit the solutions of (5) is to plot the right hand side of this equation as a function of ϕ for various parameter values. The system has stationary states where $\dot{\phi}$ crosses the ϕ-axis. These are stable states if the slope of $\dot{\phi}$ is negative and are unstable states if the slope is positive at this point. The arrows in figure 3 indicate the direction of the flow. For $\Delta\omega < a$ and $\Delta\omega < b$, two stable states exist, one close to in phase and one close to antiphase (fig. 3a). If, for example, b is decreased, the amount of attraction to these stable states decreases until one (fig. 3b) and ultimately both (fig. 3c) cease to be stationary solutions. When stationary solutions are no longer found, "running" solutions (ϕ ever increasing or decreasing), or, if the periodicity convention to relative phase is applied, "wrapping" solutions occur. There is no longer any phase or frequency-locking, a condition called loss of entrainment.

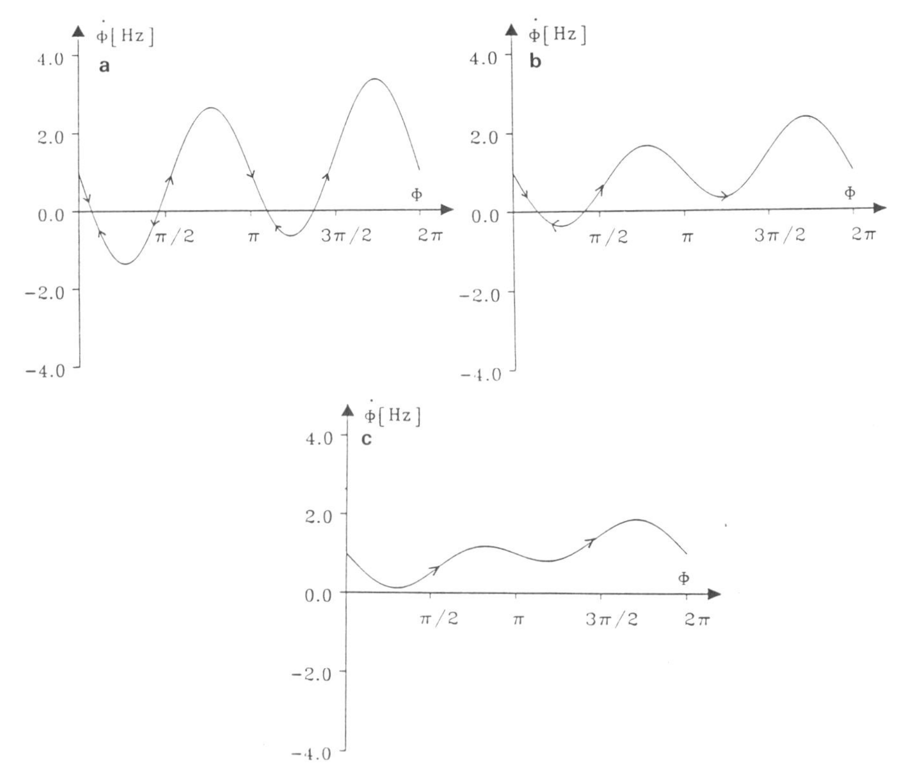

Figure 3 The right hand side of Eq. 5 plotted as a function of relative phase for a = .5Hz., $\Delta\omega$ = 1 Hz and b = 1.0 Hz (left), b = .5 Hz (center) and b = .25 Hz. (right). The arrows indicate the flow of $\dot{\phi}$. Zeros of $\dot{\phi}$ are stationary solutions that are stable if the slope is negative and unstable if the slope is positive at the stationary state (see text for discussion).

One can see intuitively in figure 3 how loss of stability occurs as a transition is approached. The slope of $\dot{\phi}$ becomes flatter and flatter as the critical point is approached, indicative of so-called *critical slowing down*. The corresponding broadening of the attractor lay out corresponds to *enhancement of relative phase fluctuations*. Such theoretically predicted features have been experimentally observed in several experimental systems (e.g. Kelso & Scholz, 1985; Kelso, Scholz & Schöner, 1986; Kelso et al. 1990; Schmidt, Carello & Turvey, 1990; Scholz, Kelso & Schöner, 1987).

The running solutions of figure 3 have a fine structure that we believe is important. Note the system spends more time at those relative phase values at which the force $\dot{\phi}$ is minimal. Numerical simulation of equation (5) in this 'running' parameter régime shows plateaus in the relative phase trajectory that exhibit the tendency to maintain certain preferred phase relationships (see figure 4). This is the phenomenon of *relative coordination* observed first by von Holst (1937/73) in his studies of locomotion in several species. There is attraction to certain phase relations even though the relative phase itself is unstable (see also figure 3(c) where there is phase attraction but the collective variable is not itself a stationary state). Several experimental examples are shown in Kelso et al.'s (1990) studies on synchronizing movement to periodic environmental signals and in Schmidt et al's (1990) work on visual coordination between people. We believe this ordered, but nonstationary behavior to be a crucial feature of neurobiological dynamical systems whose individual component periods and amplitudes are never quite the same. Rigid mode-locking (absolute coordination) which reflects the system's asymptotic approach to a well-defined frequency and phase relation state may well be the exception. The problem of discrepancies between ideal models using oscillators that can "only do it right" all the time versus the real biological system that does it "occasionally" and still function may be more than finding the right oscillator. In Section 5 below we make a connection between *intermittency* in maps, i.e. abrupt slips away from a previously stable fixed point (see fig. 3) and the phase slippage characteristic of biological systems whose intrinsic component frequencies are subject to variation. The idea is that relative coordination and the generic feature of intermittency in dynamical systems near tangent bifurcations have a common origin.

4. *The Phase Attractive Circle Map: Multifrequency coordination*

Biological coordination is often characterized by some invariant temporal relationships which hold only at some points of the components' trajectories. In between such points, the trajectories may adjust themselves both spatially (e.g. characteristic amplitudes) or temporally (e.g. characteristic periods). An example from experimental studies of rhythmic multi-limb coordination in humans illustrates this point (Kelso & Jeka, to appear). Figure 5 represents 4-limb patterns on a 3-D torus for a 'pace' (homologous limbs cycling antiphase, ipsilateral limbs cycling in phase) and a 'trot' (ipsilateral limbs cycling antiphase). An unfolded topologically equivalent 2-D version of the same tori is shown below in Figure 5 for each pattern. Note that the limbs converge only at certain points in their trajectories suggesting that the essential information for coordination is confined or localized to a discrete region in relative phase space. These convergent points have a unique value for each pattern, thereby allowing patterns to be distinguished using a single graphical representation. Thus, not only does relative phase prove to be an adequate collective variable for multicomponent systems (see also Schöner, Jiang & Kelso, 1990) but the discrete estimate of relative phase is arguably a more relevant metric for coordination than the continuous variable (see also Kelso, et al., 1986).

That the individual components may be coupled only at certain points also comes out as a dominant feature of experiments in which the frequency relations between rhythmically moving components are experimentally manipulated. In the first experiments (Kelso & DeGuzman, 1988; DeGuzman & Kelso, 1989/to appear) the frequency and amplitude of one component (left hand) were determined passively by a torque motor while the other component (the right hand) actively maintained a base rhythm. Six different frequency ratios were studied: f(L):f(R)=4:3, 3:2, 5:3, 2:1, 5:2 and 3:1 over long (90 sec.) time series. In the second study (Jiang, Kelso & DeGuzman, to appear) the frequency ratio was scanned in ascending and descending order from just below 1:1 to just above 2:1 and *vice*

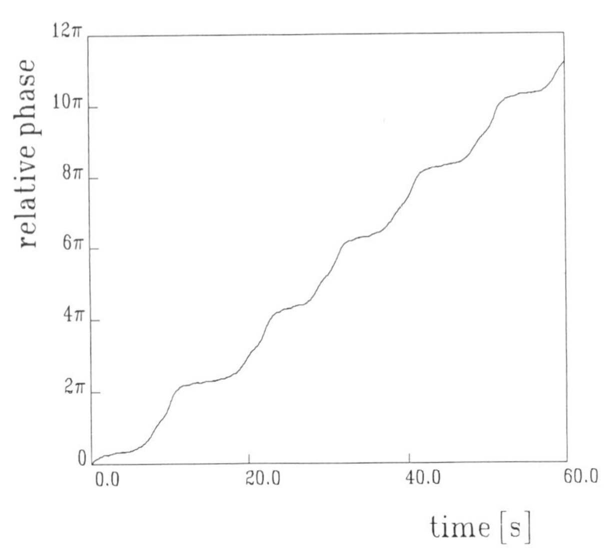

Figure 4 Numerical integration of Eq. (5) at $\Delta\omega = 1.0$ Hz, a = .5 Hz, b = .25 Hz., i.e. in the régime where only "running" solutions exist. Note, however, that there is still a tendency to maintain certain preferred phase relations.

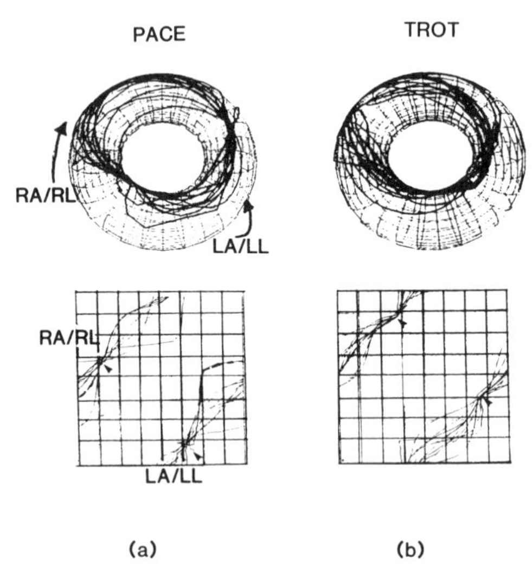

Figure 5 Torus plots of two 4-limb patterns, 'pace' and 'trot'. A centroid is calculated from an average position of all four limbs. Two sets of angular data are then calculated from the centroid for the right arm (RA) vs. the left arm (LA) and for the right leg (RL) vs. the left leg (LL). These two angular sets are plotted orthogonal to one another to produce a torus, RARL on the small circle and LALL on the large circle. Below each torus is an 'unfolded' flat representation with arrows pointing to the coordination points.

versa in steps of .1 Hz. In neither case was the initial phase between the components fixed; in fact, it was arbitrary, yet phase attraction still occurred.

Collectively, the experiments revealed the following features: 1) phase attraction to in-phase and anti-phase occurred even in situations when absolute coordination is not easily attained; 2) multifrequency patterns were differentially stable (as measured by the variability of the frequency-ratio) depending on the frequency ratio (2:1, 3:1, and 3:2 being least variable); 3) short-term spontaneous jumps occurred from less stable to more stable frequency ratios, as revealed through analysis of power spectra; 4) transitions between in-phase and anti-phase modes occurred *within* a frequency ratio (e.g. 2:1); and 5) the amplitude of the driven (LH) component was modulated and occasional extra cycles of LH were inserted apparently to maintain an in-phase relation when the frequency ratio was not 1:1. This modulation arose when the driver was allowed some flexibility (by lowering the torque from the driving motor) indicating that a two-way interaction between the components was present, of sufficient strength that an otherwise perfect harmonic trajectory was modified. These modulated cycles appeared rarely and randomly near 1:1 and more frequently near 2:1 where sometimes even trains of alternate low and high amplitudes were seen (see figure 6). Remember, the modulation appears to be due to the cross-coupling between the components causing a back reaction of the LH on the torque signal.

To characterize coordination in multifrequency situations in which the components synchronize at only certain event times and may not have a well defined frequency relationship, we need a local measure that characterizes time relationships between events. An often quoted measure of coordination is the relative frequency with which these events occur such as the frequency ratio. But frequency related measures are with respect to events in components taken individually, not between them, and are therefore inadequate. A better measure is the *point estimate of relative phase* calculated experimentally as

$$\phi_n = 2\pi \frac{\tau_n}{T_n} \qquad \textbf{(6)}$$

where T_n is the peak-to-peak period of component 1 beginning at time t_o, and τ_n is the interval between that event and the peak onset of component 2.

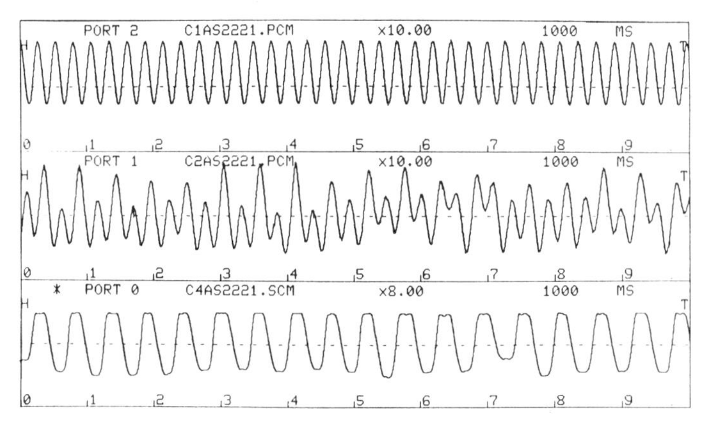

Figure 6 Modulated trajectory near a 2:1 ratio in Jiang, Kelso & DeGuzman's experiment. (Top) Driver input signal; (Middle) Driver output signal, i.e. the LH trajectory.; Note isolated cycles or groups of cycles with either depressed or enhanced amplitudes compared to the overall amplitudes surrounding them; (Bottom) RH output signal.

To study how the components coordinate in time, one examines the evolution of the phases $\{\phi_n\}$. The utility of (6) in expressing coordinated behavior in terms of the evolution of ϕ_n at discrete times τ_n depends on the actual dynamics of the system. Our phase transition experiments and consequent theoretical modeling (Section 2) show that the relative phase dynamics, with observed fixed point attractors (in-phase and anti-phase), arises from nonlinearly coupled autonomous nonlinear oscillators. We can readily extend this continuous form of the phase dynamics to a discrete circle map of order 1 which has the property that in the 1:1 case not only are $\phi=0$ and $\phi=½$ fixed points but also that either one or both are stable (angle ϕ is now normalized to the interval [0,1]). Putting the map in the form

$$\phi_{n+1} = \Omega + f(\phi_n) \tag{7a}$$

$$= \phi_n + \Omega + g(2\pi\phi_n)\ , \ \text{mod } 1 \tag{7b}$$

where Ω is an added constant term corresponding to the initial frequency ratio and g is a smooth periodic function of period 1 in ϕ (for complete details, see DeGuzman & Kelso, 1989, to appear). If in-phase and anti-phase patterns are locally stable fixed points in the 1:1 case, then for $\Omega=1$, and for some parameters of g

$$g(2\pi\phi^*) = 0 \tag{8}$$

$$-1 < 1 + g'(2\pi\phi^*) < 1 \tag{9}$$

where $\phi^*=0,½$. If the map is bistable, then both 0 and ½ are locally stable. If it is monostable, either 0 or ½ , but not both, is stable. Doing a fourier series expansion of g, the lowest order expansion that can simultaneously satisfy (8) and (9) for some values of the coefficients is

$$g(2\pi\phi) = a_1 \sin(2\pi\phi) + a_2 \sin(4\pi\phi) \tag{10}$$

Transforming (10) to the new parameters (K,A) by letting $a_1 = -K/2\pi$, $a_2 = -KA/2\pi$ and using (10) in (7b) we obtain the *phase attractive circle map*

$$\phi_{n+1} = \pi_n + \Omega - \frac{K}{2\pi}(1 + A\cos 2\pi\phi_n)\sin 2\pi\phi_n \quad \text{mod } 1 \tag{11}$$

For the standard circle or sine circle map which characterizes a bewildering array of coupled multifrequency systems in nature (Glazier & Libchaber, 1988 for review) K is a measure of the amplitude strength of an external driving force, or, in general a nonlinear coupling parameter. Also for a general g (x) in (7b), the dominant parameter that scales the overall magnitude of g may be given this interpretation. The "intrinsic parameter" A expresses the bistability of relative phase and behaves as the ratio 4 b/a in the original Haken, Kelso and Bunz (1985) model (Eq. 3). Since A is the ratio of the coefficients for the sin ϕ and sin 2ϕ terms in the fourier expansion of g, A is roughly inversely proportional to the basic frequency of oscillation. For a given frequency ratio, either both in-phase and antiphase affect the overall performance, or only the in-phase is important depending on how fast the basic frequency is. Thus the parameter A is a measure of the relative importance of the intrinsic phase states 0 and π.

The relative strength of A, representing the intrinsic phase attractive dynamics and K, the nonlinear coupling term determines many of the observed experimental features of multifrequency behavior mentioned above. The mechanism for pattern *selection*, in fact, may be seen as a *competition* between these terms. For example, it is easy to see why only a few, low-integer frequency lockings are seen in biological systems and why the simple frequency ratios in our experiments are the most stable (least variable). The reason is that the widest mode-locked regions or *Arnold Tongues* (e.g. 1:1, 2:1, 2:3 etc.) are the most stable and attractive. If the experimental noise level is low, it may be possible to observe other locked states, but the presence of noise can easily kick the system into nearby, more stable attractors (e.g 3:2 to 2:1, see figure 7). In our case an additional mechanism for inducing transitions comes from the intrinsic phase attractive dynamics itself. In the 1:1 case there is a shift from bistability to monostability as A is decreased below A=1, for a reasonable value of K.

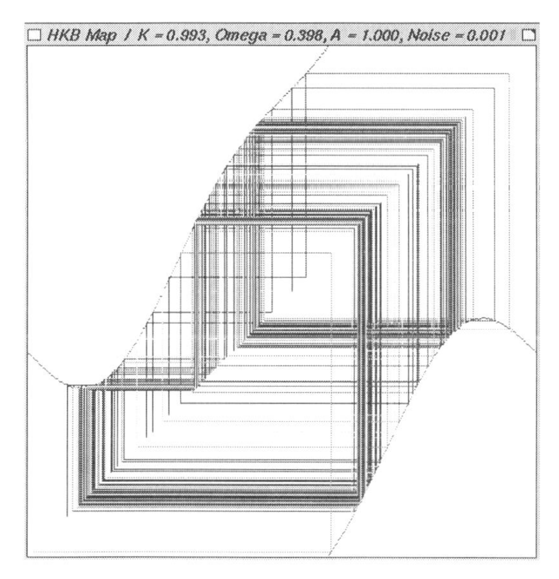

Figure 7 Noise induced transitions (3:2 to 2:1). Parameter values are $A = 1.0$, $K = 0.993$, $\Omega = 0.398$, with a small gaussian noise (amplitude of 0.001) added to the phase. Behavior jumps back and forth from one orbit (period 3) to another (period 2).

The map is invertible if the function $f(\phi)$ is monotonic or the slope $f'(\phi)$ is never zero. The boundary in the parameter space separating invertible and non-invertible regions, is called the *critical surface*. Mode-locked regions, i.e. parameter regions of *absolute coordination* in our language, are found below the critical surface of the map. For the sine circle map this is the region where the nonlinearity parameter K is below 1. For the phase attractive map, the critical surface is obviously also a function of the intrinsic parameter A. For $A>0$ the critical surface is depressed toward smaller K. For $A<0$ the surface at first shifts up (beyond $K=1$) and then is depressed again. Calculations of this critical surface (DeGuzman & Kelso, 1989/to appear) are shown in figure 8 (see also Section 6).

Computation of the winding number ρ defined by

$$\rho = \lim_{n \to \infty} \frac{|\phi_n - \phi_o|}{n} \tag{12}$$

where $\phi_n = f^{(n)}(\phi_o)$ is the nth iterate of the initial phase ϕ_o, for $0 \leq K \leq 1$, $0 \leq \Omega \leq 1$ and representative values of Λ reveals dramatic effects of the phase attractive (intrinsic) dynamics on the widths of the tongues. The approach to the critical surface (relative to K=1 in the sine circle map) is either delayed or accelerated. For example, at A=1.0 the critical surface is depressed toward smaller K, the tongues are broader and therefore overlap at smaller values of the driving term. Thus, relative to the sine circle map (Eq. 10 without the phase attraction parameter A), it is easier to stay in *some* mode-locked region beyond the critical surface as K is increased. On the other hand, for A = -.5 the tongues are thinner but the critical region is reached later. The drawback in this régime is that a very small amount of noise can easily kick the system out of these narrow stability regions inducing quasiperiodicity and chaos. Representative examples of these tongues for different A values are shown in figure 9.

In short, our experimental observations *cum* theoretical modeling reveal that generic or universal features of a system's ability to generate multi-frequency behavior are governed by the differential stability of mode-locked states as seen through the width of Arnold tongues. In our map, behavioral patterns arise from the competition between extrinsic and intrinsic, phase attractive dynamics. Pattern complexity is related to the hierarchy of the frequency ratios (the so-called Farey tree, see also Beek, 1989) and is inversely proportional to tongue width. This rationalizes the degree to which different tasks are "hard" or "easy" to perform, as well as the frequency with which different patterns occur in a large variety of experimental preparations, at many levels of observation. Phase attraction sits inside these multifrequency patterns acting as an ordering mechanism even when absolute coordination is not easily attained or possible.

5.0 *Relative Coordination Revisited: Intermittency and Anticipation*

If experimental systems can be prepared so as to correspond neatly to a pair of low dimensional coupled oscillators, coordination can be nicely captured in terms of mode-locked states (see also Glass & Mackey, 1988). However, as mentioned earlier, asymptotic convergence to mode-locked orbits may well be the exception in biological systems or at best an, albeit useful, idealization. Within the time scale of observation what is more typical are *tendencies* toward certain frequency ratio values and *tendencies* toward in-phase or anti-phase attraction. This occurs when the individual components are allowed to express their intrinsic variation, i.e. when special efforts are not made to fix them into a mode-locked state (see e.g. figure 6). The intuition is that the system retains its *flexibility* by living near mode lockings, but not in them. Beek (1989) has suggested that the skill of juggling might be characterized this way as well. Our view is motivated by: a) observations that the trajectories of components play a minor role compared to the temporal coordination of certain spatial events; and b) that living near the edge of Arnold tongues is crucial if the system is to maintain a vital mix of *stability* and *change*.

As an analogy, consider the example of a father walking along with a small child. Since their intrinsic frequencies are different, synchronization is difficult unless either one or both adjust their periods. If the mean frequencies are constant for time scales not too large

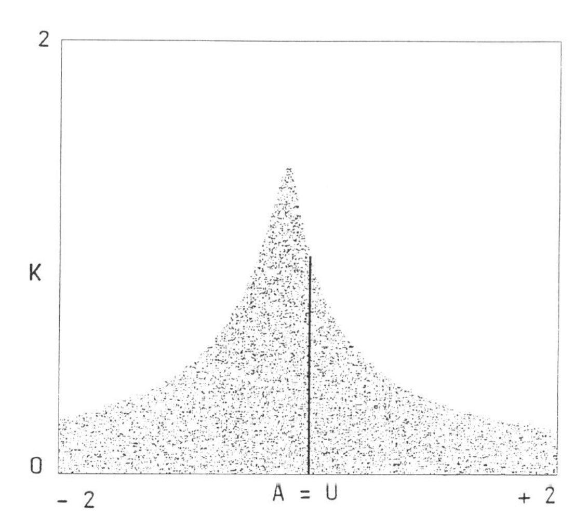

Figure 8 Subdivision of the (A,K) parameter space into invertible and noninvertible regions. The vertical strip at A = 0 defines the region of K-values below the critical region for the sine circle map. For nonzero A, the boundaries shift, stretch or shrink depending on the A parameter. The onset of chaotic behavior is either suppressed or facilitated.

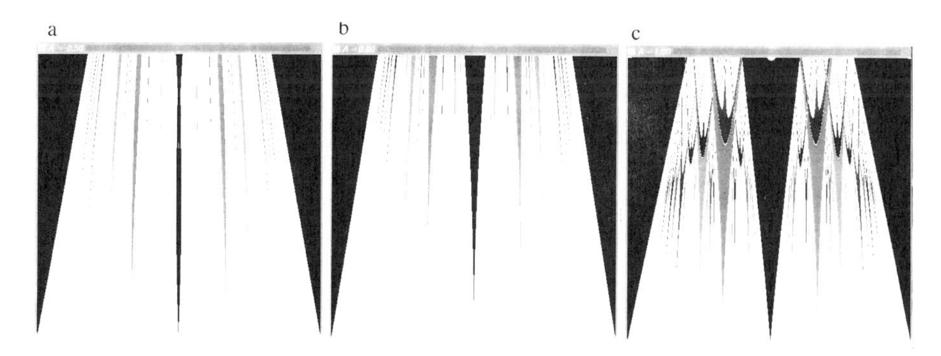

Figure 9 Arnold tongues for various strengths of the intrinsic parameter A relative to K. K, plotted between 0 and 1 is on the Y-axis, Ω between 0 and 1 is on the X-axis (a) A = 0 gives well-known result for the sine circle map. (b) A = 1.0 yields widened tongues and earlier overlappings in the direction of increasing K. (c) A = -.5 shrinks the tongue widths and delays overlapping.

compared to the larger of the two periods (the father's) a less rigid form of coordination may be observed. Like our data (cf. Fig. 6 and von Holst, 1939/1973 for many more examples), in order to keep pace with each other, the child sometimes skips a step or the father slows down and we observe *phase slippage*, followed by more or less regular cycles again.

A generic or universal feature of dynamical systems (and, of course, of our map) near a tangent bifurcation is the phenomenon of *intermittency*, i.e. abrupt slips away from a previously stable fixed point. Phase slippage in a circle map occurs because of an unstable phase direction while return to more regular cycles is due to the stable direction. That is, the order parameter relative phase might best be conceived of as *hyperbolic*, attraction taking place in one direction and divergence in another. The surface containing convergent trajectories is called the *stable manifold* while the surface containing the divergent trajectories is called the *unstable manifold*. Figure 10 shows how the phenomenon of intermittency occurs in a circle map, i.e. a diffeomorphisim of a circle (after Mannville & Pomeau who refer to it as Type 1 intermittency, see Berǵe, Pomeau & Vidal, 1984). In figure 10(a) a pair of fixed points are shown, one stable (slope <1) the other unstable, i.e. there is frequency-locking. When the slope of the function is exactly 1 the graph is just tangent to the diagonal, a single fixed point subsists. When the graph leaves the diagonal, the iterates pass through the channel, go round the circle and enter from the other side. Phase slippage occurs between the two quasiregular oscillatory periods and there is no frequency-locking (Figure 10b) because the fixed points have disappeared.

The appearance of phase slippage means that between two channel crossings, one of the oscillators gains a period with respect to the other -- exactly the phenomenon of *relative coordination*. Note that close to the tangent bifurcation the phase concentrates (*phase attraction persists*) and slows. The reason, of course, is because the iterations are compressed in the channel. We note that the appearance of phase gathering near mode-locked solutions has an *anticipatory* quality about it. A good way to see this is through the corresponding bifurcation diagram shown in Fig. 11. In Fig. 11(a) the fuzzy area reflects quasiperiodic motion because the frequency ratio between the components is irrational.

Notice, however, that the progressive darkening "anticipates" the upcoming stable solution (the single line) which indicates the system is trapped or mode-locked 1:1. For $A=0$ and $K<1$ the map is still invertible. Much richer dynamics, of course, is possible beyond the critical surface where the map loses invertibility. Fig. 11(b) shows a bifurcation diagram as K is increased *along* the 1:1 Arnold Tongue, i.e. a constant distance C from the tongue, instead of a straight cut through the $[K,\Omega]$ parameter space as in Fig. 11(a). Virtually no stable solutions exist along the tongue except for very narrow bands of overlap. Nevertheless, even though the phase is unstable, a good deal of orderly structure remains as the dark band shows.

Our main points are: (i) that relative coordination -- a ubiquitous, but largely unrecognized phenomenon in biological systems -- is to be understood as a generic property of maps called intermittency. One may speak of a "dwell" time, the average time the system stays near the same point (or the same orbit as in a non 1:1 case) before running away. Because the frequency ratio among the components is not necessarily 1:1, the fixed point states of the intrinsic dynamics need not be realized. An optimal synchronization would then be measured by how close the system gets, overall, to these intrinsic phase points; and (ii) the growth of phase densities and the resulting slowing is *predictive* of upcoming critical points, what one might call an *anticipatory dynamical system* (Kelso, in press). Anticipatory dynamical systems stay in contact with the future by living *near* but not *in* mode-locked regions, thereby retaining an optimal mixture of flexibility and stability.

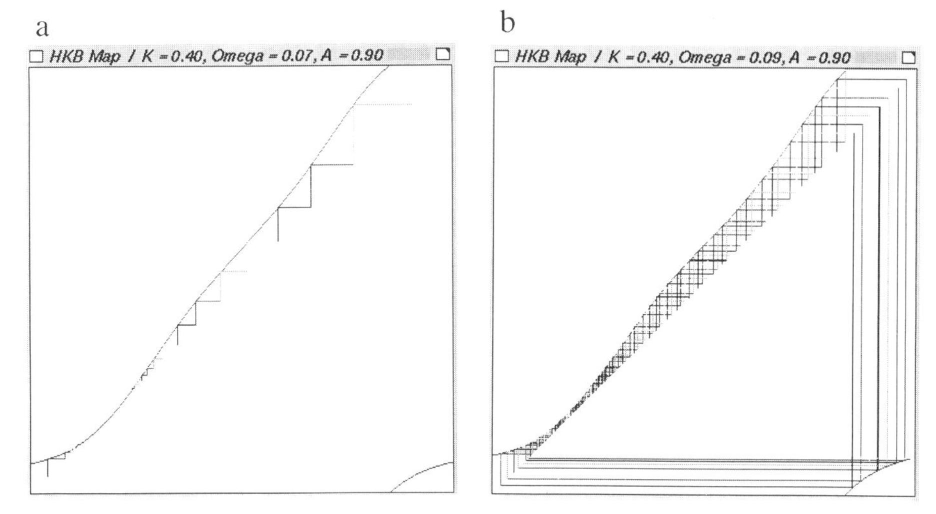

Figure 10 The phenomenon of intermittency in the phase attractive circle map. Note that the discontinuity of the function (Eq. 11) is not real but due to rectangular coordinates parameterized in the interval [0,1]. In (a) an initial condition near the unstable fixed point moves up and to the right until it wraps around and falls into the stable fixed point (lower left). In (b) the stable and unstable fixed points have disappeared, leaving a channel through which iterates can pass causing phase slippage.

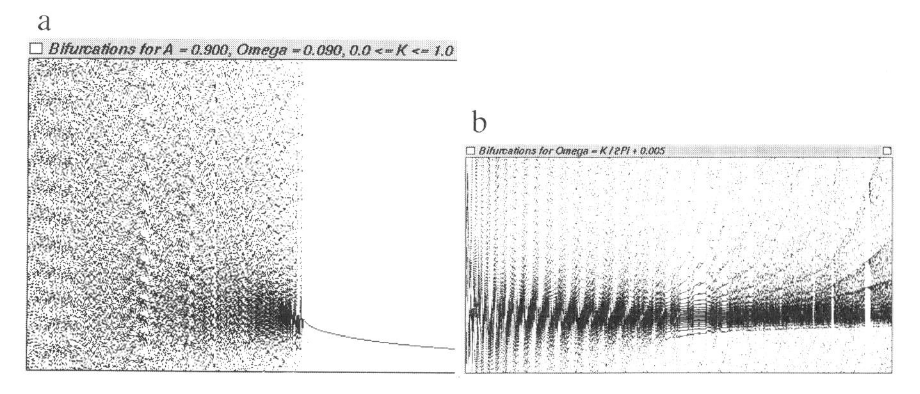

Figure 11 Bifurcation diagrams of the map. K is the bifurcation parameter, relative phase, ϕ is on the y-axis. (a) Darkening region "anticipates" upcoming stable solution (single line) which indicates the system is trapped or mode-locked 1:1. (b) Travelling along the Arnold tongue 1:1 for $\Omega = K/2\pi$ at a constant distance c = .005.

6.0 *Further theoretical features of the Phase Attractive Circle Map*

6.1 *Chaotic-Chaotic transitions*

Obviously it is not possible to make categorical statements regarding the mode-locking behavior in our experiments with the same rigor as e.g. in electronic or hydrodynamic experiments (see Bak, et al. 1984 for brief review). Certainly physiological noise is present, and earlier we established empirically and theoretically that noise-induced short term transitions from one frequency-ratio to another (e.g. 4:3 to 1:1, 5:2 to 2:1 etc.) were possible (see fig. 7). It is well known that the addition of noise can throw the system out of one attractor and into the basin of attraction of a different attractor. Here we draw attention to switching behavior that is not noise-induced but deterministic.

Often, as parameters are varied, the attractors of a dynamical system undergo sudden drastic changes, such as bifurcations or crises. A crisis (Grebogi, Ott & Yorke, 1987) occurs when an attractor comes into contact with a region of the state space (a portal) through which the system can escape from the attractor. An example of this is the quadratic map, $f(x) = rx(1-x)$, with $r = 4$. At this point the chaotic attractor is the unit interval. If r is increased slightly above 4, an initial condition in the unit interval follows a chaotic transient for a finite time until it escapes from the unit interval and diverges to negative infinity.

It can also happen that when an orbit escapes from one attractor due to a crisis, it falls into the basin of attraction of another attractor. An example of this occurs in our map (Eq. 11) with A = 0, Ω = 0.5, and K near 4.35. At K = 4.35, the system possesses two chaotic attractors, with independent basins of attraction. A random initial condition falls into one of the two attractors and stays there. At K = 4.36, the system possesses a single two-lobed attractor composed of the two original attractors connected through small portals in the state space.

An orbit of this system follows a chaotic transient inside one of the lobes until it escapes and falls into the other lobe (it can also fall back into the same lobe). Thus the system switches randomly back and forth between two states. These switches are deterministic and not the result of noise.

6.2 *Dimension calculations and self-overlapping tongues*

In circle maps the transition from quasiperiodicity to chaotic behavior occurs as the map develops an inflection point. In the quasiperiodic regime the map is monotone increasing. As K is raised above a critical value, the map develops a 'hump' and becomes non-invertible. Of course, in the phase attractive map the critical surface also depends on A, and is given by the following formula:

$$K = \begin{cases} -8A(1 + 8A^2)^{-1} & , A \leq -\frac{1}{4} \\ (1 + A)^{-1} & , A > -\frac{1}{4} \end{cases} \tag{13}$$

The curve, shown in fig. 8, has a maximum at $A=-1/\sqrt{8}$ where $K=\sqrt{2}$ and decreases as $|A|^{-1}$ on both sides of the maximum for $|A|>1$. At the critical surface, the Arnold tongues completely fill the Ω axis, except for a set of measure zero. The fractal dimension of this set was measured, using a two-dimensional Newton's method, as described in Jensen, et al. (1984). The widths of all the tongues up to period 100 were found to an accuracy of 10^{-10}. Using the widths of the tongues with winding numbers of the form 1/Q

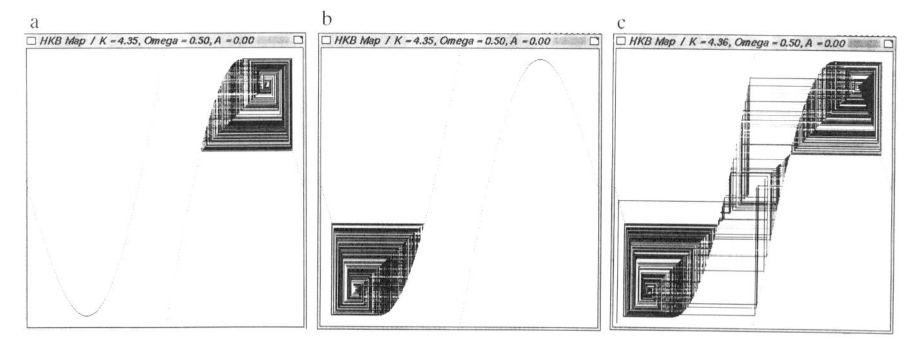

Figure 12 (a) shows a chaotic attractor in one region of the state space, (b) shows a different chaotic attractor. In (c) a connection exists between the two regions, creating a two-lobed attractor.

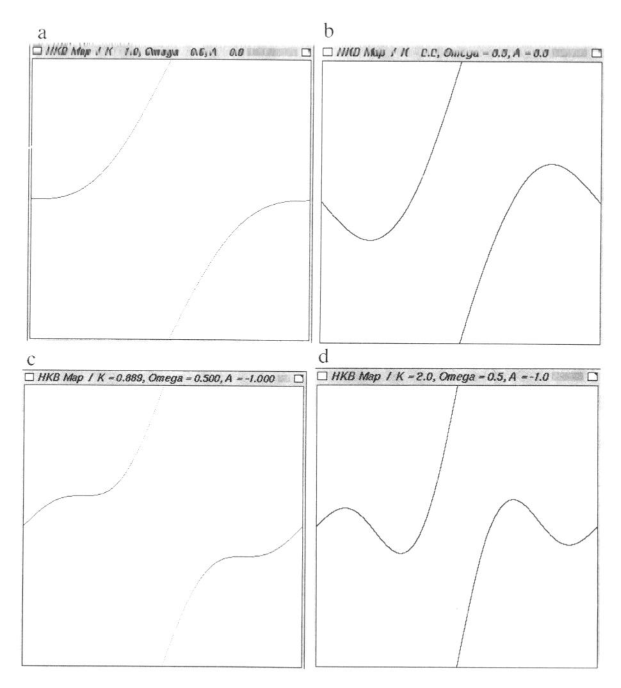

Figure 13 (a) the phase attractive circle map (Eq. 11), ϕ_{n+1} vs ϕ_n at the critical surface with A = 0. A single inflection point exists at ϕ = 0. (b) Above the critical surface, showing the single 'humps' on either side of the inflection point which make the map non-invertible. (c) At the critical surface with A < -0.8. There are now two inflection points which form two pairs of 'humps' above the critical surface (d).

as rulers, the total length of all the tongues at the critical line was measured, and subtracted from 1, giving the length of the gaps between the tongues. This number was found to scale with the ruler length, according to a linear fit of log (1/ruler length) vs. the log of the total length of the gaps, giving the fractal dimension D of the set. For A ≥ 0, D = ~ 0.87, which is in agreement with the value for the sine circle map found in Jensen et al. This shows that our map with A > 0 is in the same universality class as the sine circle map (A = 0). For A<-0.25, however, D = ~ 0.81 a value that suggests a different universality class. This class is characterized by the behavior of the inflection point along the critical surface as A and K are varied. As shown in Fig. 13, for A$\geq$0, only a single inflection point exists, and thus a single 'hump' when K is above the critical surface. For A < -0.25, the inflection point splits into two inflection points, which move away from each other as A becomes more negative, forming a double 'hump' when K is above the critical surface. Calculation of dimension as a function of A is graphically depicted in Figure 14.

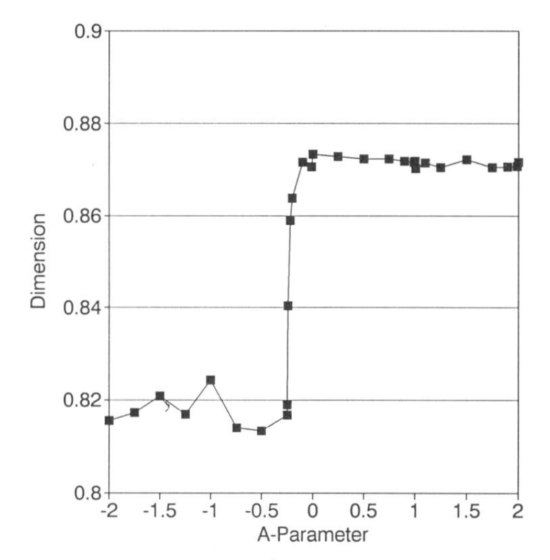

Figure 14 Dimensions at the critical surface as a function of the A-parameter. The sudden drop in dimension value coincides with the appearance of the 'swallows' or 'tridents' in the Arnol'd tongues.

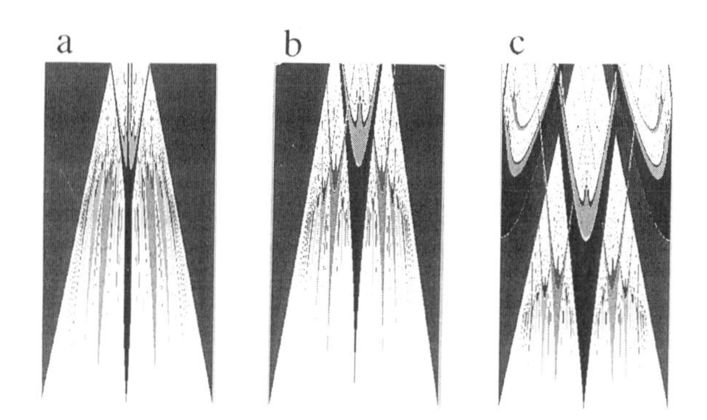

Figure 15 A series of Arnold tongues at (a) A = -0.8, (b) A = 0, (c) A = 1. (a) shows how the tongues (note the period two tongue, (a) center) have split in two and overlap with themselves.

The presence of two inflection points affects the Arnold tongues as well. The tongues also split into two when the inflection point splits, forming self-overlapping (i.e. multi-stable) mode-locked regions below the critical surface. Above the critical surface, the tongues remain self-overlapping, forming very complicated regions of multistability (see Fig. 15). As mentioned in Jensen et al., the inflection points are a local feature of the map, but determine the global scaling behavior and bifurcation structure (see also Shenker, 1982).

7.0 *Summary and Conclusions*

Experimental observations of ordered spatiotemporal patterns, multistability, switching, intermittency (phase slippage) and so forth seen on a number of scales of observation reflect generic, universal properties of dissipative dynamical systems. Variables of temporal ordering prove to be adequate function-specific collective variables or order parameters. A key concept concerns the stability of a collective state which is most clearly defined at phase transitions. At such transitions, self organization becomes apparent: different patterns arise as phase attractive states due to nonlinear coupling among interacting components. In a number of cases, the cooperative coupling can be calculated, e.g. using measures of fluctuations, local relaxation times, etc., and the pattern dynamics derived. Although dynamical laws of coordinated action are expressible in continuous differential form, evidence from multifrequency experiments in animals and humans indicates that a discrete map may be more appropriate. In particular, the trajectories of individual components may vary widely in space and time, yet nevertheless converge at discrete relative phase values. A modified circle map (a Poincaré transformation of the flow on a torus) with built-in phase attraction captures a number of experimentally observed features of coordination, including multistability and switching from one phase relation to another within frequency-ratios.

In this Phase Attractive Circle Map, the presence of phase attraction influences the width of Arnold tongues (stable mode-locked regions) and (depending on its strength) may either suppress or accelerate irregular behavior, including chaos. The map rationalizes the phenomenon of *absolute coordination* (von Holst, 1939/1973) in terms of frequency- and phase-locked patterns that correspond to attractor states. We believe that absolute coordination in which mode-lockings are asymptotic limits of an iterative process may be the exception, rather than the rule. Once the individual components are allowed to express their inherent variability, much less less rigid, more flexible forms of *relative coordination* emerge. In relative coordination, a tendency for phase attraction still remains, but occasional skips and jumps occur as the individual components adjust themselves both spatially and temporally. We establish the connection between relative coordination and phase slippage in the phase attractive map which is due to *intermittency* , a generic feature of dynamical systems near tangent bifurcations.

In the intermittent regime of the phase dynamics close to critical points the system is endowed with a predictive property; it acts as an *anticipatory dynamical system.* We believe these properties of coordination and their theoretical interpretation may reflect quite fundamental design features of biological systems. Similar features may be found in areas of research as diverse as the patterns generated by invertebrate and vertebrate neuronal networks (e.g. Cohen, Rossignol & Grillner, 1988), phase and frequency synchronization in visual cortex (e.g. Eckhorn, this volume; Gray et al., 1989) as well as hippocampus, where the phasing among neural signals appears to play a key role in cellular 'learning' *qua* synaptic modification (Stanton & Sejnowski, 1989). There is the strong impression that even in complex biological systems many of the degrees of freedom are suppressed and only a few (but not too few) contribute to behavior. If so, there is the hope that the dynamical laws governing brain and behavior may be found. We conjecture that phase attraction, but not necessarily phase-locking, will be a prominent feature of such laws, engendering, as it does, the vital mix of stability and adaptability.

Acknowledgments

Much of the research discussed herein is supported by NIMH (Neuroscience Research Branch) Grant MH42900 ONR contract N00014-88-J1191 and BRS Grant RR07258. We thank Arnold Mandell for discussion and encouragement.

References

Bak, P., Bohr, T., Jensen, M.H., 1984, Mode-locking and transition to chaos in dissipative systems. *Phys. Scripta, T9*, 50-58.

Beek, P.J. 1989. *Juggling dynamics*. Free University Press, Amsterdam.

Bergé, P., Pomeau, Y., Vidal, C., 1984, *Order within chaos: Towards a deterministic approach to turbulence.* John Wiley, New York.

Bressler, S.L.,1990, The gamma wave: a cortical information center? TINS, *13*, 161-162.

DeGuzman, G.C. & Kelso, J.A.S., 1989, Multifrequency behavioral patterns and the phase attractive circle map. Center for Complex Systems, FAU preprint (June)

Grebogi, C., Ott, E., & Yorke, J.A., 1987, Chaos, strange attractors and fractal basin boundaries in nonlinear dynamics. *Science, 238,* 632-638.

Glass, L. & Mackey, M.C., 1988, From clocks to chaos - the rhythms of life. Princeton, N.J.: Princeton University.

Glazier, J.A., Libchaber, A., 1988, Quasi-periodicity and dynamical systems, *IEEE Transactions on Circuits and Systems, 35*, 790-809.

Gray, C.M., Konig, P., Engel, A.K., & Singer, W., 1989, Oscillatory responses in cat visual cortex exhibit inter-columnar synchronization which reflects global stimulus properties. *Nature, 338* 334-337.

Haken, H., 1977/1983, *Synergetics, an introduction: Non-equilibrium phase transitions and self-organization in physics, chemistry and biology.* Berlin: Springer-Verlag, 3rd edition.

Haken, H., Kelso, J.A.S., & Bunz, H., 1985, A theoretical model of phase transitions in human hand movements. *Biological Cybernetics, 39* 139-156.

Jeka, J.J. & Kelso, J.A.S., 1989, The dynamic pattern approach to coordinated behavior: A tutorial review. In S.A. Wallace (Ed.), *Perspectives on the coordination of movement* (pp. 3-45). Amsterdam: North Holland.

Jensen, M.H., Bak, P. & Bohr, T., 1984, *Phys. Rev. A, 30,* 1960

Jiang, W., Kelso, J.A.S., & DeGuzman, G.C. (to appear). Phase attraction in relative coordination, Center for Complex Systems, FAU preprint, July, 1990.

Kay, B.A., Kelso, J.A.S., Saltzman, E.L., & Schoner, G., 1987, The space-time behavior of single and bimanual movements: data and model. *Journal of Experimental Psychology: Human Perception and Performance, 13,* 178-192.

Kay, B.A., Slatzman, E.L., & Kelso (in press). Steady-state and perturbed rhythmical movements: dynamical models using a variety of analytical tools. *J. Exp. Psych: Human Perc. & Perf.*

Kelso, J.A.S. (in press). Anticipatory dynamical systems, intrinsic pattern dynamics and skill learning. *Human Move. Sci.*

Kelso, J.A.S. & Jeka, J.J., (to appear). Dynamic patterns and direction-specific phase transitions in human multilimb coordination.

Kelso, J.A.S., Southard, D.L., & Goodman, D., 1979, On the nature of human interlimb coordination. *Science, 203*, 1029-1031.

Kelso, J.A.S., 1981, On the oscillatory basis of movement. *Bulletin of the Psychonomic Society, 18*, 63.

Kelso, J.A.S., 1984, Phase transitions and critical behavior in human bimanual coordination. *American Journal of Physiology: Regulatory, Integrative and Comparative Physiology, 15,* R1000-R10004.

Kelso, J.A.S., & Scholz, J.P., 1985, Cooperative phenomena in biological motion. In H. Haken (Ed.), *Complex Systems: operational approaches in neurobiology, physical systems and computers* (pp. 124-149). Berlin: Springer-Verlag.

Kelso, J.A.S., Scholz, J.P. & Schöner, G., 1986, Non-equilibrium phase transitions in coordinated biological motion: critical fluctuations. *Physics Letters, A118*, 279-284.

Kelso, J.A.S., & DeGuzman, G.C., 1988, How the cooperation between the hands inform the design of the brain. In: H.Haken (ed.), *Neural and Synergetic Computers.* Springer, Berlin, 180.

Kelso, J.A.S., & Schöner, G., 1988, Dynamic patterns of sensorimotor behavior: Biological control structures and constraints. *Neural Networks, 1,* (Suppl. 1), 343.

Kelso, J.A.S., 1990, Phase transitions: foundations of behavior. In H. Haken, M. Stadtler (eds.). *Synergetics of Cognition.* Springer, Berlin, Heidelberg, New York, pp. 249-268.

Kelso, J.A.S., & Schöner, G., 1987, Toward a physical (synergetic) theory of biological coordination. In R. Graham and A. Wunderlin, (Eds.), *Lasers and Synergetics, Springer Proceedings in Physics, 19*, 224-237.

Kelso, J.A.S., & DeGuzman, G., 1988, Order in time: how cooperation betwen the hands informs the design of the brain. In H. Haken (ed.), *Neural and synergetic computers.* Berlin: Springer-Verlag.

Kelso, J.A.S., Delcolle, J.D., & Schöner, G., 1990, Action-perception as a pattern formation process. In M. Jeannerod (Ed.), *Attention and Performance XIII* (pp. 139-169). Hillsdale, N.J.: Erlbaum.

Kristan, W.B., Jnr., 1980, In *Information Processing in the Nervous System.* H.M. Pinsker & W.D. Willis, Jnr., (Eds.). Raven Press, New York, pp. 285-312.

Matsumoto, G., Aihara, K., Hanyu, Y, Takahashi, N., Yoshizawa, S., Nagumo, J., 1987, Chaos and phase locking in normal squid axons. *Physics Letters A, 123*, 162-166.

Nicolis, G. & Prigogine, I., 1977, *Self-Organization in nonequilibrium systems.* New York, J. Wiley & Sons.

Nicolis, G. & Prigogine, I., 1989, *Exploring complexity - an introduction.* W.H. Freeman and Co., New York.

Schmidt, R.C., Carello, C. & Turvey, M.T., 1990, Phase transitions and critical fluctuations in visually coupled oscillators. *J. Exp. Psychol. Human Perc. & Perf.,16*,227-247.

Scholz, J.P., Kelso, J.A.S., & Schöner, G., 1987, Non-equilibrium phase transitions in coordinated biological motion: critical slowing down and switching time. *Physics Letters* A, 123, 390-394.

Schöner, G. (in press). A dynamic theory of coordination of discrete movement. *Biol. Cybern.*

Schöner, G, 1989, Learning and recall in a Dynamic Theory of Coordination Patterns. *Biol. Cybern, 62*, 39-54.

Schöner, G., Haken, H., & Kelso, J.A.S., 1986, A stochastic theory of phase transitions in human hand movement. *Biological Cybernetics, 53,* 442-452.

Schöner, G., & Kelso, J.A.S., 1988a, Dynamic pattern generation in behavioral and neural systems. *Science, 239,* 1513-1520. Reprinted in *From Molecules to Models: Advances in the Neurosciences.*

Schöner, G., Jiang, W.Y., & Kelso, J.A.S., 1990, A synergetic theory of quadrupedal gaits and gait transitions. *Journal of Theoretical Biology, 142,* pp. 359-391.

Schrödinger, E., 1944, *What is life?* Cambridge University Press.

Selverston, A.I., 1988, Switching among functional states by means of neuromodulators in the lobster stomatogastric ganglion. *Experientia 44,* 376-383.

Shenker, S., 1982, Scaling behavior in a map of a circle onto itself: Empirical results. *Physica 5D,* 4005-411.

Stanton, P.K., Sejnowski, T.J., 1989, Associative long-term depression in the

Tuller, B. & Kelso, J.A.S., 1990, Phase transitions in speech production and their perceptual consequences. In M. Jeannerod (Ed.), *Attention and Performance X111,* pp. 429-452. Hillsdale, N.J.: Erlbaum.

von Holst, E., 1939/1973, Relative coordination as a phenomenon and as a method of analysis of central nervous function. Reprinted in: *The collected papers of Erich von Holst*: Univ. of Miami Press, Coral Gables, FL.

Zanone, P.G., & Kelso, J.A.S. (to appear). The evolution of behavioral attractors with learning: Nonequilibrium phase transitions.

POSSIBLE NEURAL MECHANISMS OF FEATURE LINKING IN THE VISUAL SYSTEM: STIMULUS-LOCKED AND STIMULUS-INDUCED SYNCHRONIZATIONS

Reinhard Eckhorn and Thomas Schanze

Philipps-University, Biophysics Department
Renthof 7, D-3550 Marburg, Federal Republic of Germany

ABSTRACT

We could recently show that stimulus-specific oscillatory activities of 35-80 Hz are synchronized among distributed cell assemblies of the same and of different areas of cat visual cortex. Such synchronizations are proposed to support pre-attentive feature grouping: cells that are activated by the same visual object synchronize their activities and by this means the object is labeled as a coherent entity. In the present investigation signal dynamics and interactions of stimulus-locked and stimulus-induced oscillatory synchronization processes are analyzed. Further indications were obtained how and when oscillatory activities are generated, synchronized, desynchronized and inhibited under stimulus control. Such processes are analyzed in neural network models and their relevance for natural vision is discussed.

1 INTRODUCTION

Simultaneously occurring visual stimuli with coherent features such as a sudden shift of a partly occluded object are linked by our visual system: the object is completed and "pops out" perceptually against the background. The visual system obviously detects the coherent properties of the object's local stimulus features and is able to link, intensify and isolate them. Central to these unsolved problems of preattentive visual processing is the question, how a "perceptual object" is "neurally defined". Object definition requires, however, previous separation of the object's region from ground. Such capabilities of the visual system for feature linking, grouping, region marking and figure-ground separation have extensively been analyzed by psychophysical methods, but the underlying neuronal mechanisms are still unknown.

We follow the hypothesis that synchronization forms the neural basis for the definition of perceptual relations: receptive field (RF) properties of visual neurons are linked into a perceptual unit when the neurons' activities are synchronized. The linking-by-synchronization hypothesis is supported by recently discovered synchronization phenomena in cat visual cortex: Stimulus-forced (event-locked) responses and stimulus-induced oscillatory activities (35-80 Hz) are intensified and synchronized among cell assemblies of the same and of different cortical areas if common stimulus features drive the assemblies simultaneously (e.g. Eckhorn et al. 1988abc; Gray et al. 1989). After a brief description of our recording and data evaluation methods a short review on

Self-Organization, Emerging Properties, and Learning
Edited by A. Babloyantz, Plenum Press, New York, 1991

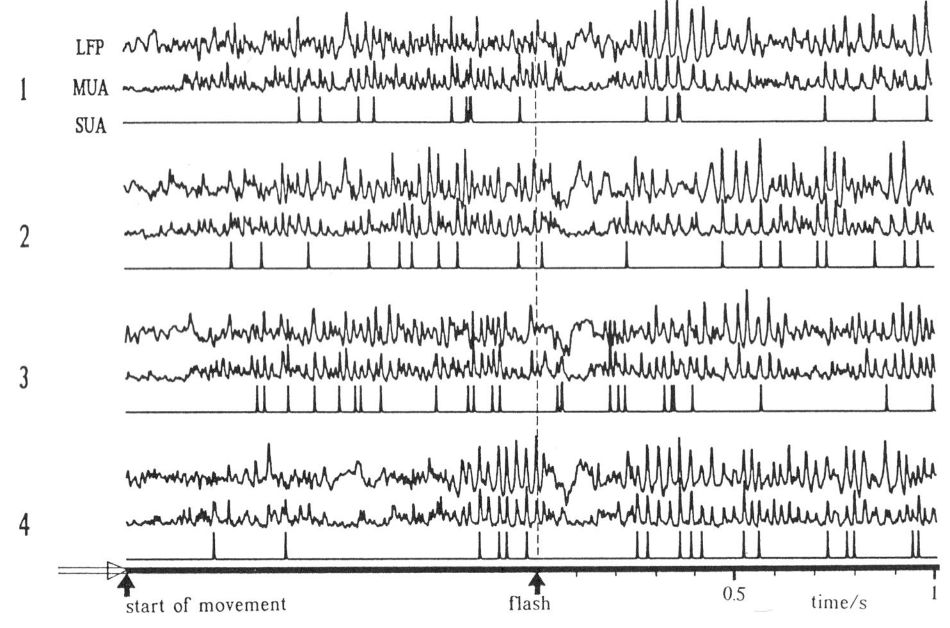

Fig. 1. Oscillatory signals in cat visual cortex induced by a moving stimulus. Responses to 4 identical stimulus repetitions of three types of neuronal signals recorded in parallel with the same microelectrode: local field potentials (**LFPs**; 12-120Hz); multiple unit activities (**MUA**; derived by filtering (1-10 kHz) and subsequent envelope demodulation (0-120Hz)); single unit spike activity (**SUA**). Stimulus: grating of 0.25 cycl/deg, moving at v = 4.5 deg/s. Arrows below mark beginning of stimulus movement (left) and instance of a light flash (middle) that was given while the grating continued to move. Note the 200ms post flash depressions of LFP high frequency components and of MUA and single unit spike discharges.

stimulus-induced synchronizations in cat visual cortex is given. Moreover, we will present additional details of stimulus related synchronizations that favor the "linking-by-synchronization" hypothesis. For this purpose we analyzed statistical parameters of oscillation "spindles" during dynamic interactions among stimulus-locked synchronized responses (as they are evoked by sudden object movements or saccades of the eyes), and oscillatory cortical synchronization processes (as they are induced during fixation periods with slow retinal image shifts). Finally, we will discuss the discovered suppression and postinhibitory facilitation of stimulus-induced oscillatory synchronizations by stimulus-locked responses and explain the effects by a neural network simulation.

1.1 Recording Synchronized Neural Signals at Inputs and Outputs of Multiple Cortical Assemblies

We developed *recording and signal analysis methods* suitable for investigations of global sensory processing. With multiple microelectrode techniques we can record single cell and mass signals from several areas of cat visual cortex in parallel (Figure 1; Reitboeck et al. 1981; Reitboeck 1983abc; Eckhorn and Reitboeck 1988). We are able to monitor with each of the singly movable fiber electrodes the mass input signals of local cell assemblies near the lectrode tip (recorded as local slow wave field potentials (**LFPs**, 13-120 Hz)), the mass output signals of local assemblies (multiple unit activities, **MUA**, 1-10 kHz), and the action potentials of single neurons (single unit activities (**SUA**); Eckhorn et al. 1989a; 1990a).

The visual cortex is functionally compartmentalized in "blobs", "patches", columns, hypercolumns, and layers. These structures delimit groups of neurons (local assemblies) that have common receptive field properties (e.g., Gilbert and Wiesel 1983). Cells in local cortical assemblies therefore generally respond in "concert" to appropriate stimuli. The simultaneous activation of many similar neurons gives us the opportunity of recording just their synchronized components because neural signals superimpose linearly in extracellular space. Synchronized components of neural mass signals add up to high local amplitudes whereas statistically independent signals average out. Multiple electrode recordings of extracellular mass signals in parallel with single neuron spike trains can, therefore, give us valuable insights into the degree of synchronization in local and among remote cortical assemblies.

1.2 Previous Work on Stimulus-Induced Oscillations

It has been shown in *cat visual cortex that signal oscillations* of 35-80 Hz can be induced by specific visual stimulation in local assemblies (Gray and Singer 1987ab; Eckhorn et al. 1988abc) and that oscillations in local assemblies at topographically remote positions of the same cortex area can be synchronized by appropriate global aspects of a stimulus (Eckhorn et al. 1988abc; Gray et al. 1989). Stimulus-response characteristics of oscillatory signal components (LFPs and MUA) were found to have specific correlations with classical single cell receptive field properties. This means that the generation of oscillatory signals is specifically linked to the tuning properties of local cell assemblies (Gray and Singer 1987ab; Eckhorn et al. 1988abc). Large LFP and MUA oscillation amplitudes are preferentially generated with sustained binocular stimuli, extending far beyond the limits of the classical receptive fields (Eckhorn et al. 1988c; Bauer et al. 1990). A considerable fraction of the investigated cell population did not exhibit oscillatory grouped spike responses with the stimuli we tested, although they could well be driven.

We discoverd, in addition, *stimulus-specific synchronizations among assemblies in different cortical areas* of the cat (A17 and A18; Eckhorn et al. 1988abc, 1989e, 1990a), and recently also of A17 and A18 with a third visual area, A19 (Kruse et al. 1990; Eckhorn et al. 1990b). Interareal synchronizations were investigated in some detail: at corresponding positions of the visual representations in different cortical areas (where receptive fields overlap), synchronized oscillatory MUAs and LFPs of high amplitudes were induced, even if the cells' preferred stimulus orientations and directions differed markedly. At non-corresponding positions, phase-locking of MUA oscillations only occurred with those assemblies that prefer similar stimulus aspects (like movement directions).

We recently proposed that *two different types of stimulus-specific synchronizations* are utilized in the CNS for the definition of feature relations (Eckhorn et al. 1989b, 1990c). a) *Stimulus-forced synchronizations*; these can occur with a high degree of coherence in primary responses of cortical neurons to transient stimuli. b) *Stimulus-induced oscillatory synchronizations*; these occur with a considerably higher degree of temporal jitter relative to the stimulus phases, and their "momentary frequencies" are highly variant. Both types of stimulus-specific synchronizations can be identified to contribute to Figure 2, for which local postsynaptic mass activities (LFPs) were recorded. Signal components due to stimulus-forced synchronization are contained in the averaged visually evoked cortical potentials (VECPs; Fig. 2B). VECPs are revealed by stimulus-locked averaging, whereas oscillatory components generally are averaged out. Fig. 2 also shows that oscillations do not last throughout visual stimulation. Instead, we found that oscillation spindles with durations of about 50-800 ms do occur, separated by pauses with irregular activity of lower amplitudes and less coherence. The latencies of the spindles' occurrences were found to be significantly longer than the latencies of the VECPs. This means that stimulus-forced (-locked) synchronization is a faster process in response to a stimulus than stimulus-induced (oscillatory) synchronization.

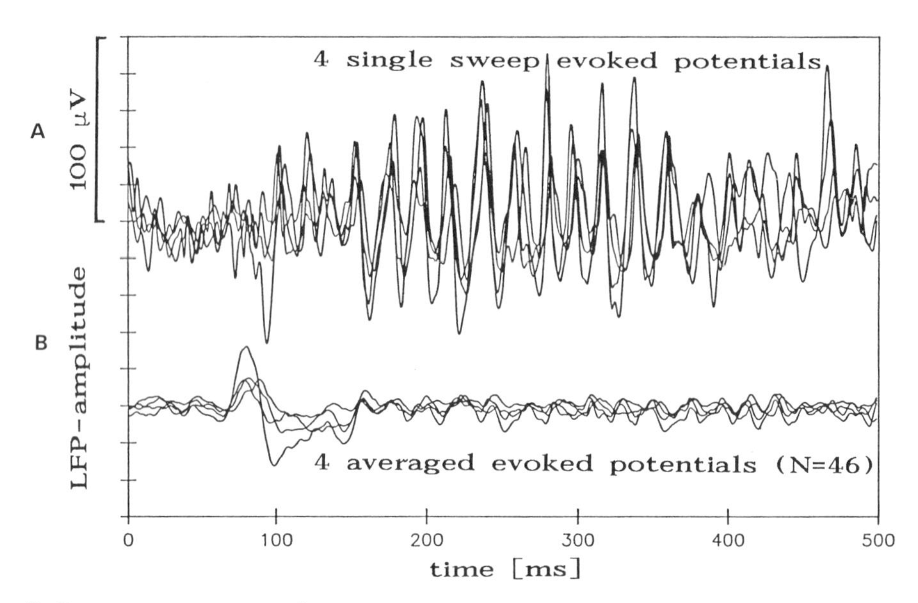

Fig. 2. Superimposograms of 4 simultaneously recorded LFPs from two different visual cortex areas (A17 and A18) in response to a drifting grating (0.7 cycles/deg moved at 8 deg/s in preferred direction; movement starts at t=0 and continues for 2s). **A:** single-sweep LFP response epochs; 3 recordings are from A17, one from A18 (overlapping RFs). Note the low amplitude uncorrelated activity during the first 70ms, in contrast to the synchronized, high amplitude oscillatory signals (spindles) beginning about 140ms after stimulus onset. **B:** averages of 46 responses to identical stimuli. Note the suppression of the oscillatory components due to stimulus-locked ensemble averaging relative to the amplitudes of the event-locked potentials at 70-130 ms that are evoked by the movement onset of the stimulus. (Fig. from Eckhorn et al. 1989d).

2 DETAILS OF STIMULUS-INDUCED OSCILLATIONS IN THE VISUAL CORTEX

2.1 Detection and Analysis of "Oscillation Spindles"

In order to obtain quantitative estimates of the observed stimulus-induced oscillatory events ("spindles") we developed an algorithm with high detection sensitivity and high temporal resolution of individual spindles (Schanze et al. 1990). For theoretical and practical reasons compromises had to be made between the contradictory demands of high temporal and spectral resolution, and between sufficient signal to noise ratio and sensitivity (Harris 1978). A two-step approach proved to be sensitive and efficient. In the first step oscillation spindles are detected with high reliability by using a sliding window short-epoch Fourier transform in combination with special conditions for the spindles' signal-to-noise ratio, duration, and the frequency and phase variations in successive overlapping data epochs. The duration of the sliding window is taken two times as long as the duration necessary for an "optimal" compromise between frequency and time resolutions. The longer window gives a doubled frequency resolution and better signal (spindle) to noise ratios. Once a spindle is detected by the above procedure, we can more than compensate for the coarse temporal resolution by a second step of analysis. After a symmetric band-pass filter is applied to the short-epoch spectrum, centered on the spindle's spectral maximum, the filtered spectral components are inversely Fourier transformed into the time domain. This spindle "extraction and cleaning" procedure enables us to apply detailed temporal analyses on single detected spindles, including time of onset, duration, number of cycles, times of zero crossings, times of each maximum and minimum, and the time courses of spindles' amplitude envelopes and "momentary frequencies". A more detailed description of these methods will be given elsewhere (Schanze et al., in preparation).

Table 1: Oscillation spindle statistics of local field potentials (average ± standart deviation). **A18:** recording from visual cortex area 18; **A17**: recording from area 17; **A18 ★ 17**: values of "correlation spindles", calculated from the simultaneously recorded A17 and A18 data. **Stimulation: A**: whole-field grating of 0.25 cycl/deg, **a:** drifting at v = 8.5 deg/s in "preferred" direction, or **b**: remaining stationary; **B**: stimulation as in a, with two additional stimulus jerks of half a spatial cycle of the grating; one jerk in, the other against the stimulus drifting direction (inter-jerk interval: 750 ms). Amplitudes are half peak-to-peak values. For more details of the spindle-search algorithm see text. Receptive fields in the two A17 and A18 recording positions: adjacent, non-overlapping, same preferences of orientations and stimulus movement directions.

		a			b		
		drifting	– Stimulus	–	stationary		
		A18	A18★A17	A17	A18	A18★A17	A17
A	Amplitude/μV	28±6.4	29±6.8	31±7.8	11±0.9	12±1.3	15±2.2
	Duration/ms	125±93	138±110	119±94	78±31	81±40	81±41
	Frequency/Hz	60±11	60±9.7	60±10	53±13	54±12	55±14
	Probability	0.72	0.74	0.72	0.23	0.17	0.24
		two superimposed	stimulus jerks				
B	Amplitude/μV	23±5.6	25±6.1	27±7.6	21±5.5	23±5.9	27±9.1
	Duration/ms	88±53	93±60	89±54	104±57	115±68	119±83
	Frequency/Hz	68±15	68±15	67±14	57±11	59±9.1	60±9.0
	Probability	0.48	0.48	0.53	0.66	0.70	0.81

With only minor changes, the spindle search algorithms could also be applied for the evaluation of correlations among spindles that were recorded in separated cortical positions. "Correlation spindles" can simply be found by calculating short-epoch cross-spectra among two recording channels instead of calculating spectra from a single channel. After appropriate normalization the same procedures reveal the statistical parameters of correlation spindles in the time domain.

2.2 Properties of Stimulus-Induced Oscillation Spindles Under Different Stimulus Conditions

Oscillation Spindle Characteristics. In Table 1 some statistical parameters of LFP oscillation spindles are listed. The data were recorded simultaneously in two different visual cortex areas (A17 and A18) while a whole-field grating with "optimal" orientation for the neurons in the two recording positions was used.

Table 1A shows that during stimulation with the drifting grating (1Aa) considerably higher values of oscillation spindle parameters occurred than during stimulation with the stationary grating (1Ab): we found a factor of 2.3 for amplitudes, 1.6 for durations, 1.1 for frequencies and 3.4 for probabilities of spindles in this typical example.

Part B of Table 1 shows spindle parameters that were obtained from recordings at the same positions under the same stimulus conditions as in Table 1A, except that the grating stimulus made two sudden jerks. These jerks were superimposed on the constant velocity drift (1Ba) or were given in isolation

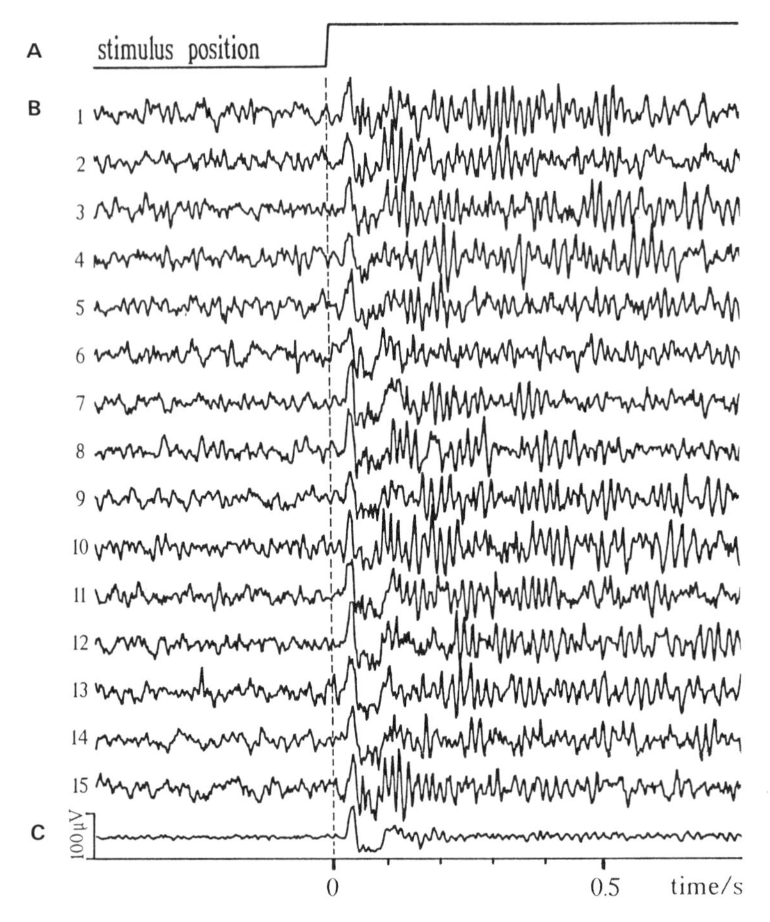

Fig. 3. Jerk-induced oscillatory LFPs in visual cortex (A18). **A**: Att = 0 the previously stationary grating jerks by half a spatial cycle; orientation and jerk direction correspond with the preferred movement direction of the neurons. **B**: 15 consecutively recorded LFP epochs from the same electrode in response to identical jerk stimuli (upward is negative). **C**: Stimulus-locked ensemble average of LFP epochs in A. Amplitudes in B and C are calibrated to the same scale. Note the high amplitude oscillation spindles after the primary "jerk potential" in B. Note also the suppression of oscillatory components in C due to stimulus-locked averaging.

(1Bb). The jerks of half a spatial cycle were directed into and against the drifting direction (740ms interval). We used this type of stimulation, because it is typical for conditions of natural vision where sudden retinal image shifts are followed by slow velocity drifts due to eye or object movements. The actions of image jerks on the generation of oscillation spindles by a slowly drifting grating are considerabe: Superimposed jerks led to a reduction of average spindle amplitudes (17%) and spindle durations (41%), to an increase in average spindle oscillation frequency (13%), and to a lower oscillation probability (46%).

When the jerks were applied to the otherwise stationary grating we found another remarkable effect (see Figure 3): after a short stimulus-locked burst and an inhibitory pause oscillation spindles were generated over several hundred milliseconds. The spindle parameters under these conditions were slightly smaller (amplitudes and durations) or nearly equal (frequency and probability) compared with those of drifting stimulus conditions. (Apossible significance of these findings for feature linking in natural vision is discussed below).

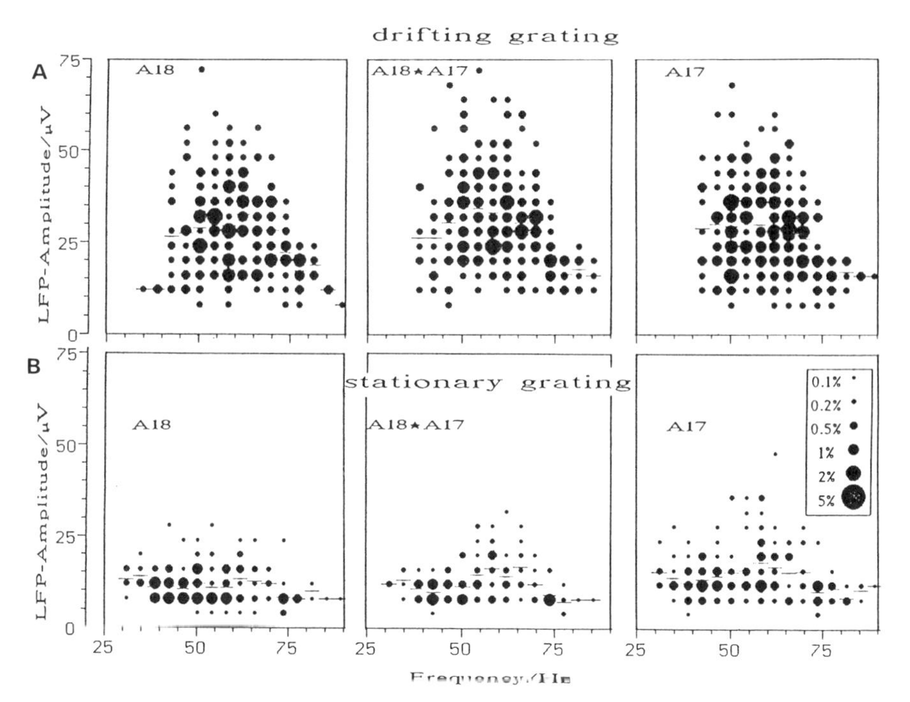

Fig. 4. Distributions of amplitudes over oscillation frequencies of LFP-spindles, obtained from simultaneously recorded data in two different visual cortex areas (A17 and A18; same data as used for Tab. 1A). The probability of presence of LFP-oscillations is shown for data recorded during stimulation with a drifting grating (**A**) and while the grating remained stationary (**B**). **A18** and **A17** denote in the outer panels that single channel data were analyzed, while **A18★A17** in the center panels denotes "correlation spindles" of the two channels (for details see text). Note the wide "spread" in the direction of large oscillation amplitudes due to stimulus drift compared with the stationary state. During stationary stimuli lower frequency oscillations are present that are absent during drift stimulation.

Oscillatory correlations among remote assemblies. Correlations between the visual areas A17 and A18 were analyzed by applying the above descibed spindle detection algorithm to the short-epoch cross-spectra of recording pairs (center columns of Table 1 and Figure 4). This method enables us to analyze correlations among single spindle events without major deteriorations by other correlated and uncorrelated signal components. Our results (Tab. 1 and Fig. 4) show that the spindle events in the two areas are highly correlated and that their generation is stimulus dependent. Such high degree of correlation can only be explained, if spindles that occur simultaneously in two locations, follow highly similar temporal courses in their "momentaneous" amplitudes and frequencies. This could indeed be proved (see also Figure 5).

3 DYNAMICS OF SPINDLE GENERATION AND SUPPRESSION

3.1 Interactions Between Stimulus-Locked Responses and Stimulus-Induced Oscillations in the Visual Cortex

Stimulus-locked responses of sufficient strength show a typical temporal course in slow wave visually evoked cortical potentials (VECPs; review: Basar 1980) as well as in spike activities of cortical neurons (see Figures 1,2 and 3).

Beginning at about 50 ms after short transient stimuli, VECPs show a succession of some negative and positive waves that do depend in amplitudes and sign on the cortical layers but are relatively invariable of the type of stimulation. Such layer-specific VECPs are well analyzed (for a review see Mitzdorf 1987) and we will discuss only those aspects that seem interesting with respect to the "linking-by-synchronization" hypothesis.

VECPs of the above mentioned type are generated by groups of neurons that are synchronized in their activities to some extent. Synchronized signal components can, therefore, easily be separated from other components by stimulus-locked ensemble averaging of responses to identical stimulus repetitions. In the next two sections we will describe in some detail the influence of the primary response (duration of peak about 50 ms), of the succeeding inhibitory phase (duration about 100 ms), and of the following phase of increased excitability (duration about 300 ms) on the generation process of stimulus-induced oscillations. In a third paragraph some remarks will be given on implications of these interactions for feature integration during saccade and fixation periods.

Delayed Oscillatory State After Fast Stimulus Transients. Efficient "jerk" stimuli (short and fast movements; Fig. 3) or flashes (Fig. 1) generally evoke a short precisely timed primary response peak in cortical mass activities and also in many single neurons (standart deviation of LFP response delays is typically ± 2 ms), that is followed by a silent phase (50-150ms) and a long lasting period of moderate activation (200-700ms), during which spindles occur with enhanced probabilities. The primary response is known to be due to direct afferent input from lateral geniculate afferents that activate a considerable number of cortical neurons shortly. The silent phase might partly be due to a discharge pause in the geniculate input but it is obviously more influenced by intra cortical inhibition (Creutzfeldt et al. 1966). During the inhibitory phase LFP components of higher frequencies and nearly all spike discharges (MUA) are suppressed, they "recover" gradually to the end of the inhibitory period (Figs. 1 and 3).

In the following long post-inhibitory activation phase non stimulus-locked oscillation spindles of considerable amplitudes were found to be generated, in the shown example with latencies of 90 ± 55ms after the primary VECP peak (mean ± standard deviation; N=55 response sweeps; Tab. 1,Bb and Fig. 3). The observed activation is probably a post-inhibitory rebound effect during which spike encoders of cortical cells are highly sensitive due to their previous discharge pause.

Suppression of Stimulus-Specific Oscillations by Fast Stimulus Transients. Oscillatory cortical synchronizations that are, for example, induced by a drifting stimulus, can immediately be suppressed by strong stimulus-locked cortical responses, like those after a flash (Fig. 1) or a stimulus jerk (Fig. 5). Strong transient stimuli lead to a sequence of two prominent response periods: complete or partial suppressions of oscillation spindles and a following period during which stimulus-induced oscillations occur with reduced amplitudes and lower probabilities, and during which they have a broader spectrum of oscillation frequencies (increased average frequency and standart deviation) compared with oscillation spindles during constant velocity stimulation alone (see Table 1,Ba). In our opinion two effects can explain the observed suppressions of LFP-spindles. The first is a desynchronization due to a dominance of afferent activations via geniculate fibers, the other is due to intracortical inhibition.

Transient visual activation dominates cortical activities and thereby generates essentially different spike patterns in individual cortical neurons. This heterogenous response behavior is probably one of the reasons that excludes the generation of mutual oscillatory synchronizations. Such different behavior is physiologically plausible, because each neuron has different visual response properties, including different response delays, because it is connected with a different neural network, and has, therefore, input from different visual and non-visual sources. After vigorous transient visual activations sufficient "relaxation" time is required for the dominating stimulus influence to wear off until a process of self organization can lead to oscillatory synchronizations. This means that oscillatory cortical synchronization probably needs release from dominating stimulus influences. Release from dominating visual drive is even more necessary for synchronization, if we make the highly probable assumption that individual intra- and inter-cortical linking connections are relatively weak.

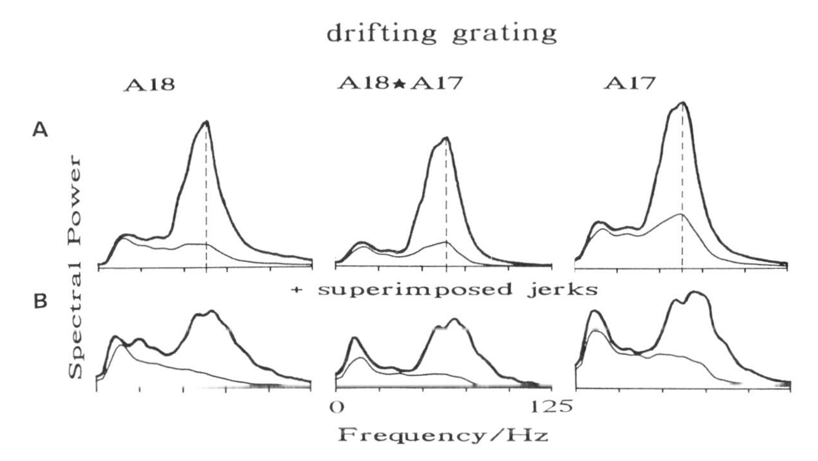

Fig. 5. Spectral amplitude distributions of parallel LFP recordings from two visual cortex areas (same data as for Tab.1 and Fig.4). **A:** data recorded while a grating stimulus drifted across the RFs, and **B**: while two fast stimulus jerks were superimposed on the drift movement of the stimulus (2 deg jerk amplitude; 740 ms inter jerk interval). **A18** and **A17** denote in the outer panels that single channel data were analyzed, while **A18 ★ A17** in the center panels denotes cross-spectra of the two channels. Thick curves: spectral amplitudes during stimulus movement, thin curves: during stationary grating (stimulation pauses). Note the lower and broader spectral distributions during superimposed jerks (B) compared with spectra obtained during drifts alone (A). Note also the high degree of correlation among both visual cortex areas. (RFs were adjacent and cells at the recording locations in both areas had similar directional preferences).

We might guess, therefore, that the more differently the neurons are activated by a stimulus, the less probable is the formation of a common phase-locked oscillatory state. On the contraray, the more similar their discharge patterns are and the nearer they are to the preferred oscillatory frequency, the higher the probability that the neurons will join a common oscillatory state, even though they are very loosely coupled. Weak linking connections are advantageous in sensory systems that have to provide many different coupling configurations with the same neural elements in order to represent various sensory situations. Direct stimulus activations via the afferent visual feeding connections should, however, be strong in order to provide fast and sensitive reactions of the organism to unexpected visual events.

Summarizing we can say, that "reactive" (stimulus-dominated) processing modes frequently change into more "constructive" oscillatory modes (and vice versa). In the latter state the internal properties and influences of the system dominate visual processing, allowing more sustained visual activity to be integrated into the ongoing synchronized cortical processing.

3.2 Successive Periods of Stimulus-Locked and Oscillatory Stimulus-Induced Processing Under Natural Conditions of Vision

Short stimulus shifts are often followed by phases with more stationary retinal images, for example during saccade-fixation sequences or when a visual object suddenly moves and stops again. In both visual situations the primary stimulus-locked responses, that occur synchronized in many stimulated visual neurons, might help to signal relatively crude but fast the "When", "Where" and "What" of the current visual events. The following inhibitory phase may play a role for perceptual suppression of the fast retinal image shifts during saccades, or/and it may provide the system transitorily with a fast but relatively insensitive state during which stimulus-locked signals can precisely be transmitted with high signal-to-noise ratios, while internally generated signals, including cortical oscillatory signals, are mainly suppressed.

Post inhibitory rebound activations after sudden shifts of an object or after saccades, might be useful for the sensitive generation of oscillatory synchronizations in just those parts of the cortical network in which the linking connections are still sensitized by the fast stimulus-locked response components (Eckhorn et al. 1990c, and Section 4.3). On the basis of the linking hypothesis we might further speculate, that during a prolonged activation state sophisticated iterative interactions could be carried out among many different structures of the visual system, including interactions from "bottom down" as they might occur during iterative interactions with memory via a "focal attention processor" (see also the concepts of Grossberg (1983), Damasio (1989), Reitboeck (1989), Sheer (1989), Sporns et al. (1989)). Oscillatory signal processing, however, is time consuming (20 ms per cycle at 50 Hz; average spindle duration (130 ms). In temporally critical situations it is therefore advantageous, if stimulus-induced oscillatory states can immmediately be suppressed by sudden stimulus transients.

3.3 Possible Relations Among Stimulus-Induced Oscillations and 40-Hz-EEG-Rhythms During Focal Attention

Synchronization, and thus, extracellular oscillation spindles are assumed to occur preferentially if facilitatorily coupled neurons are simultaneously activated at a certain similar level (e.g., Basar 1980; 1983). Oscillations might primarily be due to (bottom up) visual activations, but also "higher" mental processes, like visualization and focal attention may induce (top down) oscillatory states in visual cortex. During focal attention indeed "40-Hz-rhythms" (30-50 Hz) were found in EEG recordings of man, monkey, cat and rabbit. Significant 40Hz EEG amplitudes were recorded inter alia at locations above cortical areas of those sensory modalities where focal attention had been guided on (for a review on 40Hz rhythms during focal attention see Sheer 1989). 40-Hz-rhythms seem to occur mainly during "difficult" tasks, while the rhythms have much less amplitude or are even absent during executions of simple or previously trained tasks. During learning, amplitudes are high at the beginning and they become progressively smaller when the task is executed faster and more automatically. We expect direct relations between stimulus-induced synchronizations and 40-Hz-EEG-rhythms during focal attention, i.e. we expect both phenomena to occur synchronized and that they establish, by this means, transient relations among bottom-up and top down mechanisms of visual processing. During learning processes, when tasks are executed gradually faster, the slow processes of oscillatory synchronizations are expected to be gradually replaced by the fast processes of stimulus-locked (non-oscillatory) synchronizations.

4 STRUCTURES AND MECHANISMS THAT MIGHT SUPPORT STIMULUS-RELATED SYNCHRONIZATIONS

4.1 Levels of Organization

It is not clear to date how stimulus-induced oscillations are generated. There are, however, several clues from neurophysiological observations and from models of neural networks that allow to confine the hughe variety of possibilities to a few probable mechanisms. Different structural levels have to be considered for their involvement in oscillatory synchronizations, including the following:

On the ***level of synapses and dendrites*** we expect active bandpass properties in the preferred spindle range of 45-65 Hz in those cortical pyramide cells that participate actively in the generation of oscillatory synchronizations. We assume further, that in cortical pyramide cells the apical dendrites are the locations where oscillatory signals are injected, amplified and possibly also synchronized with other pyramide cells (see below).

On the ***single neuron level*** oscillatory and synchronizing properties might be due to the combined and coupled properties of several processes. The interactions between synaptic, dendritic and somatic signalling properties can, together with the spike encoders' dynamic threshold mechanism, lead to repetitive bursting (at 35-80 Hz) under appropriate sustained drive, even though the membrane potential does not contain oscillatory signal components in the spindles' frequency range. Cortical pyramide cells with the latter properties were identified by intracellular current injections in slice preparations (e.g., Chagnac-Amitai and Connors 1989).

On the ***level of local assemblies*** coupled cortical neurons have similar receptive field properties so that they are activated "in concert" by the same stimulus features. On this level oscillatory synchronization might be accomplished by short synaptic feedback links via local interneurons in such a way that spindle oscillation frequencies of 35-80 Hz are intensified or even generated via the feedback. Dendro-dendritic coupling was also suspected to be involved in the generation of oscillatory synchronizations in local assemblies. This notion was reinforced by the finding of vertical bundles of closely packed apical dendrites of layer-III and -V pyramide cells which can support direct dendro-dendritic coupling among the cells, even if oscillatory dendritic potentials are subthreshold (for cortical dendritic bundles in cat cortex, see Fleischhauer (1974)).

On the ***level of remote coupled assemblies*** the neurons are located in the same or in different cortical areas and they differ in their RF properties in at least one aspect (e.g., RF-position for neurons in the same area, or RF-width for different areas). We assume that association fiber systems with linking synapses provide the mutual coupling that is necessary for oscillatory synchronization. Arguments for that are given below.

4.2 Oscillations are Probably Generated in Local Cortical Units

Our ***experimental observations*** support the view, that stimulus-induced oscillations are basically generated in relatively small cortical local "units" that can generate stimulus-induced oscillatory activities by their own without the necessity for linking connections with more remote assemblies. We collected arguments in favor of local cortical oscillator units:

First, oscillation spindles of high coherence are induced by stimuli to which cortical cells respond highly specific. Second, we and others did not yet find oscillatory activities in the lateral geniculate nucleus (LGNd)(Eckhorn et al. 1989b; Gray et al. 1989). Oscillations where not present in the LGNd, although we could record oscillations simultaneously in the visual cortex (A17 and A18). Third, stimulus-induced oscillations are probably not "injected" into the cortex from thalamic or other subcortical structures via the afferent inputs in layer IV (and VI). Otherwise a polarity reversal of oscillatory LFPs should be present between upper and lower cortical layers, which was never observed by us in numerous penetrations. Such a polarity reversal is caused by layer-IV afferent inputs, because they form "dipole sources" when they are activated by synchronous activities (for a review see Mitzdorf 1987). Furthermore, we did not yet find systematic changes of oscillation phases between stationary reference recording positions in upper or lower layers and a second electrode that was systematically advanced in steps of 50 µm, either perpendicular through the cortical layers or tangentially through the same or a different cortical area. The main (average) phase shift was zero, provided the coupling strengths of the oscillatory activities were significant. We take this as an indication that cortical oscillator units are not spatially dispersed (above the resolution of our recordings), and that they neither form chain-like couplings among one another (Kammen et al. 1989) nor that they are entrained by a remote "central oscillator". The latter case seems improbable because the entraining connections should have the same fixed latencies to all cortical positions in order to guarantee zero phase lag between remote oscillation spindles.

However, the above arguments do not provide direct proof of local cortical oscillation units. In addition, phase differences between 0-180 degrees were repeatedly observed by us in LFP- and also in MUA-correlations, but rules for underlying structures or mechanisms that would argue against local oscillators could not yet be extracted from our data. Finally, we do not want to exclude the presence of systematic phase differences between oscillatory signals in pyramidal cells and those in small local interneurones that are part of a common feedback circuit. The small neurons are assumed to contribute only little to the recorded mass signals (MUAs and LFPs) because of their high generator impedances. We therefore probably did not detect their influences on MUA and LFP recordings.

From our experimental observations we derived simplified "rules" of how we assume that stimulus induced oscillatory synchronizations and stimulus-locked synchronizations are generated and how they interact. In order to explain the main properties observed in cat visual cortex we developed some simple neural networks models (Eckhorn et al. 1989cd, 1990c).

4.3 Modelling of Cortical Synchronization Processes

We modeled visual feature linking in interacting "cortical areas" using two-layered, one dimensional networks of model neurons (Eckhorn et al. 1990c). Our models have four important features: 1. Leaky integrators in the "synapses" and "spike encoders" do support repetitive bursts. A short time constant (about 18ms) provides fast response properties important for rapid synchronization and for intra-burst summation, and a slow time constant (about 80ms) provides inter-burst facilitation and longer term summation. 2. Convergent feeding connections can directly activate the model neurons' spike discharges in response to "visual" inputs. Feeding inputs specify its "receptive field" properties. 3. In our networks divergent linking connections (forward, backward, lateral) do permit the formation of synchronized activity of neurons of the same and different layers and of remote assemblies. 4. Within an "interaction time window" that is specified by the dynamic properties of synapses and spike encoder, the modulatory action of linking inputs onto feeding inputs provides mutual enhancement and synchronization of stimulus-specific responses.

Feeding connections establish in our models locally the relatively invariant "receptive field" properties of "visual neurons" by convergence of axons from (generally) "lower order" neurons. They form the afferent "visual" pathways and can activate the cells' spike outputs "directly" when the "preferred" stimuli are applied. The global arrangement of feeding connections establishes the "visuotopic" maps in the stages of the model.

Linking connections act modulatory on the single model neuron's feeding inputs. By this means they can mediate transitory facilitation and synchronization among neurons that are coupled by linking connections. In contrast to feeding connections linking connections are less strictly arranged in "visuotopic" order. Instead, they also project to "non-visuotopic" targets and can, thereby, potentially synchronize neurons that differ in one or several of their "receptive field" properties. Linking connections are, therefore, preferentially in diverging forward, backward and lateral directions. Our simplifying view of two general types of connections in the visual system is schematically illustrated in Figure 6.

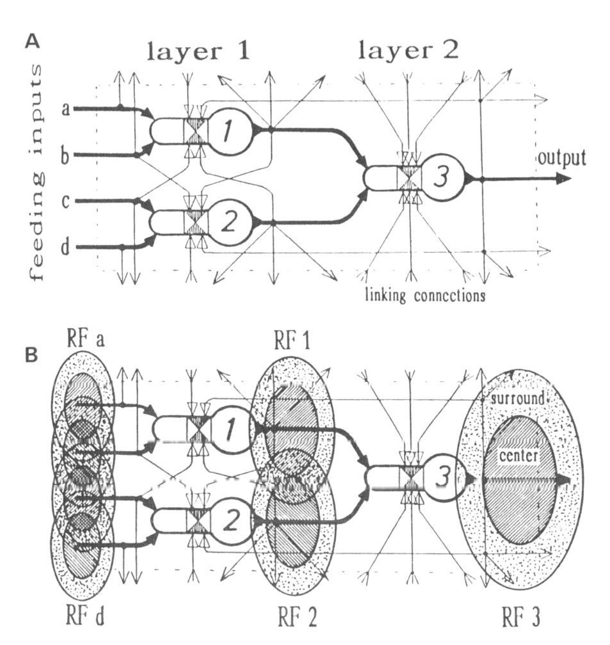

Fig. 6. Schema of two different types of connections in sensory systems. 1, 2 and 3 denote three schematic neurons with feeding and linking inputs. Feeding connections (thick lines) converge in retinotopic order from lower onto higher layer neurons. They directly activate the neurons and thereby superimpose their receptive fields (**RF**, with **center** and **surround**, shown as overlay on the circuit diagram in **B**). Linking connections (thin lines) are much more numerous; they project in forward, lateral and backward directions (mainly non-topographically) and terminate on linking synapses that facilitate the feeding signals and provide mutual synchronizations among the neurons.

Figure 7 shows *simulation results* with a two-layered network of model neurons. The primary bursts of spikes occur synchronized in response to the onset of the stimulus in both activated subregions, but they are not oscillatory. Repetitive bursting ("oscillation spindles of MUA") appear only after a considerable relaxation time of less correlated discharges. During this interval the model neurons are released from the dominating stimulus so that they can synchronize their activities via the weak linking connections into oscillatory synchronized bursts of spikes. This simulation mimics in part our experimental results from cat visual cortex: there we found that primary stimulus-locked responses suppress spindle formation over a certain interval, and that spindle suppression was followed by a phase of enhanced probability for the occurrence of synchronized oscillations (see Figs. 1, 3 and 5).

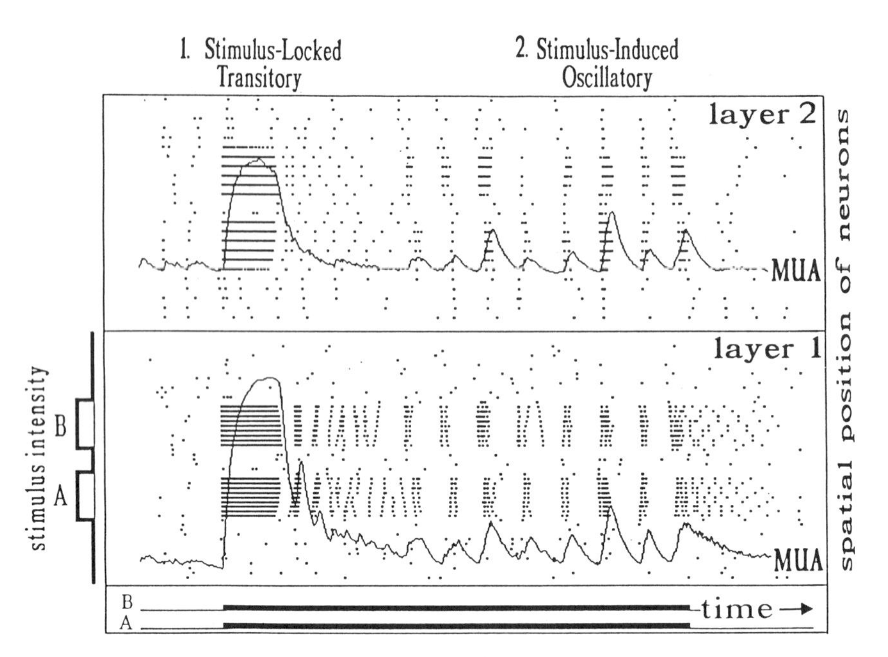

Fig. 7. Neural network simulation of stimulus-locked and stimulus-induced synchronizations among two "stimulated patches" **A** and **B** of the lower (one-dimensional) layer 1 (50 neurons) mediated via feedback from the higher layer 2 (23 neurons). A neural network was used with lateral linking connections to each four neighbors of laterally declining strength in both layers and feed forward feeding connections with convergence from 4 layer 1 neurons onto a single layer 2 neuron. Feedback linking connections from layer 2 to layer 1 are diverging by a factor of 4. Occurence times of "action potentials" are marked by dashes. Independent analog noise was continuously added to all feeding inputs. Black horizontal bars indicate stimulus duration for "patch" **A** resp. **B**. **MUA:** denotes the continuous traces that were derived by spatial summation of spikes from many neurons and by subsequent lowpass filtering, in order to obtain a similar signal like in cortical MUA recordings. Note the vigorous synchronized bursts in both layers and both patches after stimulus onset and the following period with only slightly synchronized spikes. Note the "inter-patch" burst synchronizations in layers 1 and 2, and the precise separation of synchronized activity patches in layer 1, while those in layer 2 bridge ("fill-in") the stimulus gap. (Simulation for this Fig. was made in our department by Peter Dicke; details of the model are given in Eckhorn et al. 1990c).

Synchronizations from the higher to the lower level play a substantial role in bridging the gap between the two stimulated subregions in our model network. A similar feedback synchronization might also be responsible for the synchronizations we observed in cat visual cortex during stimulation with coarse gratings. We found not only induced synchronized oscillations in A17 within the region stimulated by a single stripe, but also among A17-positions that were stimulated by other, neighboring stripes. Parallel recordings from neurons in A18 that had overlapping receptive fields with the A17 neurons (and that were stimulated by neighboring stripes), showed oscillatory inter-areal synchronizations, just as in our neural network simulations.

5 SUMMARY OF RESULTS AND CONCLUSIONS

The following questions were adressed: 1. What are the properties of synchronized neural events and what are the underlying neural mechanisms? 2. Where are oscillatory signals induced and in what visual situations? 3. What are the possible roles of stimulus-related synchronizations in visual sensory processing?

We could answer these questions partly:

* Large amplitude LFP and MUA oscillations ("spindles", 40-60 Hz) are preferentially generated with sustained binocular stimuli that extend far beyond the limits of the classical receptive fields.
* Spindles occurred with considerably larger amplitudes, higher probabilities and oscillation frequencies, and longer durations during stimulations than in "stimulation pauses".
* During continuous "optimal" stimulation (with a moving whole field pattern) oscillations do not last throughout stimulation. Instead, oscillation spindles of variable durations occur (120 ± 100ms), separated by pauses of lower amplitudes and less coherence.
* After onset of a continuing stimulation, oscillation spindles occur with relatively long delays and high temporal jitter compared with primary (non-oscillatory) responses.
* Oscillatory activities are synchronized among distributed cell assemblies of the same and of different cortical areas under stimulus control.
* Phase differences between oscillatory activities of the same type were on average, zero. Larger differences (up to 180 deg) were frequently observed, but systematic phase changes with recording distance and position could not be established.
* Oscillation spindles of LFPs that occur simultaneously in different cortical areas show a high degree of coherence due to similarities of their amplitude-envelopes and momentaneous frequencies.
* Fast stimulus-locked responses of sufficient amplitudes lead to a suppression of stimulus-induced oscillatory synchronizations. Suppression is followed by a post-inhibitory rebound period of enhanced sensitivity with high probabilities of spindle occurrence.
* Oscillatory cortical synchronizations need for their occurrence the release from dominating transient stimulus influences.
* We collected evidence that stimulus-induced oscillations are generated in local cortical units and that such units are coupled via special linking connections.
* Our model neurons and their connectivity, therefore, do support our concepts of feeding and linking synapses as a basis for both, stimulus-locked synchronizations and oscillatory stimulus-induced synchronizations.

Summarizing we can conclude that oscillation spindles are generated, synchronized, desynchronized and shut off under stimulus control. In natural vision, stimulus dominated ("reactive") modes of processing alternate with modes of ("constructive") oscillatory synchronizations. In addition, our physiological results and the network simulations support the hypothesis, that association fiber systems in the visual cortex essentially act as linking networks that can provide facilitation and synchronization among stimulus-driven assemblies in the same and between different cortical areas. Finally, the results explain the neural mechanisms of perceptual feature linking phenomena, such as mutual enhancement among similar, spatially and temporally dispersed features and figure-ground discrimination. This is in agreement with our proposal that mutual enhancement and synchronization of cell activities are general principles of temporal coding within and among sensory systems: *Event-locked synchronizations* support crude instantaneous "preattentive percepts", and *stimulus-induced oscillatory synchronizations* support more complex, "attentive percepts" that require iterative interactions among different processing levels and memory.

Acknowledgements: We thank our colleagues Dr. R. Bauer, H. Baumgarten, M. Brosch, W. Kruse, and M. Munk for their help in experiments and data processing, and our technicians W. Lenz, U. Thomas, and J. H. Wagner for their expert support. It is greatly acknowledged that P. Dicke simulated our cortical findings in the neural network model for Figure 7. We are also thankful to Prof. H.J. Reitboeck, the head of our department, for supporting this investigation. Our project was sponsored by Deutsche Forschungsgemeinschaft Re 547/2-1 and Ec 53/4-1 and by Stiftung Volkswagenwerk I/64605.

REFERENCES

Basar E (1980) EEG-Brain Dynamics. Elsevier, North-Holland Biomedical Press, Amsterdam New-York Oxford

Basar E (1983) Synergetics of neuronal populations. A survey on experiments. In Synergetics of the Brain, E Basar, H Flohr, H Haken, A Mandell eds. Springer Verlag, Berlin Heidelberg New York, pp 183-200

Bauer R, Brosch M, Eckhorn R (1990) The spatial distribution of stimuli evoking oscillations of neural responses in the visual cortex of the cat. In: Brain and Perception, N Elsner, G Roth (eds), Thieme, Stuttgart New York, p 238

Chagnac-Amitai Y, Connors BW (1989) Horizontal spread of snchronized activity in neocortex and its control by GABA-mediated inhibition. J Neurophysiol 62:1149-1162

Creutzfeldt OD, Watanabe S, Lux HD (1966) Relation between EEG-phenomena and potentials of single cells. Part I and II. Electroenceph Clin Neurophysiol 20: 1-37

Damasio AR (1989) The brain binds entities and events by multiregional activation from convergence zones. Neural Computation 1:121-129

Damasio AR (1989) Time-locked multiregional retroactivation: A systems-level proposal for the neural substrates of recall and recognition. Cognition 33:25-62

Eckhorn R, Reitboeck HJ (1988) Assessment of cooperative firing in groups of neurons. In: Springer Series in Brain Dynamics, E Basar, edit., Vol 1, Springer Verlag, Berlin Heidelberg, pp 219-227

Eckhorn R, Bauer R, Jordan W, Brosch M, Kruse W, Munk M, Reitboeck HJ (1988a) Are form- and motion- aspects linked in visual cortex by stimulus-evoked resonances? Multiple electrode and cross-correlation analysis in cat visual cortex. EBBS-Workshop on Visual Processing of Form and Motion, Tübingen, Confer Vol, p 7

Eckhorn R, Bauer R, Brosch M, Jordan W, Kruse W, Munk M (1988b) Functionally related modules of cat visual cortex show stimulus-evoked coherent oscillations: A multiple electrode study. Invest Ophthalmol Vis Sci 29: 331,12

Eckhorn R, Bauer R, Jordan W, Brosch M, Kruse W, Munk M, Reitboeck HJ (1988c) Coherent oscillations: A mechanism of feature linking in the visual cortex? Multiple electrode and correlation analysis in the cat. Biol Cybernetics 60: 121-130

Eckhorn R, Bauer R, Reitboeck HJ (1989a) Discontinuities in visual cortex and possible functional implications: Relating cortical structure and function with multi-electrode/correlation techniques. In: Basar E (ed), Springer Series in Brain Dynamics, Vol 2 , Springer Verlag, Berlin Heidelberg New York, pp 267-278

Eckhorn R, Reitboeck HJ, Arndt M, Dicke P (1989c) Feature linking via stimulus-evoked oscillations: experimental results from cat visual cortex and functional implications from a network model. Proceed Int Joint Conf Neural Networks, Washington. IEEE TAB Neural Network Comm, San Diego, Vol I: 723-730

Eckhorn R, Reitboeck HJ, Arndt M, Dicke P (1989d) A neural network for feature linking via synchronous activity: results from cat visual cortex and from simulations. In: Models of Brain Function. Cotterill RMJ (ed). Cambridge Univ Press, pp 255-272

Eckhorn R, Munk M, Kruse W, Brosch M, Jordan W, Schanze T, Bauer R(1989e) Stimulus-evoked synchronizations between visual cortex areas of the cat: A feature linking mechanism. In: Elsner N, Singer W (eds) Dynamics and Plasticity in Neural Systems. Thieme, Stuttgart New York, p 349

Eckhorn R (1990; in press) Stimulus-specific synchronizations in the visual cortex: linking of local features into global figures? In: Neuronal Cooperativity. Krüger J (ed). Springer Verlag, Berlin Heidelberg New York

Eckhorn R, Brosch M, Salem W, Bauer R (1990a) Cooperativity between cat area 17 and 18 revealed with signal correlations and HRP. In: Brain and Perception, Elsner N, Roth G (eds), Thieme, Stuttgart New York, p 237

Eckhorn R, Reitboeck HJ, Dicke P, Arndt M, Kruse W (1990b) Feature-linking across cortical maps via synchronization. In: Proceedings Intern Conf Parallel Processing in Neural Systems and Computers, Düsseldorf (FRG), Eckmiller R et al (eds), North-Holland, Amsterdam pp 101-104

Eckhorn R, Reitboeck HJ, Arndt M, Dicke P (1990c) Feature linking among distributed assemblies: Simulations and results from cat visual cortex. Neural Computation 3: 293-307

Fleischhauer K (1974) On different patterns of dendritic bundling int the cerebral cortex of the cat. Z anat Entw Gesch 143: 115-126

Gilbert CD, Wiesel TN (1983) Clustered intrinsic connections in cat visual cortex. J Neurosci 3: 1116-1133

Gray CM, Singer W (1987a) Stimulus-dependent neuronal oscillations in the cat visual cortex area 17. 2nd IBRO-Congrs, Neurosci Suppl, 1301P

Gray CM, Singer W (1987b) Stimulus specific neuronal oscillations in the cat visual cortex: a cortical functional unit. Soc Neurosci (Abstr) 404.3

Gray CM, König P, Engel AK, Singer W (1989) Oscillatory responses in cat visual cortex exhibit inter-columnar synchronization which reflects global stimulus properties. Nature 338: 334-337

Grossberg S (1983) Neural substrates of binocular form perception: filtering, matching, diffusion and resonance. In Basar E, Flohr H, Haken H, Mandell AJ (eds), Synergetics of the Brain, Springer-Verlag, Berlin Heidelberg New York Tokyo, pp 274-298

Harris FJ (1978) On the use of windows for harmonic analysis with the discrete fourier transform. Proc IEEE 66:51-83

Johannesma P, Aertsen A, van den Boogaard H, Eggermont J, Epping W (1986) From synchrony to harmony: ideas on the function of neural assemblies and on the interpretation of neural synchrony. In: Brain Theory, Palm G, Aertsen A (ed), Springer-Verlag, Berlin Heidelberg, pp 25-47

Kammen DM, Holmes PJ, Koch C (1989) Cortical architecture and oscillations in neuronal networks: feedback versus local coupling. In: Models of Brain Function. Cotterill RMJ (ed). Cambridge Univ Press, Cambridge New York Melbourne, pp 273-284

Kruse W, Eckhorn R, Bauer R, Jordan W, Brosch M, Reitboeck HJ (1990) Stimulus-induced synchronization among three visual cortical areas of the cat. Abstr Vol ECVP-Conf, Paris, Perception (in press)

Mitzdorf U (1987) Properties of the evoked potential generators: current source-density analysis of visually evoked potentials in the cat cortex. Intern J Neurosci 33: 33-59

Reitboeck HJ, Adamczak W, Eckhorn R, Muth P, Thielmann R, Thomas U (1981) Multiple single-unit recording: Design and test of a 19-channel micro-manipulator and appropriate fiber electrodes. Neurosci Letters, Suppl 7: S148

Reitboeck HJ (1983a) A 19-channel matrix drive with individually controllable fiber microelectrodes for neurophysiological applications. IEEE SMC 13:676-682

Reitboeck HJ (1983b) Fiber microelectrodes for electrophysiological recordings. J Neurosci Meth 8: 249-262

Reitboeck HJ (1983c) A multi-electrode matrix for studies of temporal signal correlations within neural assemblies. In Basar E, Flohr H, Haken H, Mandell A (eds), Synergetics of the Brain. Springer-Verlag, Berlin Heidelberg New York, pp 174-182

Reitboeck HJ (1989) Neuronal mechanisms of pattern recognition. In: Sensory Processing in the Mammalian Brain. JS Lund (ed), Oxford Univ Press, New York, pp 307-330

Schanze T, Eckhorn R, Baumgarten H (1990) Properties of stimulus-induced oscillatory events in cat visual cortex. In: Brain and Perception. Elsner N, Roth G (eds), Thieme, Stuttgart New York, p 238

Sheer DE (1989) Sensory and cognitive 40-Hz event-related potentials: behavioral correlates, brain function, and clinical application. In: Springer Series in Brain Dynamics 2, Basar E, Bullock TH (eds), Springer-Verlag, Berlin Heidelberg New York, pp 339-374

Sporns O, Gally JA, Reeke GN, Edelman GM (1989) Reentrant signaling among simulated neuronal groups leads to coherency in their oscillatory activity. Proc Natl Acad Sci USA, 86: 7265-7269

A Model for Synchronous Activity in the Visual Cortex

Christian Kurrer, Benno Nieswand, and Klaus Schulten

Beckman Institute and Department of Physics
The University of Illinois at Urbana-Champaign
405 N. Mathews Ave., Urbana Il 61801, U.S.A.

Abstract

We investigated the problem of figure-ground separation or *binding problem* of image processing in the brain. Recent experiments by Singer et al. have shown in the visual cortex of cat synchronous firing activity among neurons coding similar features. The observations suggest that synchronization may be an important coding principle for information processing in the brain.

The investigations reported here are based on a dynamical description of single neurons as excitable elements with stochastic activity. We demonstrate that sets of weakly coupled neurons of this type can readily develop synchronous activity when subject to coherent excitation. We provide a mathematical analysis of the dynamical model chosen as well aspresent numerical simulations illustrating how synchronous firing can be used in the visual cortex to segment images.

1 Introduction

Most present neural network models, e.g. back-propagation or Hopfield neural nets, use a single state variable to describe a neuron. This state variable represents the firing activity of a physiological neuron. In spite of important advances using these kind of neural nets, many problems concerning information processing through neural networks are left unsolved. A most important unsolved problem is the socalled *binding problem*[1], which addresses the question how the brain segments images into objects and which neurons correspond to different objects. An example is the figure-ground separation task, i.e. the task to determine the parts of the visual cortex involved in

Self-Organization, Emerging Properties, and Learning
Edited by A. Babloyantz, Plenum Press, New York, 1991

"seeing" the figure and the parts involved in "seeing" the background. This separation is a nontrivial task since both figure and background can contain very similar optical properties such as color, luminosity, texture. Also the figure often does not have a clearly marked outline.

Whereas formerly the *firing rate* of a neuron was assumed to contain all the information transmitted to the brain by sensory neurons, it recently became apparent that in the cortex also the *timing of the firing* of one neuron relative to other neurons in the same cortical area contains essential information. For example, recordings in the olfactory cortex performed by Freeman [2, 3] revealed the occurrence of specific spatio-temporal patterns as soon as a specific odor is identified. However, the problems connected with the measurement of olfactory input seemed to be a major obstacle in establishing quantitative input-output relations between the spatio-temporal patterns and the excitation of the olfactory sensory neurons. Thus it was not easy to discover the mechanisms causing the formation of spatio-temporal patterns and the importance of these for the subsequent steps in information processing.

Important progress in clarifying the role of spatio-temporal patterns in the firing activity of cortical neurons was achieved through recent experiments performed on the visual cortex, initiated by Gray and Singer as well as by Eckhorn et al. [4, 5]. These experiments, for the first time, related the firing correlation of two cortical neurons to the fact that these cortical neurons processed information originating from the same object, such as an illuminated bar shown to the retina. When two neurons processed information originating from the same bar, a situation which can be checked by determining their receptive field beforehand, their firing activity was observed to be synchronized. This result supports the conjecture by von der Malsburg [6] that the "binding" of different stimuli to the same object is achieved through synchronization of the corresponding neural activity in the cortex.

Models for Synchronously Firing Neurons

The new findings on the temporal correlation properties of interacting neurons prompted an investigation of neuron models that incorporate the temporal aspects of the firing of neurons. Some approaches [7, 8, 9, 10] employed oscillator models for a single neuron, and introduced rules that govern the synchronization of these oscillators. These approaches implicitly assume that the normal state of a neuron is described by oscillators which adjust their phases when subjected to certain kinds of input.

Previous research on coupled nonlinear oscillators [11, 12, 13, 14, 15] elicited essentially three ways of enhancing the synchronization of coupled nonlinear oscillators: (1)increasing *coupling strength* between the oscillators, (2)reducing *spread of the intrinsic frequencies*, and (3) reducing *random pertubations* of the phases of the nonlinear oscillators. Oscillator models for neural networks addressing the synchronization issue, thus, were restricted to stating rules how the excitations of the sensors influences one of these three properties of the population of neurons. With the aid of these rules, some of the experimental results could be reproduced[8, 16].

However, it seems to be hard to translate the rules back into physiological context. This limits the stimulus such models might provide for further experiments, and one may doubt whether the mechanim which are relevant in the brain have actually been described by these rules.

Moreover, periodic oscillations of single isolated neurons have been observed very rarely in experimental recordings. Other models, the ones discussed in [17], employ a more detailed representation of the single neuron incorporating a firing threshold, relaxation and refractory behavior. Their models, however, nevertheless describe single neurons only in a rather abstract fashion.

2 Nonlinear Dynamics Models

To achieve synchronous oscillations in the cortex, our approach employs single neurons which are *excitable element* (EE) [18] rather than oscillators. An EE is a dynamical system with a stable state to which the system relaxes *directly* after *small* perturbations. When the strength of the perturbation exceeds a certain *threshold*, the equilibrium state will only be regained after the system passes through a series of *specific states* which significantly deviate from the stationary state.

In the case of a neuron this corresponds to the phenomenon that small variation of the transmembrane voltage result in a direct relaxation back to the -65 mV resting potential, whereas larger variations cause the cell first to increase its transmembrane potential up to +40 mV before the resting potential is reestablished.

As a mathematical model for such an EE we employ a set of equations known as the Bonhoeffer-van-der-Pol (BvP) or Fitzhugh-Nagumo (FN) equations [19, 20]. These equations, on the one hand, contain only the minimal number of nonlinear terms necessary to yield an EE behavior and, therefore, are mathematically relatively simple; on the other hand, the equations are closely related to the Hodgkin–Huxley description of neurons [21], i.e. to a quantative description of certain neurons (squid giant axon), as shown by Fitzhugh. Hence, the dynamical variables involved can be interpreted in terms of physiological observables.

2.1 The Bonhoeffer-van-der-Pol or Fitzhugh-Nagumo Equations

The dynamics of a neuron in the BvP model is given by the set of equations

$$\begin{aligned} \dot{x}_1 &= F_1(x_1, x_2) &= c(x_1 - x_1^3/3 + x_2 + z) \\ \dot{x}_2 &= F_2(x_1, x_2) &= (a - x_1 - bx_2)/c. \end{aligned} \tag{1}$$

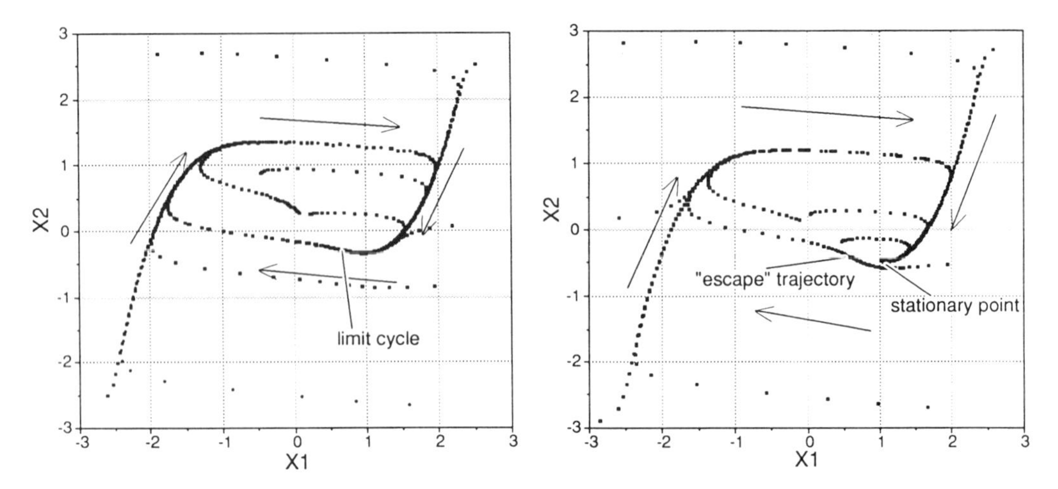

Figure 1. (left side) Some typical phase space trajectories for $z = -0.4$: All trajectories eventually lead into a stable limit cycle. Represented is a stroboscopic view of trajectories for 9 different initial states.

Figure 2. (right side) Some typical phase space trajectories for $z = 0$: All trajectories lead to the stable stationary point at (1.1,-0.5). Note that a trajectory which passes very near by the stationary point will lead the phase point back to its stationary state only after a long path through the phase space.

According to Fitzhugh's derivation of the BvP equations, x_1 represents the negative transmembrane voltage and x_2 is closely related to the potassium conductivity. The nonlinear term in the first equation reproduces the effect of the voltage-dependent sodium channels. The equations reproduce the response of a single neuron for the choice of control parameters $a = 0.7$, $b = 0.8$, and $c = 3.0$. The dynamical character of the solutions of this system of equations is determined by the parameter z, which corresponds to the excitation current I in the Hodgkin-Huxley equations. Within the physiological range $2.0 > z > -0.6$, the dynamical behavior depends on whether z is larger or smaller than $z_{crit.} \approx -0.34$. For $z > z_{crit.}$ the variables (x_1, x_2) asymptotically reach a stable fixed point, whereas for $z < z_{crit.}$ the solutions are periodic in time. The latter solution corresponds to a periodic firing of neurons. The phase portrait for these two dynamical modes is shown in Figs. 1 and 2.

Whereas the BvP model describes a neuron which at z_{cr} abruptly changes its behaviour from complete silence to firing at a fairly constant frequency, Treutlein and Schulten [22] showed that the model can be made more realistic by adding noise, which opens the possibility of varying the firing frequency smoothly from zero to its maximal value. The resulting dynamics of the neuron is described by

$$\begin{aligned} \dot{x_1} &= F_1(x_1, x_2) + \eta_1(t) &= c(x_1 - x_1^3/3 + x_2 + z) + \eta_1(t) \\ \dot{x_2} &= F_2(x_1, x_2) + \eta_2(t) &= (a - x_1 - bx_2)/c + \eta_2(t) \end{aligned} \quad (2)$$

where $\eta_i(t), (i = 1, 2)$ represents Gaussian white noise with amplitude $\sigma = \sqrt{2\beta^{-1}}$ charactarized through the relations

$$\langle \eta_i(t) \rangle = 0, \qquad \langle \eta_i(t_1)\eta_j(t_2) \rangle = \beta^{-1}\, \delta(t_1 - t_2)\, \delta_{ij} \ . \tag{3}$$

These equations were investigated for the range of z leading to limit cycle dynamics in [23]. Here, we will focus on a range of z where these equations describe an EE (corresponding to the case shown in Fig. 2). Because of the noise added to the BvP dynamics, the phase point will not be trapped at the stationary state, but rather exhibits diffusion-like behavior in the phase space. Eventually it will reach an *escape trajectory*, shown in Fig. 2, which attracts the phase point to negative x_1 values, and in this way action potentials are generated. The influence of the noise thus leads to trajectories that follow a *stochastic limit cycle*.

The rate by which a *stochastic excitable element* (SEE) releases action potentials depends on the amplitude of the noise as well as on the distance between the stationary point and the closest escape trajectory. This distance determines the excitability of the EE and depends on the parameter z. A change in z will thus influence the average firing rate of the SEE, as shown in Fig. 3.

A change in the excitation parameter z is thus the key physiological mechanism by which the behavior of an individual neuron is controlled in our model. In the remaining part of this text, we will investigate how this parameter influences the synchronicity of firing of coupled neurons, and how this firing synchronicity can serve visual information processing.

3 Dynamics of Coupled Neurons

To describe the dynamics of coupled neurons, we expanded the BvP-equations to include an interaction term between neurons

$$\begin{aligned} \dot{x}_{1,i} &= c(x_{1,i} - x_{1,i}^3/3 + x_{2,i} + z) + \eta_1(t) + \textstyle\sum_j W_{j\to i}(t) \\ \dot{x}_{2,i} &= (a - x_{1,i} - bx_{2,i})/c + \eta_2(t). \end{aligned} \tag{4}$$

The coupling of neuron j to neuron i is described by

$$W_{j\to i}(t) = \theta(-x_{1,j}(t)) \cdot (x_{1,j}(t) - x_{1,i}(t)) \cdot w_{ji}, \quad i,j \in \{1, 2, \ldots N\}. \tag{5}$$

This interaction term is to be interpreted as follows : (1) neuron j only excites neuron i when neuron j is momentarily firing, i.e when the step function θ is 1; (2) the interaction strength is proportional to the difference in the voltages of the neurons, i.e. an excited cell cannot be further excited effectively by another excited cell; (3) the interaction strength is scaled by "synaptic weights" w_{ji}.

A Measure for Synchronicity

In the following we provide a measure for the time correlation of the firing activity of populations of coupled neurons. In case individual neurons are described by nonlinear oscillators, the synchronicity of N oscillators can be determined in terms of the phases $\phi_i(t)$ of the individual oscillators as follows:

$$C_{lc}(t) = \frac{1}{N(N-1)} \sum_{i \neq j}^{N} cos(\phi_j(t) - \phi_i(t)). \quad (6)$$

The definition of a phase $\phi_i(t)$ of the individual oscillators can be expanded to stochastic nonlinear oscillators in close proximity to the limit cycle [11]. In general, however, it is not easy to find a meaningful definition of the phase of an SEE. Since each SEE will stay an undetermined amount of time in the vicinity of the stationary point, the knowledge that two systems are close to each other on the stochastic limit cycle does not allow a prediction of the time when these two systems will start their next turn on the stochastic limit cycle. Therefore, the relative location of SEE's in the phase space is not a good basis for defining their synchronicity. For this reason we define the synchronicity of coupled SEE in terms of the function $t_i(t)$ for neuron i; this function gives at the instant t the time t_i at which neuron i started to fire last, the beginning of firing being defined as the time when x_1 crosses the x_2-axis from $x_1 > 0$ to $x_1 < 0$. The synchronicity C_{SEE} for SEE's was then

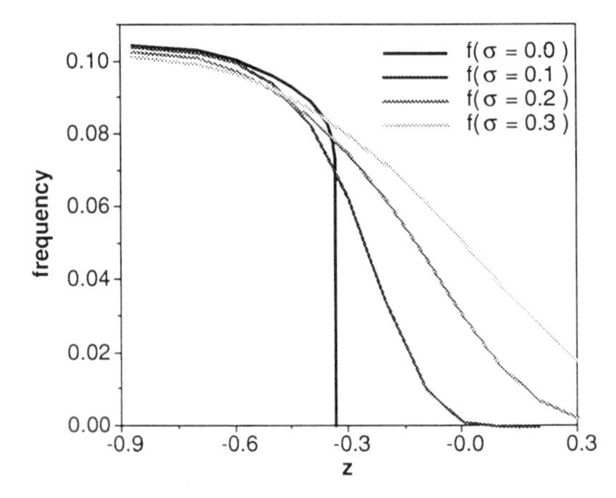

Figure 3. Dependence of the average frequency of action potentials on the input parameter z in the BvP–model (Eq. 2) for different noise amplitudes σ. The frequencies were obtained by integrating the stochastic equations (Eq. 2) over a long time and counting the number of action potentials released.

defined accordingly as

$$C_{SEE}(t) = \frac{1}{N(N-1)} \sum_{i \neq j}^{N} cos(2\pi \frac{(t_j(t) - t_i(t))}{T}). \quad (7)$$

where T is the average firing frequency of the population of SEEs. This definition becomes equivalent to the definition in Eq. (6) when z is changed from $z > -0.34$ to $z < -0.34$, i.e. when the SEE's adopt a limit cycle which does not require noise, as shown in Fig. 1.

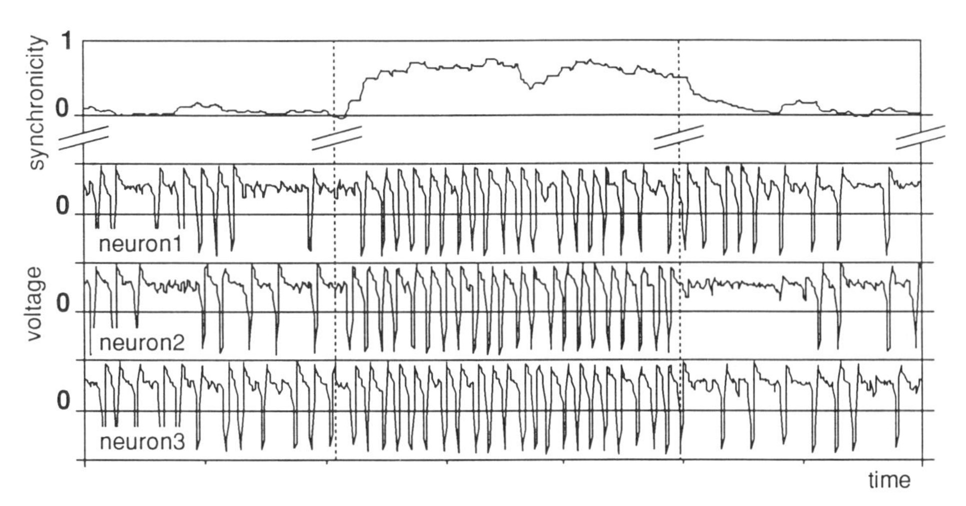

Figure 4. Simulation of the firing activity of 50 coupled BvP neurons. At the time indicated by the vertical dashed lines, the excitation z is first increased from $z = -0.16$ to $z = -0.24$ and then restored to the initial value $z = -0.16$. The top trace shows the correlation function $C_{SEE}(t)$. The synchronicity C_{SEE} responds to the change in z with a delay of about two firing cycles. The lower three traces represent the transmembrane voltage signal $x_1(t)$ of three neurons. (noise amplitude is $\sigma = 0.1$, synaptic weights are all $w_{i,j} = 0.01$).

3.1 Dynamical Properties of Coupled SEEs

To study the dynamics of a population of coupled SEE's, we numerically integrated the dynamics of 50 uniformly coupled SEE's described by Eqs. (4).

Figure 4 shows the result of a typical simulation. In the initial time interval the neurons were only slightly excited ($z = -0.16$). At the instant indicated by the first dashed line, the excitation value is suddenly increased ($z = -0.24$) for a time period lasting to the instant indicated by the second dashed line, at which time the value of $z = -0.16$ is restored. The uppermost graph shows how the correlation function C_{SEE} defined in Eq. 7 responds to the changes of z. One can see, that the correlation function responds within approximately two periods to changes of z. This fast synchronization and desynchronization is a key feature of our model and is in good agreement with experimental data [4]. The three lower traces in Fig. 4 show the voltage signal $x_1(t)$ of three of the 50 neurons. At low excitation values ($z = -0.16$), the neurons fire at a relatively low frequency. As the excitation of the neurons increase ($z = -0.24$), the firing frequency rises and at the same time, the firing pattern of the coupled neurons becomes synchronized.

In Fig. 5 we present histograms showing the time correlation of action potentials for large populations of neurons for different values of the excitation parameter z. The histograms show the firing probability of neuron i at a time δt after neuron j has fired. For the first two simulations with $z = -0.12$ and $z = -0.16$ the firing is essentially asynchronous. There is only a small correlation due to the finite size of the neuronal population ($N = 100$). For $z = -0.20$ and $z = -0.24$ the firing is synchronous. The synchronicity becomes more pronounced as the parameter z is further lowered below the value of $z_{crit} \approx -0.185$, for which the system would show a synchronization phase transition with order parameter C_{SEE} in the limit of $N \to \infty$. (A detailed discussion of this phase transition will be published elsewhere.)

Figure 6 shows the the firing frequencies of coupled and uncoupled neurons and order parameter C_{SEE} as a function of z, obtained from simulations of $N = 500$ neurons. It is worth noting that this transition in the firing pattern already occurs at quite low excitation values, much lower than necessary for an individual neuron to become continuously firing.

Experiments by Gross and Kowalsky [24] have shown that the synchronous firing activity in populations of randomly coupled neurons is usually connected with intense firing activity. This relation is naturally reproduced by our model (see Fig. 6).

4 Artificial Cortical Geometry

We simulated the dynamics of the neural activity in a system where the coupling strengths between any two neurons are no longer uniform, but mimic the coupling scheme of neurons in the visual cortex. The aim was to re-

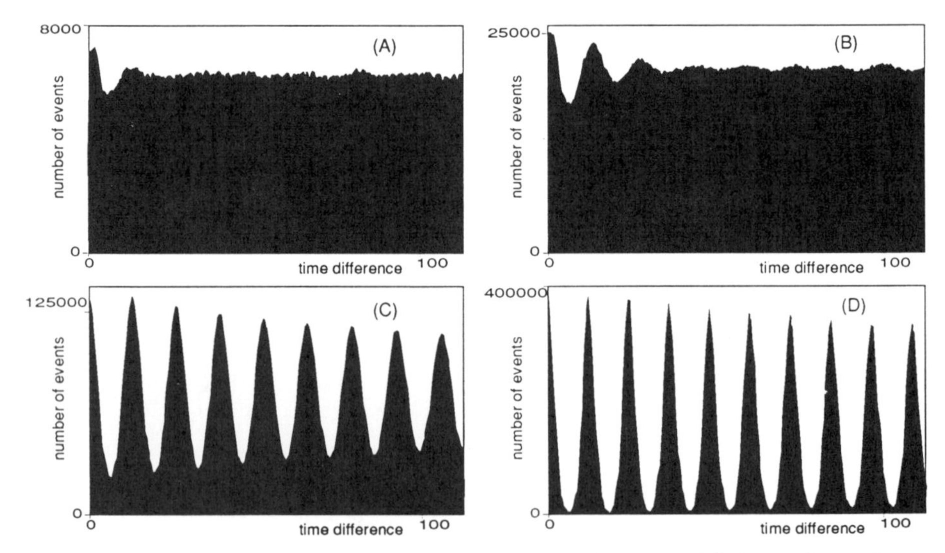

Figure 5. Histograms showing the time correlation of firing of 100 homogeneously coupled neurons. Plotted are the number of times that two firing events have been recorded with a time difference of δt in a simulation involving 100 neurons. The values of z are -0.12 for (A), -0.16 for (B), -0.20 for (C), and -0.24 for (D). The time span over which firing events are correlated increases with lowered z-value (noise amplitude is $\sigma = 0.1$, synaptic weights are all $w_{i,j} = .005$).

produce experiments by Gray and Singer [4] in which oriented light bars presented to the retina of a cat evoked synchronous activity in area 17 of the visual cortex.

Extensive neuroanatomical studies based on Hubel and Wiesel's pioneering works [25, 26] have resulted in information on the structure of the visuo-cortical pathways: activity in the retina projects onto the visual cortex in such a way, that signals from nearby areas on the retina project to nearby areas of the visual cortex. On a lower hierarchal level one finds cells that respond best to stimuli of a specific orientation. Cells representing all orientation prefences are grouped in the socalled hypercolumns of the primary visual cortex.

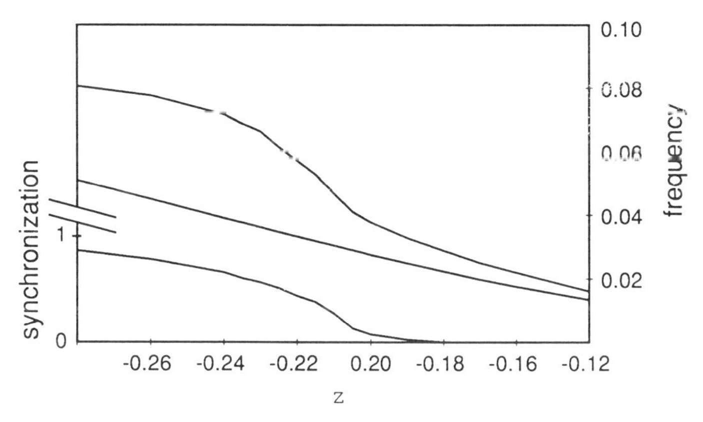

Figure 6. Correlation and firing frequency as a function of the excitation parameter z. The graph shows the dependence on z of the frequency of coupled neurons (upper graph) compared to the frequency of uncoupled neurons (middle graph). The bottommost graph gives the corresponding C_{SEE} for the coupled system which is simulated using 500 neurons coupled by $w_{i,j} = 0.001$ and with noise $\sigma = 0.1$.

Accordingly, we distribute the neurons in our simulation on a two dimensional grid labeling them by their horizontal and vertical position (i, j) and assigning them four different orientation preferences $\theta_{i,j} = 0°, 45°, 90°, 135°$ relative to the x-axis according to the scheme displayed in Figure 7. This distribution scheme ensures that four neurons on any 2×2 area of the artificial cortex assume all four orientation preferences. The scheme is a caricature of the observed orientation homogeneity in the hypercolumns of the cortex [27, 28].

The coupling between two neurons (i, j) and (i', j') was described as in Eq. 5. The weights $w_{i,j}$ are chosen

$$w_{i,j;i',j'} = w_{\max} \cdot w^{(1)}_{i,j;i',j'} \cdot w^{(2)}_{i,j;i',j'} \cdot w^{(3)}_{i,j;i',j'} \tag{8}$$

where

$$w^{(1)}_{i,j;i',j'} = \cos^2(\theta_{i,j} - \theta_{i',j'}) \tag{9}$$

$$w^{(2)}_{i,j;i',j'} = \exp\left(-\frac{\sqrt{(i-i')^2 - (j-j')^2}}{r_0}\right) \tag{10}$$

$$w^{(3)}_{i,j;i',j'} = \frac{\cos^2(\frac{\theta_{i,j}+\theta_{i',j'}}{2} - \xi_{i,j;i',j'}) + 1}{2}. \tag{11}$$

and $\xi_{i,j;i',j'}$ is the orientation of the connection vector between neurons (i, j) and (i', j'). The weights are scaled by the global parameter $w_{\max}$. The three factors in this formula describe how the coupling strength depends (1) on the relative orientation preference, (2) on the distance, and (3) on the alignment of the orientation preferences of neighboring neurons, respectively.

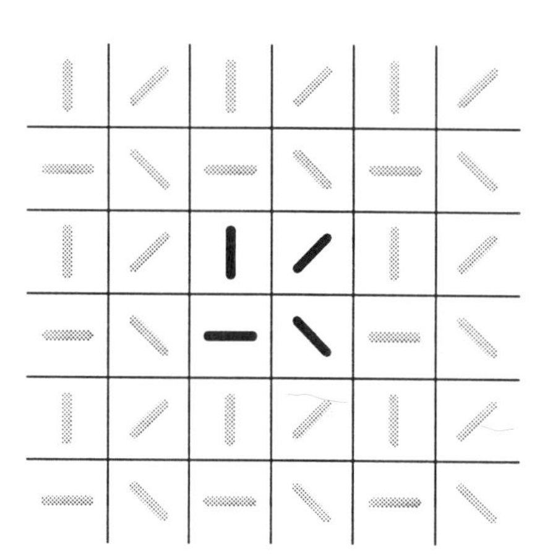

Figure 7. Distribution of orientation preferences of neurons in the artificial cortex. The center part of the figure shows a quadruple of neurons which represent an orientation column.

Rotating Bar on the Artificial Cortex

We use our artificial cortex to simulate the synchronous activity evoked in the visual cortex by a rotating bar presented to the retina. This bar is shown schematically in Fig. 8. In these simulations, cortical neurons the receptive fields of which overlap with the light bar receive an additional excitation of strength Δz

$$\begin{aligned} \dot{x}_{1,i} &= c(x_{1,i} - x^3_{1,i}/3 + x_{2,i} + (z + \Delta z_{i,j}) + \eta + \textstyle\sum_j W_{j\to i} \\ \dot{x}_{2,i} &= (a - x_{1,i} - b x_{2,i})/c + \eta(t), \end{aligned} \tag{12}$$

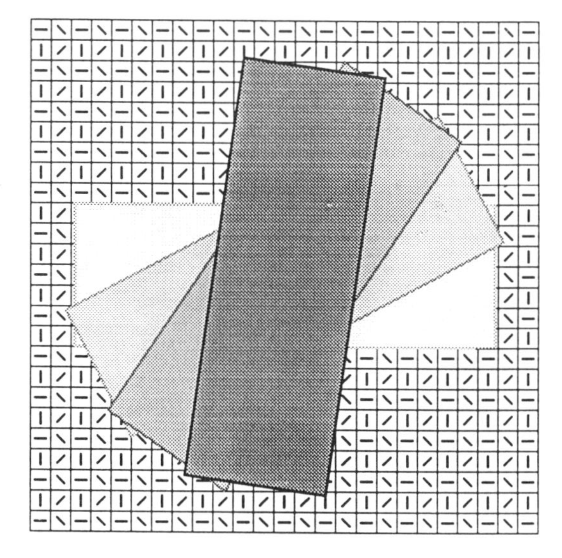

Figure 8. Scheme showing the geometry of a rotating light bar, providing excitation to the receptive fields of 25 × 25 neurons.

where

$$\Delta z = \begin{cases} \Delta z_{\max} \cdot \cos^2(\theta_{i,j} - \theta_0) & : \text{for neurons receiving input from the light bar} \\ 0 & : \text{for all other neurons} \end{cases} \tag{13}$$

with θ_0 being the orientation of the bar. The other parameters are $z = -0.16$, $\Delta z_{\max} = -0.16$, $\sigma = 0.1$, and $w_{\max} = 0.04$.

We simulated a light bar presented through the retina to an array of 25 × 25 neurons. The light bar initially covered the central portion of the retina which is connected to 7 × 21 neurons of the cortex then rotated around its center with a constant angular velocity (see Fig. 8). In these simulations, the neurons covered by the light bar and having an orientation preference approximately parallel to the bar engaged in synchronous, periodic oscillation. Neurons covered by the light bar, but having an orientation preference approximately perpendicular to the orientation of the bar, did not fire periodically and had a much lower frequency; their firing spikes however were correlated with the firing spikes of the neurons with correct orientation preference. Figure 9 presents a histogram of time differences between spikes for the two populations of neurons, those covered by the bar (1) and those not covered by the bar (2). One can see that firing correlation is maintained over a long time for population (1). The population (2) does not show long time correlation; the small wiggle in the correlogram is caused by the influence of the neurons of (1), which s excite adjacent neurons of population (2) .

As the bar rotates, the population (1) of excited neurons constantly changed due to the rotation of the bar, with new neurons added and other neurons leaving the population. The ability of these neurons to respond quickly with their firing activity, i.e. quickly synchronize or desynchronize

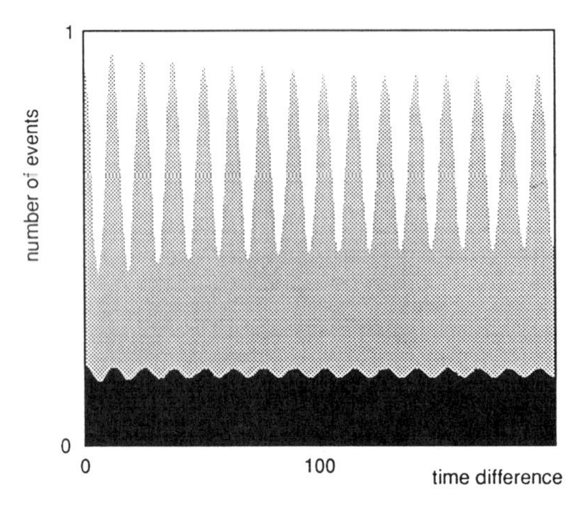

Figure 9. Histograms showing the firing time correlation of different populations of the cortical neurons. Plotted are the number of times that two firing events of neurons of the same population have been recorded with a time difference of δt. The upper curve shows the counts for neurons (1) covered by the bar, the lower curve shows the counts for neurons (2) not covered by the bar. Neurons (1) under the bar fire at a higher frequency and with larger correlation. ($z = -0.16$, $\Delta z_{\text{max}} = -0.16$, $\sigma = 0.1$, and w_{max}=0.04).

with the rest of the population, was thereby important for the coding of the location of the bar in terms of synchronous firing activity. Although different neurons participated in the synchronous activity during the rotation of the bar, the phase of the oscillations was preserved over a time period of a full turn of the object.

5 Conclusion

Our investigations demonstrated that synchronous periodic oscillations appear in populations of SEE's with sufficiently large external excitation z. Both synchronization and desynchronization occur within few firing periods after modification of z. We believe that this behaviour is relevant to understand the occurrence of synchronous firing in physiological neural networks as observed in [4, 5].

The only way to achieve synchronization and desynchronization in oscillator models, rather than SEE models, is to shift the phases of single oscillators. This shift is driven by the competition of two mechanisms, coupling and noise. This shift develops only slowly, unless coupling and noise are changed drastically. SEE models show a different, and more suitable behaviour. Synchronicity is controlled by the excitability of the neurons. Periodic activity only arises as the excitability of the neurons is increased beyond a critical threshold z_{crit}. The single neurons themselves have no autonomous phase, but respond with their firing activity such that periodic activity emerges very rapidly in a population of coupled SEEs. While oscillator models have been similarly successful in describing the state of collective oscillations in neural networks, the SEE model also describes well the state of non-synchronous firing. Furthermore the model reproduces the relation between high firing frequency and synchronicity experimentally measured for randomly connected networks [24].

The model in based on a description of single neurons in terms of observable dynamic variables. The results presented demonstrate a simple mechanisms by which synchronization can be induced and exploited for solution of the binding problem.

Acknowledgement: The authors wish to thank Thomas Martinetz, Klaus Obermayer, Helge Ritter and Guenter W. Gross for useful discussion. This work has been supported by the University of Illinois at Urbana-Champaign, by a grant of the National Science Foundation DMR 89-20538 (through the Materials Research Lab of the University of Illinois), and by a grant of the National Science Foundation DIR 90-15561.

References

[1] J. E. Hummel and I. Biederman. Dynamic binding: A basis for the representation of shape by neural networks. Manuscript submited to the 12th Annual Meeting of the Cognitive Science Society, Cambridge, MA, July 1990.

[2] W. J. Freeman. *Mass Action in the Nervous System*. Academic Press, New York, 1975.

[3] C. S. Skarda and W. J. Freeman. How brains make chaos in order to make sense of the world. *Behavioural and Brain Sciences*, 10:161–195, 1987.

[4] C. M. Gray and W. Singer. Stimulus-specific neuronal oscillations in orientation columns of cat visual cortex. *PNAS*, 86:1698–1702, 1989.

[5] R. Eckhorn, R. Bauer, W. Jordan, M. Brosch, W. Kruse, M.Munk, and H. J. Reitboeck. Coherent oscillations: A mechanism of feature linking in the visual cortex? multiple electrode and correlation analysis in the cat. *Biol. Cybernetics*, 60:121–130, 1989.

[6] C. von der Malsburg. The correlation theory of brain function. Internal Report 81-2, Department of Neurobiology, Max Planck Institute for Biophysical Chemistry, Göttingen, FRG, 1981.

[7] R. Eckhorn, H. J. Reitboeck, M. Arndt, and P. Dicke. A neural network of feature linking via synchronous activity. In *Models of Brain Function*. Cambridge University Press, 1989.

[8] H. Sompolinsky, D. Golomb, and D. Kleinfeld. Global processing of visual stimuli in a neural network of coupled oscillators. *PNAS*, 87:7200–7204, 1990.

[9] H. G. Schuster and P. Wagner. A model for neuronal oscillations in the visual cortex. In *Parallel Processing in Neural Systems and Computer*, pages 143–146. North Holland, 3 1990.

[10] T. B. Schillen and P. König. Coherency detection by coupled oscillatory responses - synchronizing connections in neural oscillator layers. In *Parallel Processing in Neural Systems and Computer*, pages 139–142. North Holland, 3 1990.

[11] A. T. Winfree. *The Geometry of Time*. Springer-Verlag, Berlin Heidelberg New York, 1980.

[12] Y. Kuramoto. *Chemical Oscillations, Waves, and Turbulence*. Springer Verlag, New York, 1984.

[13] Y. Kuramoto and I. Nishikawa. Statistical macrodynamics of large dynamical systems. case of a phase transition in oscillator communities. *J. Stat. Phys.*, 49(3/4):569, 1987.

[14] L. L. Bonilla, J. M. Casado, and M. Morillo. Self-synchronization of populations of nonlinear oscillators in the thermodynamic limit. *J. Stat. Phys.*, 48(3/4):571, 1987.

[15] S. H. Strogatz and R. E. Mirollo. Collective synchronization in lattices of nonlinear oscillators with randomness. *J. Phys. A*, 21:L699–L705, 1988.

[16] D. M. Kammen, P. J. Holmes, and C. Koch. Personal communications.

[17] H. J. Reitboeck, R. Eckhorn, M. Arndt, and P. Dicke. A model for feature linking via correlated neural activity. In *Synergetics of Cognition*. Springer Verlag, 1989.

[18] A. V. Holden, M. Markus, and H. G. Othmer, editors. *Nonlinear Wave Processes in Excitable Media*. Plenum Press, London, 1990.

[19] R. Fitzhugh. Impulses and physiological states in theoretical models of nerve membranes. *BJ*, 1:445, 1961.

[20] J. Nagumo, S. Arimoto, and S. Yoshizawa. An active pulse transmission line simulating nerve axon. *Proceedings of the IRE*, 50:2061–2070, 1962.

[21] A. L. Hodgkin and A. F. Huxley. A quantitative description of membrane cuurent and its application to conduction and excitation in nerve. *J. Physiol. London*, 117:500, 1952.

[22] H. Treutlein and K. Schulten. Noise-induced neural impulses. *Eur. Biophys. J.*, 13:355–365, 1986.

[23] Ch. Kurrer and K. Schulten. Effect of noise and perturbations on limit cycle systems. *Physica D*, 1990. in press.

[24] G. W. Gross and J. M. Kowalski. Experiments and theoretical analysis of random nerve cell network dynamics. In *Neural Networks: Concepts, Applications, and Implementations.* Prentice Hall, 1990. in press.

[25] D. H. Hubel and T. N. Wiesel. Receptive fields, binocular interaction and functional architecture in the cat's visual cortex. *J. Physiol.*, 160:106–154, 1962.

[26] D. H. Hubel and T. N. Wiesel. Sequence regularity and geometry of orientation columns in the monkey striate cortex. *J. Comp. Neurol.*, 158:267–294, 1974.

[27] G. Blasdel and Salama. Voltage sensitive dyes reveal a modular organisation in monkey striate cortex. *Nature*, 321:579, 1986.

[28] K. Obermayer. A principle for the formation of the spatial structure of cortical feature maps. *Proc. Natl. Acad. Sci., USA*, in press, 1990.

SOME COLLECTIVE PHENOMENA IN THE HIPPOCAMPUS IN VITRO

Roger D. Traub[1] and Richard Miles[2]

1) IBM Research Division
Thomas J. Watson Research Center
Yorktown Heights, NY 10598, U.S.A.

2) Institut Pasteur
Laboratoire de Neurobiologie Cellulaire
28 Rue du Dr. Roux
75724 Paris Cedex 15, France

In this chapter, we shall discuss the structure of one part of the guinea-pig hippocampus, the anatomically simplest type of cortex. We shall describe how the structure of this part (the CA3 region) leads to the existence of a repertoire of different modes of population activity. Selection of a particular mode of population behavior depends on the settings of various functional parameters such as the conductances of the inhibitory synapses. Some of the population activities observed *in vitro* and in models of the slice appear to have *in vivo* analogs. We shall tabulate some of the mechanisms which may be available to the brain for setting functional parameters in the hippocampus. Finally, we shall speculate briefly on the behavioral significance of certain population modes.

Experimental preparation. Physiological recordings are made from 300 to 400μ thick slices of the hippocampus, maintained in a plexiglass chamber. This preparation is cut so that the cells remain viable, as do sufficiently many synaptic interconnections that interesting population behaviors occur. Wedges of the slice can be dissected out as well. We estimate that the smallest number of cells able to generate a synchronized discharge is about 1,000. Most experiments involve populations of several thousand to about 20,000 cells. The slice preparation allows chemical control of the bathing medium and stable intracellular recordings from pairs of neurons. The slice has the advantage of being a somewhat simplified biological system, since it is disconnected from the rest of the brain. At the same time, this very disconnection means that collective neuronal behaviors may not have precise correlates in the hippocampus *in situ*. Nor can population activities in the slice be clearly linked to normal functions of the hippocampus. This latter defect of the slice

Self-Organization, Emerging Properties, and Learning
Edited by A. Babloyantz, Plenum Press, New York, 1991

is not devastating, since the functions of the hippocampus (in the formation of new memories or in spatial performance tasks) are poorly defined in cellular terms.

Types of synaptic interactions in the hippocampus. While many types of neurotransmitters are found in the hippocampus, two transmitters - glutamate and γ-aminobutyric acid (GABA) - appear to be dominant on the time scales of greatest interest to us (hundreds of ms). Both glutamate and GABA bind to more than one receptor type, leading in turn to different postsynaptic actions. Let us consider three classes of unitary synaptic interaction: e→e, e→i and i→e. Here, "e" denotes an excitatory neuron and "i" an inhibitory neuron. i→i interactions are not well characterized.

A) e→e synapses. When a presynaptic action potential arrives at a release site, glutamate molecules enter the synaptic cleft and bind to postsynaptic receptors, including "Quis" (for quisqualate, an agonist) or "NMDA" (for N-methyl-D-aspartate, also an agonist) types. The respective receptors may be located near to one another, within the same synaptic region. Blockers for Quis and NMDA receptors exist, and NMDA-activated currents are sensitive to Mg^{+2} concentration at the resting potential, so that the two types of synaptic response can be separated from each other experimentally. The unitary Quis-activated conductance is brief (a few ms). The NMDA response is more complex. The NMDA-activated conductance has a rapid onset, but then decays with time constant about 100 ms (Forsythe and Westbrook, 1988). In addition, the conductance is not "expressed" (that is, no current flows) unless the postsynaptic membrane is depolarized or unless $[Mg^{+2}]_o$ is lowered (Mayer, Westbrook and Guthrie, 1984). e→e synapses are on dendrites, generally on spines.

B) e→i synapses. There is a Quis component (Miles, 1990), but it is not yet known if an NMDA component exists. Within the CA3 region, the postsynaptic conductance has a more rapid upstroke, and is larger, than for e→e synapses. This probably results because e→i synapses are on or near the cell body.

C) i→e synapses. GABA can bind to either "A" or "B" receptors. It seems likely, but not definitively proven, that separate $GABA_A$ and $GABA_B$ synapses exist, and that these synapses are contacted respectively by distinct populations of inhibitory neurons. $GABA_A$ synapses are located, at least largely, on or near the cell bodies of e-cells. Let us consider a unitary IPSP (inhibitory postsynaptic potential), where unitary means "evoked by a single presynaptic action potential". Unitary $GABA_A$ IPSPs within CA1 appear to be small, perhaps less than 0.5 mV. The underlying conductance decays with a time constant less than 10 ms above 30° (Collingridge, Gage and Robertson, 1984). In CA3, unitary $GABA_A$ IPSPs vary considerably in amplitude (Table 1), but all decay with a time constant close to the membrane time constant (30-50 ms), suggesting that the underlying conductance decays this fast or faster. The peak of $GABA_A$ IPSPs in CA3 is attained in less than 5 ms. $GABA_A$ receptors open Cl^- channels. Unitary $GABA_B$ IPSPs have not been clearly identified, although CA1 lacunosum-moleculare i-cells may contribute to such IPSPs

(Lacaille and Schwartzkroin, 1988b). $GABA_B$ or "slow" IPSPs are typically evoked by stimulating afferent fibers. In such cases, the onset of the conductance is delayed for more than 10 ms and may not peak for about 200 ms (Hablitz and Thalmann, 1987). The IPSP can last several hundred ms. $GABA_B$ receptors open K channels, probably via an intracellular second messenger system (Nicoll, 1988). Pharmacological blockers exist for A and for B GABA receptors.

Properties of individual hippocampal neurons. Neurons may be classified by their location, shape, axonal arborization pattern, electrical properties (including different modes of firing), synaptic actions, presence of various calcium-binding proteins or transmitter peptides, and so on. Many or all of these properties are relevant for modeling. Certain types of synaptic actions have been mentioned and we shall describe axonal arborization - that is, synaptic connectivity - below. Here, we discuss very briefly the electrical properties of some hippocampal neurons.

A neuron may be viewed as a set of interconnected leaky coaxial cables into whose surfaces are inserted various channel and receptor proteins. The channels allow passage, with greater or lesser selectivity, of ionic species across the cell membrane. Each type of channel may be regulated in its own way by any of the following: membrane potential, neurotransmitter(s), external and/or internal divalent cations, and the internal chemical milieu. The full elucidation of the processes for even one cell type is an awesome task.

For generating the electrical output of pyramidal cells, the following intrinsic conductances (each corresponding to a channel type) appear to be relevant: g_{Na}, g_{Ca}, and various types of g_K, denoted $g_{K(DR)}$, $g_{K(A)}$, $g_{K(C)}$, $g_{K(AHP)}$ and $g_{K(M)}$. Here "g" is standard notation for "conductance", and the subscript denotes the specific ion passed most readily by the channel. g_{Na} is responsible for the upstroke of the action potential. g_{Ca} can contribute to slow (tens of ms or more) depolarizations or can lead to broad action potentials. The different types of g_K contribute, in different combinations, to repolarization of Na-spikes, Ca-spikes, or bursts of spikes. They also generate spike or burst afterhyperpolarizations, which may be brief (ms or tens of ms) or long (seconds). Patch clamp techniques, applied to isolated cells (Kay and Wong, 1986) or thin slices (Edwards, Konnerth, Sakmann and Takahashi, 1989) are leading to quantitative descriptions of various channel types. Such data are essential for understanding the firing behaviors of cells. A different but related issue is to determine the density of the various channels in different regions, for such densities do not appear to be uniform, neither in hippocampal pyramidal cells, nor in many other cell types (motorneurons, Purkinje cells, inferior olivary neurons, and so on). Optical imaging of intracellular Ca^{+2} signals provides some information (Ross and Werman, 1987; Regehr, Connor and Tank, 1989) as do intradendritic electrical recordings (Wong, Prince and Basbaum, 1979) but new techniques may need to be developed.

A) *CA3 pyramidal cells.* These excitatory neurons constitute about 90% of the cells in the CA3 region. They generate several different output patterns depending on the refractoriness of the cell and how (and where) the cell is

TABLE 1

Interactions Between Neurons in CA1 and CA3 Regions of the Hippocampus in Vitro

Synapse	*Prob.*	*PSP size(mV)*	*Functional*	*Refs.*
CA3 e→e	.02	0.7±0.3	burst→burst (probability 0.2-0.5; latency 10-30 ms)	a
CA3 e→e (immature)	~.04	~2.7 mV	burst→burst	b
CA3 i→e	0.6	-0.3 to -2.6	can prevent post-synaptic burst	c
CA3 e→i	0.1	1 to 4	spike→spike (prob. 0.6, latency 2.5-5 ms)	d
CA3 e→ CA1 e	.06	.13	series of unitary EPSPs subthreshold	e
CA1 e→i (oriens)	.64	2	spike→spike (sometimes)	f
CA3 e→ CA1 i (oriens)	>0?			f
CA1 i (oriens) →e	.07	<-1 (unit.) -1 (train)	train may prevent spike	f
CA1 e→i (lacunosum)	0?			g
CA1 i (lac.) →e	.21 to .26	-0.9 (train)	delays or blocks firing of e-cell	g
CA3 e→ CA1 i (lac.)	>0?			h
CA1 e→i (pyramidale)	.28	2 to 4	spike→spike (prob. ~0.5)	i
CA3 e→ CA1 i (pyr.)	> 0			j
CA1 i (pyr.)→ CA1 e	.3	<1 (unitary) -2 to -4 (train)		i
CA1 i (oriens)→i (pyr.)	>0	-10 (train)		f
CA1 i (lacun.)→i (pyr.)	>0	-.7 (train)	delays or blocks spikes	g

References for Table 1.
a) Miles and Wong, 1986, 1987a. b) Swann, Smith and Brady, in press. c) Miles and Wong, 1984; Miles, in prep.; Traub and Miles, 1991. d) Miles, 1990. e) Sayer, Friedlander and Redman, 1990. f) Lacaille, Mueller, Kunkel and Schwartzkroin, 1987. g) Lacaille and Schwartzkroin, 1988b. h) Lacaille and Schwartzkroin, 1988a. i) Knowles and Schwartzkroin, 1981. j) Schwartzkroin and Mathers, 1978.

stimulated. For example, a brief excitatory stimulus to either soma or dendrite may evoke an *intrinsic burst*, consisting of up to 8 action potentials, 5 to 10 ms apart, riding on a depolarizing wave and followed by a long AHP. The burst may contain one or more slow calcium-mediated action potentials. A large steady stimulus to the soma evokes, in contrast, a more-or-less regular train of action potentials (Wong and Prince, 1981), a form of firing that may not occur normally but which can be seen in certain pathological situations. A typical orthodromic stimulus, consisting of overlapping excitatory and inhibitory inputs, may evoke a single action potential or just subthreshold potentials. A model of the CA3 pyramidal cell (Traub, Wong and Miles, 1990) predicts that steady excitation of the distal dendrites should lead to high frequency bursting (at about 10 Hz), a form of behavior not seen with somatic stimulation. Such stimulation is difficult to apply to real cells because of dendritic branching, but something similar may happen during one type of epileptic bursting (see below).

B) *CA1 pyramidal cells.* The voltage threshold for firing is higher in CA1 cells than in CA3 cells. Furthermore, in contrast to CA3 pyramidal cells, CA1 cells usually respond to steady depolarization of the soma by firing repetitively, not by bursting (Schwartzkroin, 1978). Nor does a brief stimulus to the soma usually generate a burst. In contrast, an excitatory pulse to a CA1 dendrite readily evokes a dendritic burst. Orthodromic mixed excitatory/inhibitory synaptic inputs generally evoke a single dendritic spike, but the same stimulus applied with $GABA_A$ inhibition blocked can evoke dendritic (and somatic) bursting (Wong, Prince and Basbaum, 1979). Likewise, synchronized firing in a population of CA3 cells can elicit bursting in CA1 cells, provided that inhibition is blocked within CA1 (Mesher and Schwartzkroin, 1980). Preliminary calculations indicate that a comparmental model reproducing CA3 firing properties (Traub, Wong and Miles, 1990) can be modified to reproduce CA1 behavior, by enhancing $g_{K(DR)}$ on and near the soma, and by reducing g_{Ca} and $g_{K(C)}$ in the dendrites.

C) *CA3 and CA1 inhibitory cells (i-cells).* While some i-cells fire in bursts (Kawaguchi and Hama, 1988), most respond to a depolarizing current by firing a steady train of narrow spikes, often with prominent AHPs. i-cells are notable for their relatively high rates of spontaneous firing (sometimes more than 10 Hz), the large size of unitary EPSPs evoked by nearby pyramidal cells (except for CA1 lacunosum/moleculare cells) (Table 1), and their low firing threshold. As a result, single pyramidal cell spikes may cause these cells to fire with latencies as brief as 2.5 ms. (Knowles and Schwartzkroin, 1981; Lacaille, Mueller, Kunkel and Schwartzkroin, 1987; Lacaille and Schwartzkroin, 1988a,b; Miles, 1990; Sayer, Friedlander and Redman, 1990). In addition, CA3 and CA1 i-cells are readily excited by stimulation of afferent pathways, generally at lower stimulus thresholds and with shorter latencies than seen in nearby pyramidal cells (Knowles and Schwartzkroin, 1981; Lacaille, Mueller, Kunkel and Schwartzkroin, 1987; Lacaille and Schwartzkroin, 1988a,b; Miles, 1990).

Functional aspects of synaptic connections. While single pyramidal cell action potentials may elicit spikes in connected i-cells, it is rare for a single CA3 pyramidal cell to elicit a spike in another pyramidal cell. Some further properties of the neuronal connections are described below.

A) CA3 pyramidal cell → CA3 pyramidal cell. While a unitary EPSP, about 1 mV, is generally subthreshold, a presynaptic burst can evoke a postsynaptic burst with probability as high as 0.5 during 0.5 Hz stimulation. The reason for this transmission of activity is that decay of an EPSP is long compared to the interspike interval during a burst, so that successive EPSPs add. Short-term facilitation (probably presynaptic in origin) and voltage-dependent currents in the postsynaptic cell may contribute to burst transmission. The latency for burst propagation is 10 to 30 ms (Miles and Wong, 1987a).

B) Disynaptic CA3 pyramidal cell excitation. When $GABA_A$ inhibition is suppressed, a burst in one CA3 pyramidal may evoke an EPSP in another one through an intercalated cell. The EPSP latency and shape are consistent with the notion that the intercalated cell is induced to burst by the stimulated pyramidal cell (Miles and Wong, 1987a).

C) CA1 pyramidal cell → CA1 pyramidal cell. Recurrent excitation appears to be rare in CA1, at least as assessed by dual intracellular recording (Knowles and Schwartzkroin, 1981). Microapplication of glutamate drops indicates that some recurrent excitation may exist within CA1, at least in longitudinal slices (Christian and Dudek, 1988).

D) CA3 recurrent inhibition. Unitary $GABA_A$ IPSPs in CA3 vary 8-fold in amplitude between different connections (Miles, in prep.). The latency to the peak of the IPSP is brief (less than 5 ms). To observe this latency in the case of small IPSPs requires averaging, because the unitary event may be less than electrode noise. A unitary disynaptic IPSP may be of sufficient amplitude to block bursting that would otherwise be elicited by a presynaptic pyramidal cell burst (Miles and Wong, 1984).

E) CA3 disynaptic inhibition is particularly effective for several reasons. First, the latency to recruit the intercalated i-cell is short (2.5 to 5 ms), and the onset of the resulting IPSP is also brief. As a result, the latency to onset of a disynaptic IPSP is less than for the onset of a monosynaptic e→e evoked burst. In addition, the divergence of connections from CA3 i-cells is large. However, this divergence does not cover the entire CA3 network. One can find excitatory connections $e_1 \rightarrow e_2$ where e_1 does not disynaptically inhibit e_2.

F) CA1 feedforward and feedback inhition. Unitary IPSPs observed to date within CA1 are quite small, so that even the summated IPSP following a presynaptic spike train may be less than 1 mV. Such summated IPSPs have been demonstrated to delay or block action potentials elicited by near-threshold current stimuli (see references in Table 1). To what extent unitary CA1 IPSPs can block stimulated activity impinging from CA3 remains to be elucidated.

Synaptic connection probabilities. Such probabilities can be estimated by recording from a large number of neuronal pairs, assuming that both neurons can be identified, by location, morphology, postsynaptic action or firing pattern. The resulting estimates of probability (Table 1) are approximate. For example, connection probability between CA3 pyramidal cells depends on intercellular distance in longitudinal slices (Miles, Traub and Wong, 1988), but collecting valid statistics on connection probability, using dual recordings, at many different intercellular distances is likely to be impractical.

Experimental estimates of connection probabilities are important parameters for constructing cellular models of hippocampal circuitry. Certain types of data, unfortunately, are not yet available. Thus, to illustrate: A) We do not have data on connections in CA3 to or from single cells mediating $GABA_B$ IPSPs. B) We do not have data on *higher order* probabilities. For example, suppose e_1, e_2 and e_3 are CA3 pyramidal cells. If $e_1 \rightarrow e_2$ and $e_2 \rightarrow e_3$, does that have any effect on the probability of $e_3 \rightarrow e_1$? Similarly, if e_4 is a CA1 pyramidal cell, does $e_1 \rightarrow e_2$ and $e_2 \rightarrow e_4$ influence the probability of $e_1 \rightarrow e_4$? It is a natural assumption, particularly for physicists who come into this field, that these probabilities are *not* independent. However, in the absence of specific data, we assume in our models a *random and independent* connectivity, and examine collective phenomena which do not (so far as we can tell) depend on correlations in the connection probabilities.

Models of the CA3 region. We have constructed two different models of the CA3 region. In each case, individual cells are simulated with a compartmental model, either with 4 types of ionic channels (Traub, 1982) or with 6 (Traub, Wong and Miles, 1990), in the latter case using kinetics constrained by voltage clamp data. With the simpler single-cell model, circuits of almost 10,000 neurons can be simulated. With the more complicated model, we have been confined to 1,100 neurons. The population phenomena in each case are qualitatively similar to each other and to experiments, except that synchronized multiple bursts (Miles, Wong and Traub, 1984) are more readily reproduced with the more complicated model (6 voltage channels per cell).

Collective phenomena: evoked events in CA3. The simplest collective phenomenon results when $GABA_A$ inhibition is blocked pharmacologically. Stimulation of a small number of cells (Wong and Traub, 1983), sometimes even of one cell (Miles and Wong, 1983) then leads to synchronized firing throughout the population. Because, in the absence of inhibition, bursting can propagate from e-cell to e-cell with non-zero probability, and since e-cell recurrent axons are divergent, this process can be viewed as a chain reaction (Traub and Wong, 1982). However, it proceeds relatively slowly, at least in its initial stages, because of the tens of ms required for bursting to propagate from one cell to another.

One can view as a time-dependent percolation the results of stimulating one cell in the presence of partial inhibition. Experimentally, this can be demonstrated by stimulating one e-cell (e_1) and recording from another (e_2) as $GABA_A$ inhibition is gradually blocked (Miles and Wong, 1987a). First, e_1 has no effect on e_2. Then a single path opens up (intermittently) leading to an EPSP in e_2 at long latency. Later, two paths may open up, leading to two EPSPs in e_2 at different latencies. Finally, percolation occurs, and e_2 bursts, in phase with e_1 itself and with a field potential indicating the simultaneous firing of a large number of cells. A similar phenomenon occurs in network models (Traub, Miles and Wong, 1987; Traub and Miles, 1991). Another way to view this percolation is to count the number of e-cells induced to fire by stimulating one e-cell at different mean strengths of the $GABA_A$ synapses. This is difficult to do experimentally, since it requires recording from all of the cells, but can be done in simulations (Traub, Miles and Wong, 1987; Traub and Miles, 1991). There are several "phases". When inhibition is above some value, say c_1, a small number of cells are induced to fire, the number being almost independent of the level of inhibition. Between c_1 and c_2 ($c_2 < c_1$), the number of cells firing rises steeply as inhibition is reduced. Below c_3 ($c_3 < c_2$), virtually all of the cells fire. As inhibition is reduced further, activity spreads through the population faster, but no new cells are recruited. The partially synchronized population bursts that occur between c_2 and c_3 resemble a normal *in vivo* phenomenon, the physiological sharp wave (SPW) (Buzsáki, 1986). During SPW, relatively synchronized firing, for tens of ms, takes place throughout both hippocampi and some interconnected structures. SPW are associated with specific behaviors such as drinking. Since not all neurons fire during a SPW, it would be intersting to know if, and how, it matters to the animal *which* neurons fire.

What is the functional significance of the excitatory connections between CA3 pyramidal cells? Some insight may be obtained from the series of simulations in Fig. 1. These simulations are of a network of 1,000 e-cells, 50 i-cells mediating $GABA_A$ inhibition and 50 i-cells mediating $GABA_B$ inhibition. e-cells are represented with the 6-current model (Traub, Wong and Miles, 1990) and i-cells are modeled with Hodgkin-Huxley-like active currents so as to fire repetitively, but not develop intrinsic bursts. Connectivity is random, and can be described by 4 parameters: the average number of excitatory inputs per e-cell and i-cell, and inhibitory inputs per e-cell and i-cell, respectively. These numbers are 40, 40, 40 and 10. Excitatory inputs to e-cells impinge on one apical and one basilar dendritic compartment, the conductance to each being $c_e t e^{-t/3}$, where c_e is a parameter. Excitatory inputs to i-cells are on the soma. $GABA_A$ inputs to e-cells are to the soma and adjacent compartments. The conductance peaks in 1 ms and is then proportional to $c_i e^{-t/7}$. Further details of this type of network will be presented subsequently, but the principles are similar to networks described previously (Miles, Traub and Wong, 1988; Traub and Miles, 1991).

In each run illustrated in Fig. 1, the same 10 cells were stimulated sufficiently to elicit bursting. This bursting constitutes the *primary response* of CA3 to a brief coherent stimulus. When $c_e = 0$ (left, with recurrent excitation to e-cells disabled), nothing more happens, except that some of the primary bursts are slightly truncated by recurrent inhibition. As recurrent excitation is "engaged" ($c_e = 2$ nS), a *secondary response* occurs as the stimulated cells excite postsynaptic e-cells. Because of recurrent inhibition, however, only a small fraction of the postsynaptic e-cells fire. As recurrent inhibition is reduced, the secondary response grows in amplitude and complexity. Growth does not just occur by adding firing cells, since an occasional cell firing at one

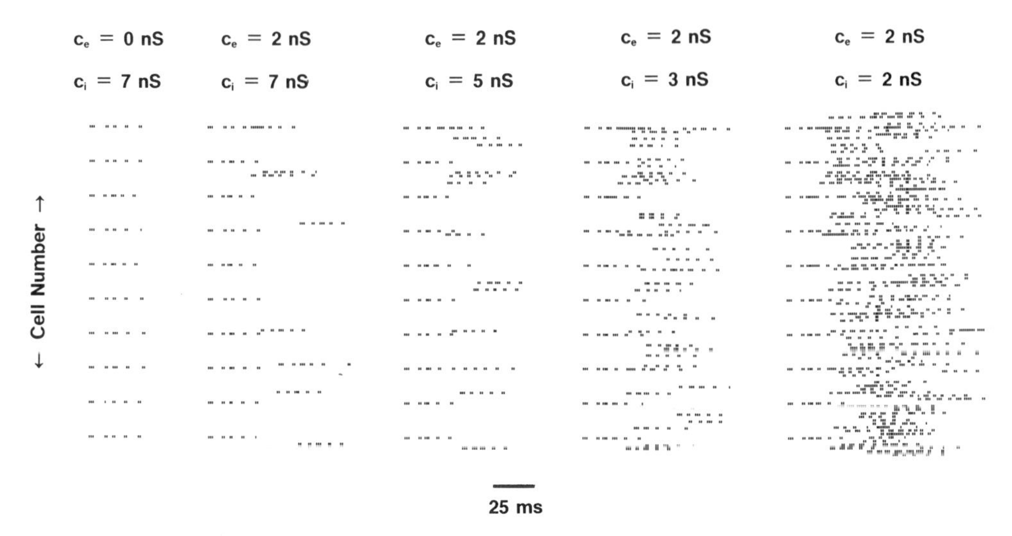

Figure 1. Primary and secondary responses in a network model of the CA3 region. 5 simulations were done in a network of 1,000 pyramidal cells (each with 6 voltage-dependent currents, 19 compartments and 209 membrane state variables) and 100 inhibitory cells. Interconnection probabilities were as described in the text. In each case, 10 cells were excited (0.3 nA for 10 ms) to elicit bursting. In each plot, a dot is inscribed at time t and ordinate i if cell i is depolarized more than 20 mV at time t. When recurrent excitation to e-cells is absent (c_e = 0) only the stimulated cells burst, giving the primary response (left). With recurrent excitation present (c_e = 2 nS) but with $GABA_A$ synapses powerful (c_i = 7 nS), a small secondary response is elicited, consisting of cells monosynaptically excited by the stimulated cells. The amplitude of the secondary response grows as inhibition is reduced further. In addition, cells begin to participate that are polysynaptically excited by the primary cells. This is especially apparent when c_i = 2 nS (right).

level of inhibition will fail to do so at a lower level of inhibition when more cells overall are firing.

Fig. 1 represent an idealization, since the network begins at rest. In the real brain, where there is always background activity, refractoriness will modify both the primary and secondary responses. The functional consequence of this is that the cellular composition of the secondary response is, at least to some extent, unpredictable.

Another method of obtaining insight into the function of CA3 circuitry is to examine autonomous activity. In the model, this is done by randomly choosing a small bias current, constant for each cell but different, in general, between cells. Under these conditions, the model again exhibits one of at least three "phases" depending upon the strength of $GABA_A$ inhibition (Traub, Miles and Wong, 1989; Traub and Miles, in press). With $GABA_A$ synapses stronger than some threshold, a series of population waves occurs at intervals of a few hundred ms. During each wave, a small fraction of the population fires (a "cluster"), but individual cells do not fire periodically, nor does the composition of the clusters repeat. The interval between waves is set in large part by the duration of $GABA_B$ IPSPs. When $GABA_A$ synapses are reduced sufficiently in strength, a new phase emerges, consisting of fully synchronized population bursts at intervals of several seconds. Each cell bursts in phase with the population, and the interburst interval is set largely by the intrinsic AHP current. With $GABA_A$ inhibition fully blocked, and provided the cells are sufficiently excitable or excitatory synapses sufficiently powerful, then synchronized multiple bursts occur, again with interevent intervals of several seconds. *Within* an event, a brief series of bursts occurs with intervals of about 60-100 ms.

Each of these phases corresponds to a type of activity observed in the hippocampal slice. Low amplitude population oscillations may be detected as small amplitude field potentials (Schneiderman, 1986; Swann, Smith and Brady, in press), or as repeating synchronized synaptic potentials in pairs of cells (Schwartzkroin and Knowles, 1984; Schwartzkroin and Haglund, 1986; Miles and Wong, 1987a,b; Traub, Miles and Wong, 1989; Swann, Smith and Brady, in press). The oscillations are observed most readily when inhibition is partially blocked, or in the presence of cholinergic agents (MacVicar and Tse, 1989; Leung and Yim, 1988). With higher doses of inhibitory blockers, fully synchronized bursts occur (Wong and Traub, 1983) that have a similar appearance to interictal EEG spikes *in vivo*. The latter events, when found in human EEGs have a high association with spontaneous epileptic seizures. Nevertheless, human EEG spikes are not generally as stereotyped or as rhythmical as those occurring in the hippocampal slice. Finally, in full does of picrotoxin (0.1 mM), synchronized multiple bursts occur (Hablitz, 1984; Miles, Wong and Traub, 1984). These events may correspond to an EEG phenomenon called polyspike and wave, an abnormal event highly correlated with clinical seizure disorders. This correspondence remains to be clearly established, however.

While synchronized bursts and multiple bursts are interesting for the insights they provide into epileptogenesis, what is the relevance of low amplitude population oscillations? It is tempting, but dangerous, to compare these oscillations with theta rhythm, an EEG pattern found in the rodent. Theta (Buzsáki, Leung and Vanderwolf, 1983) can be recorded in a number of interconnected structures, including the septum, hippocampi, entorhinal cortex and cingulate cortex (Leung and Borst, 1987). It occurs only during certain behavioral states, such as walking or during REM sleep. The frequency range of theta is similar to that seen in slices, but the mechanisms are unlikely to be identical in the two cases. First, the septum is critical for theta generation *in vivo*. Second, cellular firing is rare in CA3 during walking theta, and those cells which do fire almost never do so in bursts (G. Buzsáki, personal communication). In contrast, in the slice CA3 cells often fire in bursts.

Spontaneous low amplitude population oscillations do provide an opportunity to analyze some of the physical properties of the CA3 region. Some of the most revealing manipulations can be performed only in simulations, although the EEG correlation dimension could, in principle, be estimated for real field potential data. Let us consider first whether the activity is chaotic, that is whether there is rapid divergence after a minimal perturbation. Strictly speaking, this perhaps should be examined in the underlying phase space of all the membrane voltages and state variables. More insight is provided by considering the following two signals. First is the set of n 2-valued functions of time, $e_i(t)$, where n is the number of e-cells, and $e_i(t) = 1$ if cell i is firing at time t and is 0 otherwise. Second is the signal s(t) representing the average activity, $s(t) = \sum_{i=1}^{n} e_i(t)$.

In simulations, both signals are exquisitely sensitive to small perturbations (Traub and Miles, 1991), with divergence from the unperturbed system occurring on a time scale of hundreds of ms. This suggests that the underlying system is chaotic. On the other hand, the actual CA3 region is not a freely running autonomous system, but is driven by external inputs. It is interesting to speculate, therefore, that CA3 is "reset" periodically (theta?), with the period comparable to the time scale of divergence set by the underlying dynamics. We might speculate further that "useful" output (the secondary response?) is passed on to CA1 before the next reset, that is, on a time scale less than the time it takes for the system output to lose all relation to the most recent driving stimulus.

Some of the features of the CA3 region which contribute to its chaotic behavior (at least in simulations) are these. First is the nonlinearity of the pyramidal cell membrane, so that small differences in timing of excitatory and inhibitory inputs can determine if a cell bursts or not (Wong and Prince, 1981). Second is the high gain in the synaptic circuitry, both excitatory (onto e-cells and i-cells) and inhibitory. Third are statistical fluctuations in the connectivity, whereby some cells have more inputs (or outputs) than do others. The first two of these features are solidly established experimentally. There is indirect evidence for the third, in that stimulation of some individual cells will lead to a synchronized population burst in a disinhibited slice, whereas stimulation

TABLE 2

In Vivo Regulation of Hippocampal "System Parameters"

Substance/Action	*Mechanism*	*Effect*	*Refs.*
Repetitive stimulation	NMDA current	Enhanced e→e	a,b
Repetitive stimulation	?	First, enhanced e→i→e, then reduced e→i→e	c
Acetylcholine	stimulate i-cells	barrage of IPSPs	d
Enkephalin	hyperpolarize i-cells	reduce e→i→e	e,f
Somatostatin	? presynaptic	↓ GABA release	g
Neuropeptide Y	presynaptic inhibition	reduced CA3→CA1	h
Septal GABA release	inhibit i-cells	reduced e→i→e	i
Intrinsic GABA release	? presynaptic GABA receptors	↓ GABA release	j
$\uparrow [K^+]_o$	$\uparrow E_{IPSP(Cl)}$	↓ GABA effect	k
Intracell. energy depletion	dephosphorylate GABA receptor	↓ GABA effect	l
ACh, NE	reduce AHP current	↑ excitability	m

References for Table 2.

a) Buzsáki, 1984. b) Wigström and Gustafsson, 1983. c) Miles and Wong, 1987b. d) McCormick and Prince, 1986. e) Masukawa and Prince, 1982. f) Madison and Nicoll, 1988. g) Scharfman and Schwartzkroin, 1989. h) Colmers, Lukowiak and Pittman, 1987. i) Freund and Antal, 1988. j) Davies, Davies and Collingridge, 1990. k) Korn, Giacchino, Chamberlin and Dingledine, 1987. l) Stelzer, Kay and Wong, 1988. m) Nicoll, 1988.

Abbreviations:

ACh = acetylcholine (this compound has many additional actions)
NMDA = N-methyl-D-aspartate
NE = norepinephrine
GABA = γ-aminobutyric acid
AHP = afterhyperpolarization
e = excitatory neuron
i = inhibitory neuron

of other cells has little effect (Miles and Wong, 1983). These differences in single cell output may reflect, however, a side effect of the preparation of the slice, whereby more axonal branches of some cells might be cut than for other cells.

Much of our understanding of population behaviors in the *in vitro* CA3 region derives from experiments in which inhibition is blocked pharmacologically, either partially or fully. While these experiments provide intuition into the properties of the circuitry, could the population behaviors so revealed be directly relevant to normal brain function? It seems possible that this is the case, since certain transmitters found in the hippocampus can alter the efficacy of inhibition, as can repetitive stimulation (Table 2). Other "system parameters" are alterable as well. It remains for further work to determine whether, and how, the brain makes use of its potential ability to adjust its own operating parameters.

REFERENCES

Buzsáki G (1984) Long-term changes of hippocampal sharp-waves following high frequency afferent activation. *Brain Res.* 300: 179-182.

Buzsáki G (1986) Hippocampal sharp waves: their origin and significance. *Brain Res.* 398: 242-252.

Buzsáki G, Leung L-WS, Vanderwolf CH (1983) Cellular bases of hippocampal EEG in the behaving rat. *Brain Res. Rev.* 6: 139-171.

Christian EP, Dudek FE (1988b) Electrophysiological evidence from glutamate microapplications for local excitatory circuits in the CA1 area of rat hippocampal slices. *J. Neurophysiol.* 59: 110-123.

Collingridge GL, Gage PW, Robertson B (1984) Inhibitory post-synaptic currents in rat hippocampal CA1 neurones. *J. Physiol.* 356: 551-564.

Colmers WF, Lukowiak K, Pittman QJ (1987) Presynaptic action of neuropeptide Y in area CA1 of the rat hippocampal slice. *J. Physiol.* 383: 285-299.

Davies CH, Davies SN, Collingridge GL (1990) Paired-pulse depression of monosynaptic GABA-mediated inhibitory postsynaptic responses in rat hippocampus. *J. Physiol.* 424: 513-531.

Edwards FA, Konnerth A, Sakmann B, Takahashi T (1989) A thin slice preparation for patch clamp recordings from neurones of the mammalian central nervous system. *Pflüg. Arch.* 414: 600-612.

Forsythe ID, Westbrook GL (1988) Slow excitatory postsynaptic currents mediated by N-methyl-D aspartate receptors on cultured mouse central neurones. *J. Physiol.* 396: 515-533.

Freund TF, Antal M (1988) GABA-containing neurons in the septum control inhibitory interneurons in the hippocampus. *Nature* 336: 170-173.

Hablitz JJ (1984) Picrotoxin-induced epileptiform activity in the hippocampus: role of endogenous versus synaptic factors. *J. Neurophysiol.* 51: 1011-1027.

Hablitz JJ, Thalmann RH (1987) Conductance changes underlying a late synaptic hyperpolarization in hippocampal CA3 neurons. *J. Neurophysiol.* 58: 160-179.

Kawaguchi Y, Hama K (1988) Physiological heterogeneity of nonpyramidal cells in rat hippocampal CA1 region. *Exp. Brain Res.* 72: 494-502.

Kay AR, Wong RKS (1986) Isolation of neurons suitable for patch-clamping from adult mammalian central nervous systems. *J. Neurosci. Meth.* 16: 227-238.
Knowles WD, Schwartzkroin PA (1981) Local circuit synaptic interactions in hippocampal brain slices. *J. Neurosci.* 1: 318-322.
Korn SJ, Giacchino JL, Chamberlin NL, Dingledine R (1987) Epileptiform burst activity induced by potassium in the hippocampus and its regulation by GABA-mediated inhibition. *J. Neurophysiol.* 57, 325-340.
Lacaille J-C, Mueller AL, Kunkel DD, Schwartzkroin PA (1987) Local circuit interactions between oriens/alveus interneurons and CA1 pyramidal cells in hippocampal slices: electrophysiology and morphology. *J. Neurosci.* 7: 1979-1993.
Lacaille J-C, Schwartzkroin PA (1988a) Stratum lacunosum-moleculare interneurons of hippocampal CA1 region. I. Intracellular response characteristics, synaptic responses, and morphology. *J. Neurosci.* 8: 1400-1410.
Lacaille J-C, Schwartzkroin PA (1988b) Stratum lacunosum-moleculare interneurons of hippocampal CA1 region.II. Intrasomatic and intradendritic recordings of local circuit synaptic interactions. *J. Neurosci.* 8: 1411-1424.
Leung L-WS, Borst JGG (1987) Electrical activity of the cingulate cortex. I. Generating mechanisms and relations to behavior. *Brain Res.* 407: 68-80.
Leung L-WS, Yim CY (1988) Membrane potential osciallations in hippocampal neurons in vitro induced by carbachol or depolarizing currents. *Neurosci. Res. Comm.* 2: 159-167.
MacVicar BA, Tse FW (1989) Local neuronal circuitry underlying cholinergic rhythmical slow activity in CA3 area of rat hippocampal slices. *J. Physiol.* 417: 197-212.
Madison DV, Nicoll RA (1988) Enkephalin hyperpolarizes interneurones in the rat hippocampus. *J. Physiol.* 398: 123-130.
Masukawa LM, Prince DA (1982) Enkephalin inhibition of inhibitory input to CA1 and CA3 pyramidal neurons in the hippocampus. *Brain Res.* 249: 271-280.
Mayer ML, Westbrook GL, Guthrie PB (1984) Voltage-dependent block by Mg^{2+} of NMDA responses in spinal cord neurones. *Nature* 309: 261-263.
McCormick DA, Prince DA (1986) Mechanisms of action of acetylcholine in the guinea-pig cerebral cortex *in vitro*. *J. Physiol.* 375: 169-194.
Mesher RA, Schwartzkroin PA (1980) Can CA3 epileptiform burst discharge induce bursting in normal CA1 hippocampal neurons? *Brain Res.* 183: 472-476.
Miles R (1990) Synaptic excitation of inhibitory cells by single CA3 hippocampal pyramidal cells of the guinea-pig *in vitro*. *J. Physiol.*
Miles R (In prep.) Variation in strength of inhibitory synapses in the CA3 region of guinea-pig hippocampus *in vitro*.
Miles R, Traub RD, Wong RKS (1988) Spread of synchronous firing in longitudinal slices from the CA3 region of the hippocampus. *J. Neurophysiol.* 60: 1481-1496.
Miles R, Wong RKS (1983) Single neurones can initiate synchronized population discharge in the hippocampus. *Nature* 306: 371-373.
Miles R, Wong RKS (1984) Unitary inhibitory synaptic potentials in the guinea-pig hippocampus *in vitro*. *J. Physiol.* 356: 97-113.
Miles R, Wong RKS (1986) Excitatory synaptic interactions between CA3 neurones in the guinea-pig hippocampus. *J. Physiol.* 373: 397-418.

Miles R, Wong RKS (1987a) Inhibitory control of local excitatory circuits in the guinea-pig hippocampus. *J. Physiol.* 388: 611-629.
Miles R, Wong RKS (1987b) Latent synaptic pathways revealed after tetanic stimulation in the hippocampus. *Nature* 329: 724-726.
Miles R, Wong RKS, Traub RD (1984) Synchronized afterdischarges in the hippocampus: contribution of local synaptic interaction. *Neuroscience* 12: 1179-1189.
Nicoll RA (1988) The coupling of neurotransmitter receptors to ion channels in the brain. *Science* 241: 545-551.
Regehr WG, Connor JA, Tank DW (1989) Optical imaging of calcium accumulation in hippocampal pyramidal cells during synaptic activation. *Nature* 341: 533-536.
Ross WN, Werman R (1987) Mapping calcium transients in the dendrites of Purkinje cells from the guinea-pig cerebellum *in vitro*. *J. Physiol.* 389: 319-336.
Sayer RJ, Friedlander MJ, Redman SJ (1990) The time course and amplitude of EPSPs evoked at synapses between pairs of CA3/CA1 neurons in the hippocampal slice. *J. Neurosci.* 10: 826-836.
Scharfman HE, Schwartzkroin PA (1989) Selective depression of GABA-mediated IPSPs by somatostatin in area CA1 of rabbit hippocampal slices. *Brain Res.* 493: 205-211.
Schneiderman JH (1986) Low concentrations of penicillin reveal rhythmic, synchronous synaptic potentials in hippocampal slice. *Brain Res.* 398: 231-241.
Schwartzkroin PA (1978) Secondary range rhythmic spiking in hippocampal neurons. *Brain Res.* 149: 247-250.
Schwartzkroin PA, Haglund MM (1986) Spontaneous rhythmic synchronous activity in epileptic human and normal monkey temporal lobe. *Epilepsia* 27: 523-533.
Schwartzkroin PA, Knowles WD (1984) Intracellular study of human epileptic cortex: *in vitro* maintenance of epileptiform activity: *Science* 223: 709-712.
Schwartzkroin PA, Mathers LH (1978) Physiological and morphological identification of a nonpyramidal hippocampal cell type. *Brain Res.* 157: 1-10.
Stelzer A, Kay A, Wong RKS (1988) $GABA_A$ receptor function in hippocampal cells is maintained by phosphorylation factors. *Science* 241: 339-341.
Swann JW, Smith KL, Brady RJ (in press) Neural networks and synaptic transmission in immature hippocampus. In: Ben-Ari Y, ed., *Excitatory Amino Acids and Neuronal Plasticity. Advances in Experimental Medicine and Biology*. Plenum Press.
Traub RD, Miles R (1991) *Neuronal Networks of the Hippocampus*, Cambridge University Press.
Traub RD, Miles R, Wong RKS (1989) Model of the origin of rhythmic population oscillations in the hippocampal slice. *Science* 243: 1319-1325.
Traub RD, Wong RKS, Miles R (1990) A model of the CA3 hippocampal pyramidal cell based on voltage-clamp data. *Abstr. Soc. Neurosci.*
Traub RD, Miles R (in press) Multiple modes of neuronal population activity emerge after modifying specific synapses in a model of the CA3 region of the hippocampus. In: Wolpaw JR, Schmidt JT, eds., *Activity-driven CNS Changes in Learning and Development*, NY Acad. Sci.
Wigström H, Gustafsson B (1983) Facilitated induction of hippocampal long-lasting potentiation during blockade of inhibition. *Nature* 301: 603-604.

Wong RKS, Prince DA (1981) Afterpotential generation in hippocampal pyramidal cells, *J. Neurophysiol.* 45: 86-97.
Wong RKS, Prince DA, Basbaum AI (1979) Intradendritic recordings from hippocampal neurons. *Proc. Nat. Acad. Sci.* 76: 986-990.
Wong RKS, Traub RD (1983) Synchronized burst discharge in disinhibited hippocampal slice. I. Initiation in CA2-CA3 region. *J. Neurophysiol.* 49: 442-458.

INFORMATION THEORY AND EARLY VISUAL INFORMATION PROCESSING

D. S. Tang
Microelectronics and Computer Technology Corporation
3500 West Balcones Center Drive
Austin, TX 78759-6509
E-Mail: tang@mcc.com

INTRODUCTION

Shannon's information theory[1] is a mathematical theory of communication. It deals with the fundamental aspects of transmission of messages from place to place in the communication channels. The process itself serves the goal of making or assisting decision-makings at the receiving ends. An intelligent decision relies on how accurate, how relevant and how efficient the pieces of information are transmitted and received. One may view a natural or synthetic neural network system as an ensemble of distributed decision-making units resiting in different functional blocks linked through a massive network of communication channels operated distributively and in parallel. The immediate questions are (1) how to approach the study of a given neural network within the information-theoretic framework[2], (2) how to construct the optimal communication channels in a neural network setting and (3) how to evaluate different neural networks in terms of their information-theoretic properties. We have no ideas to answer the last two questions. We have some concrete results to answer the first question in terms of a simple synthetic neural net that captures many functionalities of the early visual information processing in retina[3,4]. The self-emerging properties of the network are derived analytically. With 2-D spatiotemporal inputs, it is found that (1) there is an emergence of statistically independent operation modes for feature detection, (2) there is an emergence of lower memory for motion detection of the temporal change of the input signals and the direction of motion and (3) the role of symmetry-breaking of parity conservation and time-reversal invariance in spatiotemporal feature detection is explained. The presentation is organized as follows. Information theory is first introduced. This is followed by the definitions of a simple feed-forward neural net model. The information-theoretic analyses of this model and its extension to image reconstruction are then presented. We conclude our discussion with a summary.

NEURAL NETWORKS AND INFORMATION THEORY

Neurons are the fundamental building blocks of a biological neural system. We know from neurophysiological experiments and anatomy[5] that [1] the site of change of connection strength in terms of the post- and pre-synaptic efficacy is at the synapse of the neurons, [2] the connection is massively parallel and distributive, [3]

Self-Organization, Emerging Properties, and Learning
Edited by A. Babloyantz, Plenum Press, New York, 1991

the system processes both spatial and temporal signals, [4] the system is a highly efficient decision engine for adaptive response to the changing environment, and [5] the sensory input signals are stochastic variables. We may configure an neural system as a black box with changeable internal parameters a in accordance with (1) and allow the parameters to be spatiotemporal dependent in conjunction with (3). At the present moment, there is no way to explain the realization of these internal parameters in terms of the chemical pathways at the molecular level in a biological substrate. Our tactic is to understand the functional aspects of the systems. If this can be done, then the next step is on the problems of realization of these functionalities. A single function may be realizable by a single neuron or a group of neurons. We can go beyond a realization in a biological substrate. As long as the technology allows, we may realize the functionalities in a silicon substrate, a GaAs substrate, and so forth. These are the general settings of our approach.

Let us denote the input signal space as S and the output decision space as D. The mapping from the signal space to the decision space is $f : S \rightarrow D$. The functional relationship between the input and the output is

$$g = f(s, a). \tag{1}$$

The assumption that s is a stochastic variable in accordance with [5] means that the decision element g is also a stochastic variable. In statistical communication theory, one of the fundamental quantities on the performance of a communication channel is the channel capacity[1]. It measures the maximum rate of transmission of information through the channel when the input is varied over all possible ensembles. In other words, one encodes the input signals with an appropriate probability distribution function (e.g. message redundancy can be introduced to obtain almost error-free transmission) which corresponds to the maximum information rate. In doing so, the communication channel is utilized with the maximum capacity. This underlies the coding theorems in information theory. In the neural system we are considering, the input signal is manipulated after it has been received. The manipulation itself is to serve the purpose of making the appropriate decision g. Let $p(s)$ be the probability distribution function of the input signals. Then the probability distribution of the decision g is

$$p(g) = \int ds \delta(g - f(s, a)) p(s). \tag{2}$$

The average information content of the input signal is given by the information entropy

$$H[s] = -\int dsp(s) \ln p(s) \tag{3}$$

and likewise, the average information content in the decision is

$$H[g] = -\int dgp(g) \ln p(g). \tag{4}$$

On the basis of a decision signal g, a measure of the uncertainty as to whether the input signal s was actually detected is given by the conditional entropy

$$H[s|g] = -\int dgdsp(g)[p(s|g) \ln p(s|g)]. \tag{5}$$

Likewise, on the basis of an input signal s, a measure of the uncertainty as to whether the decision g was actually made is given by the conditional entropy

$$H[g|s] = -\int dsdgp(s)[p(g|s) \ln p(g|s)]. \tag{6}$$

The conditional probability distribution function $p(g|s)$ is the decision rule in statistical decision theory[6]. In the usual approach, a cost function $C(g,s)$ is constructed. When the right decision is selected for a given input, the cost is low. Otherwise, the cost is high. The optimization of the risk by changing the internal parameters a produces the optimal decision rule $p(g|s)_C$ in which the supscript C refers to the optimization respect to a cost function. Now, in information theory, a measure of the net decrease in the uncertainty about the occurrence of s, before a decision g is made and after g has been made, is given by the information rate

$$R = H[g] - H[g|s] = H[s] - H[s|g] = H[s] + H[g] - H[s,g]. \tag{7}$$

Therefore, the decision rule $p(g|s)_R$ that optimizes the information rate may not be identical to the decision rule $p(g|s)_C$ that optimizes the cost function. The case that both decision rules are equal is when the cost function is a linear monotonically increasing function of R. A neural system implemented in a physical substrate is itself a physical system and therefore there are physical constraints in terms of the rate of energy consumption and heat dissipation and so forth throughout the decision making process inside the blackbox. Furthermore, the cost function depends on how relevant a decision is to the survival of the specie in a changing environment. These conditions constitute the cost function and are constraints to the information rate. The problem of finding out the optimal decision rule will become exceedingly complex if we take all these factors into account. The discussion below will be limited to a simple case that the information rate is assumed to be identical to the cost function. The question on the physical and environmental constraints remains unresolved. However, if we assume that the input-output relation in Equation (1) can be expanded in terms of a complete set of eigen-functions, then the constraints amount to limiting the expansion coefficients to definite values. This is the approach we take in this study. Below, we shall present our neural network model, then strict information-theoretic analysis is carried out.

A NEURAL NETWORK MODEL

The input signal is assumed to belong to an ensemble of 2-D pictures described by a Gaussian distribution function. This means that each pixel value at location (x,y) in the X-Y plane is Gaussian-distributed. This defines the a priori probability distribution function of the input signal. There is no obvious reason on why a Gaussian distribution is chosen except for the fact that analytic solutions of the decision rule can be derived in a rigorous manner. The picture of a human face or the picture of a string of hand written characters belong to this ensemble. The differences between them are their frequencies of occurrence and their structural distributions of the pixel values in the X-Y plane. Take for example an arbitrary picture B. The probability that this particular picture is selected randomly among all the pictures in a Gaussian ensemble is

$$p(B) = \int \Pi_{ij}[dL_{ij}\delta(f(i,j) - L_{ij})]p(L) \tag{8}$$

where the structure function $f(i,j)$ defines the grey level at location (i,j). This also shows that all pictures belong to this Gaussian ensemble. The difference between a picture in a Gaussian ensemble and the same picture, say, in an uniformly distributed ensemble is its frequencies of occurrence in these two ensembles. This is the first step in applying information theory. That is, by some means (assumptions, approximations, actual statistical analysis and etc.) the a priori probability distribution of the input signal is known. Then we can proceed to the next step of the information-theoretic study. Equation (1) is a general statement on the relation between the signal and the decision. We have to be more specific on this

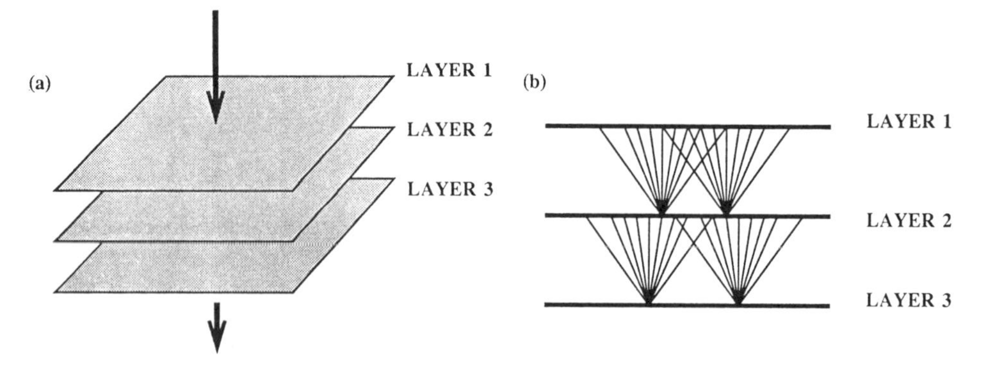

Figure 1. (a) A three dimensional view of the network model. (b) A cross-sectional view of the model.

relation in terms of a general network configuration. According to the previous discussion, the neural system is massively parallel and distributive. Below we present a simple neural network model satisfying this requirement and to show that it is capable of capturing the essential functionalities of early visual information processing.

The model introduced below is motivated by, but not a duplicate of, the gross cellular and synaptic organization of the vertebrate retina[3,4]. There are three cellular layers, the outer and inner nuclear layers and ganglion cell layers in the retina. The two-dimensional visual signal is first captured by the photoreceptors in the outer nuclear layer and the signal is processed through the bipolar cells in the inner nuclear layer and then the ganglion cell layer respectively. There are lateral connections mediated by horizontal cells and amacrine cells separately among the photoreceptor cells to bipolar cell layers and the bipolar cell to the ganglion cell layer respectively. The simplest skeleton network structure embedded in this complex network is a three-layer feed-forward network, Fig.1. Each layer is a two dimensional sheet of units. The sheets are stacked with the input layer being layer 1, the middle layer being layer 2 and the output layer being layer 3.

Layer 1 is composed of randomly distributed photoreceptors. Each unit in layer 2 is assumed to receive signal $X_j(t)$ from layer 1 according to the following relation

$$Y_i(T_2) = C_0 \sum_j \sum_{t=T_1}^{T_2} \exp(-(T-t)/\tau) X_j(t). \tag{9}$$

Here, i and j are spatial labels. T_2 is the present time. T_1 is the past time. The input units are spatially distributed according to the following Gaussian distribution function

$$\rho(\vec{R}_i - \vec{r}_j) = C_\rho \exp[\frac{-(\vec{R}_i - \vec{r}_j)^2}{2\eta^2}] \tag{10}$$

with $C_\rho = N_i/2\pi\eta$. η is the standard deviation. Equation (9) means that the units in layer 2 collect signals locally. It takes time for the signal to pass from layer 1 to layer 2. This microscopic delay in time is modeled by an exponential decay factor. From layer 2 to layer 3, our model assumes the same type of spatial summation,

$$Z_i(T_2) = \sum_{j=1}^{N} \sum_{t=T_1}^{T_2} a_{ij}(t) Y_j(t). \tag{11}$$

Here, the internal parameters $a_{ij}(t)$ are to be determined later. As discussed in the previous section, these internal parameters determine the properties of the decision rule of the model system. To find out the decision rules, Shannon's procedure to derive the information rate of the system is followed. We view the network model defined above as an active communication channel in which the internal parameters determine the operational characteristics of the channel. Ordinarily, a physical channel cannot have infinite information rate due to the presence of a certain amount of noise in the channel. If the information rate is infinite, then the input signal can be exactly recovered from the output signal[1]. Note that the information rate is a function of the internal parameters. As far as an exact recovery of the input signal from the output signal is concerned, the information rate is also a measure of how good the recovery or the degree of information conservation the channel can offer for a given set of values of the internal parameters in the presence of noise. The lower the information rate of a given decision g is, the more ignorance of what exactly the input signal is since the probability to successfully recover the input signal becomes less. Equation (1) means that the input s is transformed to a decision g. By changing the internal parameters, we are taking different transformations from the input signal to the decision. In other words, we are taking different aspects of the input signal to arrive at different decisions. Setting the internal parameters to the values that produce the maximum information rate restricts the network capability to perform a decision in one subregion in the decision space for a given input signal. In our study here, we relax this restriction and consider the full spectrum of the internal parameters which define all statistically independent subregions in the decision space. In this sense, we take the network as a multi-mode communication channel that is actively participating in signal analyses during signal transmission. In a noiseless channel, the totality of all these decisions constitutes the full picture of the input in terms of exact recovery of the input signal. To account for noise $n_j(t)$ in each of the inputs, Equation (11) becomes

$$Z_i(T_2) = \sum_{j=1}^{N} \sum_{t=T_1}^{T_2} a_{ij}(t)[Y_j(t) + n_j(t)]. \tag{12}$$

In matrix notation, the a priori probability distribution $P(X)$ of the stochastic input signals $X_i(\tau)$ is assumed to be Gaussian

$$P(X) = \frac{\exp[-(X - \tilde{X})^T \mathcal{U}(X - \tilde{X})]}{(2\pi)^{\frac{3\tilde{N}}{2}} \sqrt{\mathrm{Det}\mathcal{U}}}. \tag{13}$$

$\tilde{N}$ is the total number of space-time indices. $\tilde{X}$ is the mean. $\mathcal{U}$ is the variance, $U_{i_1 t_1, i_2 t_2} = u\delta_{i_1 i_2}\delta_{t_1 t_2}$. Assuming there is additive Gaussian noise at each of the inputs of layer 2, the information rate from layer 1 to layer 2 can be shown to be a function of the ratio of variance of the signal to the variance of the noise

$$R(1 \to 2) = \frac{1}{2}\ln(1 + \frac{u}{v}). \tag{14}$$

It depends only on the variances and does not depend on the internal parameters. As long as the signal-to-noise is high, the input information transmitted through this part of the channel is conserved to a high degree, and therefore the outputs from layer 2 very much reproduce the input signal. The information rate from layer 2 to layer 3 can be derived as

$$R(2 \to 3) = \frac{1}{2}\ln(1 + \frac{a^T W a}{v a^T a}). \tag{15}$$

The spatial-temporal correlation matrix W of the output signal from layer 2 is,

$$W_{ij}(TT') = << Y_i(T) Y_j(T') >> = s Q_{ij} G_{TT'}, \tag{16}$$

with

$$\begin{aligned} Q_{ij} &= \exp[\frac{-|\vec{r}_i - \vec{r}_j|^2}{2 r_B^2}]. \\ G_{TT'} &= \exp[\frac{-|T - T'|}{\tau}], \quad |T - T'| \leq T_2 - T_1, \\ G_{TT'} &= 0, \quad |T - T'| > T_2 - T_1. \end{aligned} \tag{17}$$

s is an irrelevant constant factor. Equation (15) shows that the information rate is a function of the normalized internal parameters and the spatiotemporal correlation matrix of the input signals from layer 2. A complete classification of the decisions in the decision space will be possible if in Equation (15) the internal parameters are chosen to be the eigenfunctions of the correlation matrix W. Furthermore, the decision rules and their information rates can be found if the explicit expressions of the internal parameters are known. In matrix notation, the eigenvalue problem is, for the kth eigenvalue λ_k,

$$W a_k = \lambda_k a_k. \tag{18}$$

The input-output relation hence becomes

$$Z_k = a_k Y. \tag{19}$$

The information rate is

$$R(2 \to 3)_k = \frac{1}{2}\ln(1 + \frac{\lambda_k}{v}). \tag{20}$$

Therefore, the ignorance of the input signal when the kth decision has been made is determined by the ratio of the kth eigenvalue λ_k to the variance of the noise. Furthermore, the kth decision is statistically independent from the k'th decision since

$$<< g_k g_{k'} >> = \delta_{kk'} \tag{21}$$

when the eigenvectors are properly normalized. The equation for the eigenvalue problem becomes the Fredholm integral equation and can easily be solved when continuous space and time variables are used. Since the correlation function is separable into the product of a spatial factor and a temporal factor, solutions to the eigenvalue value problem naturally divide themselves into spatial solutions and temporal solutions. The eigenvalue λ_k is the product of the spatial eigenvalue and the temporal eigenvalue.

Below we list the solutions of the problem.
(1) Temporal component:
(a) The antisymmetric temporal solutions:

$$B_a = \sin[\gamma_n(t - T_m)], \quad T_1 \leq t \leq T_2, \quad n = 1, 3, 5, ... \\ B_a = 0, \quad \text{otherwise} \tag{22}$$

The frequency γ_n is

$$\gamma_n = -\frac{1}{\tau}\tan(\gamma_n \Delta) \tag{23}$$

with $T_m \equiv (T_2 + T_1)/2$ and $\Delta \equiv (T_2 - T_1)/2$.
(b) The symmetric temporal solutions:

$$B_s = \cos[\gamma_n(t - T_m)], \quad T_1 \leq t \leq T_2, \quad n = 0, 2, 4, ... \\ B_s = 0, \quad \text{otherwise.} \tag{24}$$

The frequency is

$$\gamma_n = \frac{1}{\tau}\cot(\gamma_n \Delta). \tag{25}$$

In both symmetric and antisymmetric cases, the temporal eigenvalue is given by

$$\lambda_n^T = \frac{\tau}{\Delta(1 + \gamma_n^2 \tau^2)}. \tag{26}$$

It can be shown that the temporal eigenvalues satisfy the inequality $\lambda_n^T > \lambda_{n+1}^T$.
(2) Spatial component:

$$A_n^m(\vec{r}) = H_m(\frac{x}{\sqrt{2}\sigma'_\infty})H_{n-m}(\frac{y}{\sqrt{2}\sigma'_\infty})\exp(-r^2/2\sigma_\infty^2), \\ m = 0, ..., n, \quad n = 0, 1, 2, ... \tag{27}$$

The spatial eigenvalue is given by

$$\lambda_n^S = 2\pi C_\rho C_Q r_B^2 q^{n+1}, \tag{28}$$

with

$$q = \frac{1}{1 + \frac{r_B^2}{\alpha^2} + \frac{r_B^2}{\sigma_\infty^2}}, \tag{29}$$

and σ'_∞ satisfying $r_B^2/\sigma'_\infty = 1/q - q$. It can be shown[8] that $r_B^2/\alpha^2 = 2, r_B^2/\sigma_\infty^2 = 0.73205$ and $q = 1/3.73205$. Furthermore, $\lambda_n^S > \lambda_{n+1}^S$. For example, the spatial eigenvectors with the three largest spatial eigenvalues can be found to be

$$A_0^0 = \exp[\frac{-r^2}{2\sigma_\infty^2}], \tag{30}$$

$$A_1^0 = xA_0^0, \tag{31}$$

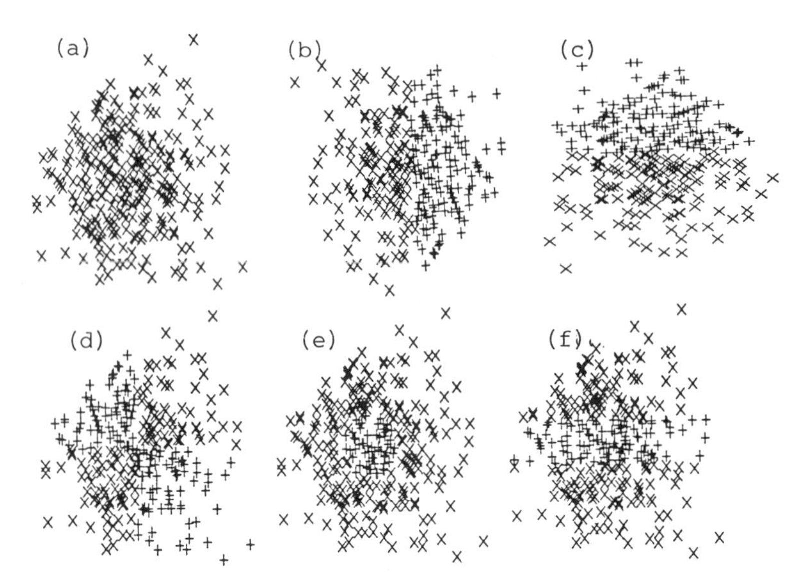

Figure 2. (a) A_0^0 morphology. (b) A_1^0 morphology. (c) A_1^1 morphology. (d) A_2^0 morphology. (e) A_2^1 morphology. (f) A_2^2 morphology. x denotes inhibitory input. + denotes excitatory input.

$$A_1^1 = yA_0^0, \tag{32}$$

$$A_2^0 = xyA_0^0, \tag{33}$$

$$A_2^1 = (r^2 - 2d)A_0^0, \tag{34}$$

$$A_2^2 = (y^2 - d)A_0^0, \tag{35}$$

$$d = \frac{1}{2[1 - \frac{1}{4r_B^4 f^2}]},$$

$$f = \frac{1}{2}[\frac{3}{2r_B^2} + \frac{1}{\sigma_\infty^2}].$$

These solutions also divide themselves into symmetric and anti-symmetric solutions with respect to spatial reflection. These spatial components are shown in Figure 2. The all excitatory morphology is given by Equation (31). The center-surround morphology is given by Equation (34). Equation (35) defines a stripe morphology.

CHARACTERISTICS OF THE DECISIONS

Equation (11) describes a local operation of the input signal both in space and time. It sums up the spatiotemporal signal discriminately according to the internal parameters $a_{ij}(t)$ which control the degree of ignorance of the input signal given that a decision has been made. The characteristics of the decisions depend on the behavior of the spatiotemporal eigenfunctions. To be specific, we focus our discussion on the eigenfunctions with the large eigenvalues. We begin with temporal processing of input signal. Let the input to layer 1 be modeled as a step function such that the signal is switched on at time T_1 everywhere in the layer. Then, according to Equation (9), the output signal everywhere in layer 2 is given by

$$\begin{aligned} Y(\vec{r}, T) &= 0, \quad T < T_1; \\ &= X_o\tau[1 - exp(-(T - T_1)/\tau]/2\Delta, \quad T_1 \geq T, \end{aligned} \tag{36}$$

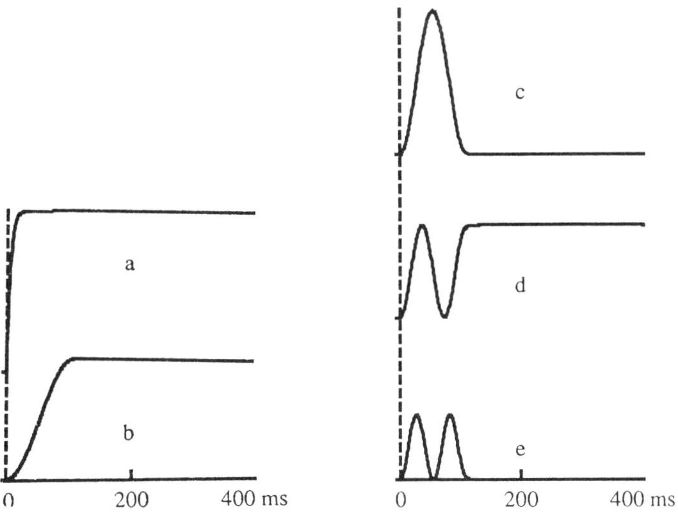

Figure 3. (a) Temporal input signal to the third layer. (b) The first mode temporal output signal. (c) The second mode temporal output signal. (d) The third mode temporal output signal. (e) The fourth mode temporal output signal.

where X_o is an arbitrary input intensity. Since the layer 1 to layer 2 channel has a band-width of order of $1/\tau$, from sampling theorem, this channel will not distort the signal with frequency less than $1/\tau$. This is the limitation on the bandwidth of the input signal impinging on layer 1. That Equation (36) is not an exact step function is a consequence of this limitation. As long as this limitation is satisfied, the output reproduces faithfully the input signal. The signal transmitted from layer 2 to layer 3 is convoluted with the internal parameters. The internal parameters are eigenfunctions and therefore the output signal from layer 3 is decomposed into different eigenmodes. Their properties are different.

Fig. 3 illustrates the output for the first four largest temporal eigenvalues. The odd numbered modes delay the temporal signal but the even numbered modes detect the changes of the temporal signal. Since the information rate of Fig. 3a is the highest, we have the least ignorance on knowing the input signal by receiving this mode of output. The ignorance increases as the mode increases. Qualitatively, the piece of information that is missed has to do with the uncertainty of what the intensity of the constant input signal is since this mode only registers the change in input signal. The information rate described by Equation (20) has an inherent invariant characteristic with respect to time-reversal operation[7] defined by changing the signs of all time-dependent variables to their opposite signs. The first symmetry-breaking internal parameter defined by Equation (22) allows the channel to operate in a mode with the ability to detect the temporal changes of the input signal. This is a temporal differentiation. Similar behavior emerges for the spatial components of the internal parameters. Inspection of the Equations (30) to (35) shows that Equations (31) and (32) are antisymmetric with respect to spatial reflection, $\vec{r} \rightarrow -\vec{r}$. The information rate remains unchanged under this operation. The operational modes defined by Equations (31) and (32) are parity-non-conserving[8,9]. The operations are spatial differentiation. Its effect on an image is shown in Fig.4.

Fig. 4a is the input image. Fig. 4b and Fig. 4c are the output images from layer 3 with internal parameters $A_2 = xA_0^0$ and $A_2 = yA_0^0$ respectively. When the

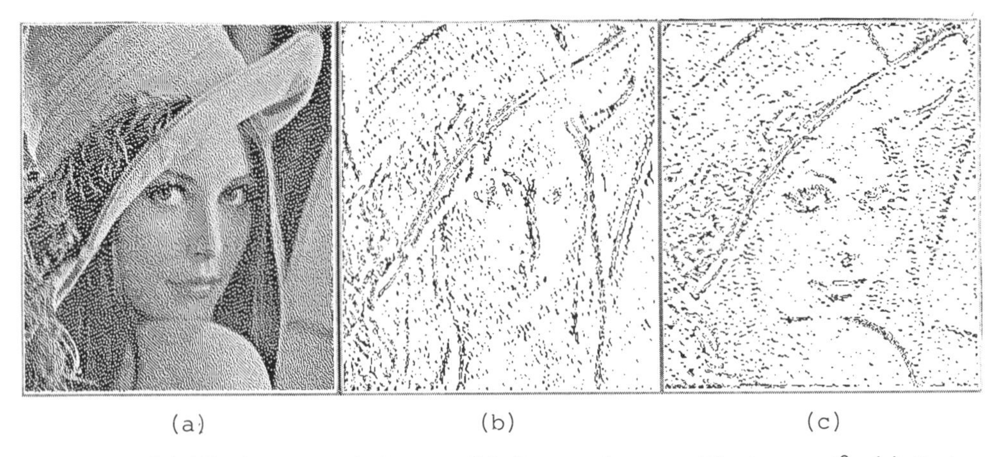

Figure 4. (a) The input static image. (b) Output image with $A_2 = xA_0^0$. (c) Output image with $A_2 = yA_0^0$.

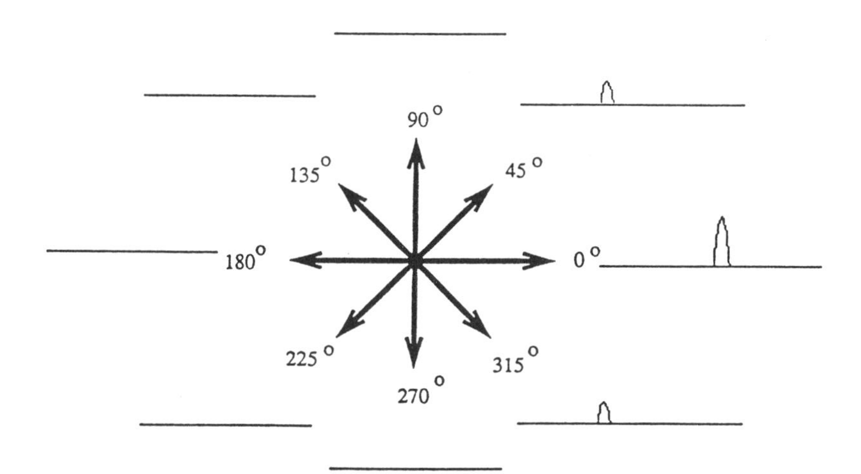

Figure 5. The temporal responses of the directionally selective unit with threshold. The arrows indicate the direction of motion of an input light spot. The optimal response direction shown here is at zero degree.

first temporal and spatial antisymmetric internal parameters are acting together on the input signal, the output from layer 3 registers both the spatiotemporal changes of the signal. The effect is motion-direction detection[10], as shown in Fig. 5. However, it can be shown that this operation does not respect contrast-invariance. This is a limitation of the linear theory. To preserve contrast-invariance in the detection of the direction of motion, a nonlinear theory has to be used. We list consecutively below the four top eigenvalues(information rate) and their functional characteristics.

Table 1. Functionalities of the highest four modes of operations

rank	eigenvalue	functional characteristics
1	$\lambda_0^S \lambda_0^T$	time-delayed relay
2	$\lambda_0^S \lambda_1^T$	temporal differentiation
3	$\lambda_1^S \lambda_0^T$	spatial differentiation
4	$\lambda_1^S \lambda_1^T$	motion-direction detection

IMAGE RECONSTRUCTION

Below we confine our discussion to static pictures and lateral connections are considered. It is sufficient to consider time-independent internal parameters. In matrix notation, the decision at the ith location with lateral connections is

$$g_i = \sum b_{ij} L_j + \sum c_{ik} g_k \tag{37}$$

Therefore,

$$[\sum (\delta_{ik} - c_{ik}) d_k = \sum b_{ij} L_j . \tag{38}$$

That is,

$$[1 - c] g = bL, \tag{39}$$

and thus

$$g = [1 - c]^{-1} bL. \tag{40}$$

This means that g is linearly dependent on L. The series expansion of this matrix equation is

$$g = bL + cbL + ccbL + cccbL + ... \tag{41}$$

in which the first term are immediate contributions, the second term is from the first neighbour, the third term is from the next-nearest neighbours and so forth. Let a denote the matrix $[1 - c]^{-1} b$. Then the decision g written in continues spatial variable is

$$g(R) = \int \rho(R - r) a(R - r) L(r) d\vec{r}. \tag{42}$$

The following completeness relation of the internal parameters can be derived

$$\delta(R - r) = \sum_k [\rho(R)\rho(r)]^{\frac{1}{2}} a_k(R) a_k(r) \tag{43}$$

from the eigenvalue problem,

$$\lambda_k a_k(r) = \int \rho(R) Q(r - R) a_k(R) dR. \tag{44}$$

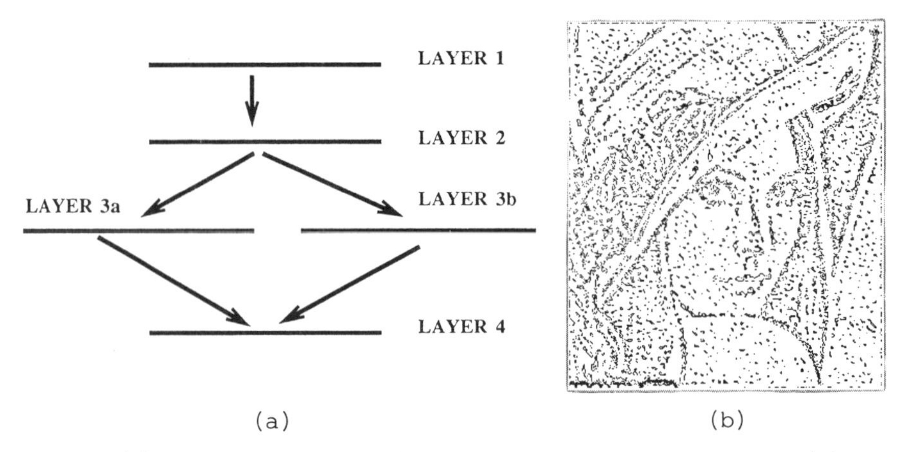

Figure 6. (a) The four-layer neural network for image reconstruction. (b) The reconstructed output image.

The decision space is subdivided into different statistically independent subregions defined by the different eigenvalues. Making use of the completeness relation, the original visual image can be reconstructed by

$$L(r) = \int \rho(R - r) \sum_k g_k(R) a_k(R - r) dR. \tag{45}$$

The information-theoretic implication of this equation is that if we collect all the decisions g_k from all subregions labelled by k, the ignorance of the input signal will be reduced to zero and therefore the original signal can be recovered exactly. Equation (45) further says that the signal recovery can be done locally at location $\vec{r}$. Based on this equation, one can add a fourth layer to the three-layer feed-forward neural network to do the image reconstruction. Below, we present the numerical results of a subclass of this network. The configuration has five layers of two dimensional networks. Layer 1 is for the photoreceptors which feed the information to layer 2. Layer 2 decomposes the information into two complementary modes in layer 3a and layer 3b. Layer 4 reconstructsthe image with inputs from layer 3a and layer 3b, Figure 6. The internal parameters are $A_2 = xA_0^0$ and $A_2 = yA_0^0$ for layer 2 to layer 3a and layer 3b respectively. Outputs from layer 3a and layer 3b give partial information on the original image. The combination of the two pieces of information produces a skeleton sketch of the original image. This is shown in Fig. 6.

SUMMARY

We have demonstrated how to approach the study of a given neural network within the information-theoretic framework. Explicit expressions have been derived for a simple three-layer feed-forward network and the functions have been identified for edge detection and motion detection. This may be the simplest synthetic neural network model that is able to capture the essential morphologies found in vertebrate retina, in terms of the center-surround morphology , Equation (34) and the direction-sensitive morphology[11-14]. Our proposal is that the network is an active multi-mode communication channel participating in signal processing for making different decisions. A few modes may be selected for their functional characteristics that are relevant to survival. Our discussion does not offer any hints on how the retinal neural network is formed and organized[15-16]. However, it suggests a possible functional explanation on how early visual information is analyzed naturally within the information-theoretic framework. It is likely that studies along this direction may shed light on how higher visual information is processed. Furthermore, the results of these studies probably suggests an important information-theoretic idea not anticipated by Shannon. Namely, with the introduction of tunable internal parameters within the system, the input messages can be analyzed and correlated with certain physical or conceptual entities. These bring out the semantic aspects of communication[1]. It appears therefore that information theory can be employed to deal with not only the statistical problems but also the semantic problems of communication.

REFERENCES

1. C. E. Shannon and W. Weaver, The mathematical theory of communication (U. of Illnois Press,1949).
2. R. Linsker, "Self-Organization in a perceptual network", Computer 21,105(1988); "Towards an organizing principle for a layered perceptual network", Neual Information Processing Systems (Denver) ,ed. D. Z. Anderson (Amer. Inst. of Physics, N. Y.),485(1987);" An application of the principle of maximum information preservation to linear systems", Advances in Neural Information Processing I(Denver), ed. D. S. Touretzky(Morgan Kaufman, San Mateo, CA),1989.
3. J. E. Dowling, The Retina (Harvard University Press, 1987).
4. J. E. Dowling, "Functional and pharmacological organization of the retina: Dopamine, Interplexiform cells, and neuromodulation",1-18, in Vision and the Brain, edited by B. Cohen and I. Bodis-Wollner(Raven Press, 1990).
5. E. R. Kandel and J. H. Schwartz, Principles of Neural Science(Elsevier, 1985).
6. A. Wald, Statistical decision functions, (John Wiley and Sons, 1950).
7. D. S. Tang and V. Menon, "Temporal differentiation and violation of time-reversal invariance in visual information processing", Neural Computation, Vol.2, no. 2(1990 in press).
8. D. S. Tang, "Information-theoretic solutions to early visual information processing: analytic results",Phys. Rev. A,40, 6626, 1989.
9. D. S. Tang, "Analytic solutions to the formation of feature-analyzing cells of a three-layer feed-forward information processing neural net", Advances in neural information processing systems 2, edited by D. S. Touretzky, Morgan Kaufmann, 160-165, 1990.
10. D. S. Tang, "Neurocomputation of image motion",IJCNN'90(San Diego) Proceedings(in press).
11. H. B. Barlow, R. M. Hill and W. R. Levick," Retinal ganglion cells responding selectively to direction and speed of image motion in the rabbit", J. Physiol, 377(1964).
12. P. L. Marchiafava, " The response of retinal ganglion cells to stationary and moving visual stimuli", Brain Research,19,1203(1079).
13. C. Koch and T. Poggio and V. Torre, "Computations in the vertebrate retina: gain enhancement, differentiation and motion discrimination", Trends in Neuro-Sciences, vol.9, no.5, 204(1986).
14. R. H. Masland, "The cholinergic amacrine cell", Trends in NeuroSciences, vol.9, no.5, 218(1986).
15. W. R. Levick and D. R. Dvorak, "The retina-from molecules to networks",Trends in NeuroSciences, vol.9, no.5, 181(1986).
16. P. Sterling, M. Freed and R. G. Smith, "Microcircuitry and functional architecture of the cat retina", Trends in NeuroSciences, vol.9, no.5, 193(1986).

Deterministic Chaos in a Model of the Thalamo–Cortical System

A. Destexhe[1] and A. Babloyantz

Université Libre de Bruxelles, CP 231 - Campus de la Plaine,
Boulevard du Triomphe, B-1050 Bruxelles, Belgium

We introduce here a simple model of the cortical tissue subject to periodic inputs which mimic thalamic activity. It is shown that in the absence of input, the dynamics of the system is turbulent and desynchronized. The onset of the pacemaker input organizes the system into a more coherent spatiotemporal behavior. The spatial coherence is characterized with the spatial autocorrelation function. Reminiscent of EEG activity, the average potential is monitored in time. The temporal coherence of this mean activity is assessed with the help of techniques from nonlinear time series analysis. The model shows deterministic chaos with dimension–frequency relationships similar to the human EEG.

Introduction

In 1875, Caton [22] measured for the first time the electrical activity of the human brain. Today, the recording of the electroencephalogram (EEG) is a common medical procedure. This electrical activity occurs as "waves" of very small amplitude (about 10 to 300 μV), with several stereotyped patterns of activity, characteristic of the behavioral state of the brain.

The electroencephalogram (EEG) of an aroused and active brain is characterized by beta waves of high frequency and low amplitude. When the eyes are closed and the subject is relaxing, spindle-like waves, called alpha waves, appear. As the subject drifts toward sleep, the brain enters into a succession of four stages where the coherence and the amplitude of the EEG gradually increase as the mean frequency of the waves decreases. In a normal brain, the fourth stage – or deep sleep – is characterized by the highest EEG amplitude, while the mean frequency of oscillations is the lowest. With the onset of rapid eye movement sleep, called REM sleep, where most of dreams occur, the EEG reverts back to beta-like activity.

[1]E-mail: R09614@bbrbfu01.bitnet

It is now largely admitted that the EEG results from the combined electrical activities of the neurons and also probably the glial cells. However the exact relationship between the EEG and neuronal activity is not yet fully understood. In 1943, Adrian and Matthews [2] showed that the large-amplitude waves result from a synchronization of neuronal activity rather than from an increase of the value of the electrical potential of individual neurons. Since this pioneering work, numerous experimental evidence of the synchronization of neurons during EEG rhythms have been provided (see for example [19, 62, 85]).

The idea that this rhythmic activity could be generated outside the cortex was first advanced by Bremer [18]. The discovery of massive interconnections between thalamus and neocortex [30] as well as the ability of the thalamus to produce periodic oscillations even when disconnected from the cortex [1] led to the "thalamic pacemaker" hypothesis for the generation of brain rhythms [4, 5, 78]. The subgroup of thalamic cells responsible for autonomous oscillations was identified later [79, 80] and has been shown to be responsible at least for spindle-type oscillations [81]. During the 8-12 Hz alpha rhythm and the 10-14 Hz sleep spindles, the membrane potential of thalamocortical (TR) cells oscillates at a frequency of the same order as the EEG. Slower activities of about 3 Hz are also observed in the thalamus during sleep episodes. These oscillations might underly the slow delta waves of the EEG [57]. Thus it seems that the behavioral states of the brain are somehow related to an oscillatory input into the cortical tissue.

The object of this paper is to introduce a simple mathematical model to investigate the influence of an oscillatory input on a cortical tissue. The cortex is described by a neural network of model neurons endowed with some of the salient qualitative and quantitative features of the biological cells. The cortical tissue is assumed to be entrained by different types of thalamic activity characterized by oscillations of various frequencies. This thalamic pacemaker is modeled by an oscillating system which may produce rhythms of various frequency as an input parameter is modified. The degree of synchrony of the network as a result of the input is investigated by evaluating the spatial autocorrelation function.

Although the molecular basis of electrical activity of the brain is reasonably well understood, the origin of the EEG remains still disputed. However from the early sixties, experimental evidence suggests that the EEG is mainly due to postsynaptic currents, while the part due to action potentials seems to be negligible [24, 25, 26, 35]. Thus the EEG appears mainly as a spatial average of cellular currents [63]. Therefore the spatially averaged membrane potentials of the network may be thought as an analogue of the EEG and is used here to compare the model with recorded EEG data.

During the last decade, it was realized that irregular behavior may be generated by relatively simple deterministic dynamics [15, 34, 72, 74]. These so-called chaotic systems have been extensively studied with the help of the new methods of nonlinear time series analysis [32, 34, 38].

The non-periodic and irregular etiology of EEG suggests that it could stem from chaotic dynamics. Therefore the EEG, considered as a time series, can be analyzed by these new methods. The algorithms apply to both experimental and model systems, therefore providing a new and important tool for comparison.

The first section contains a short overview of the analysis of EEG data with the methods of nonlinear dynamics. In the second section, we introduce a neural network model of the thalamo-cortical system. The last sections are devoted to the investigation of the average activity of the network with the help of the same methods of characterization as used for the analysis of EEG data.

1 Chaotic Dynamics of EEG

The three typical EEG rhythms considered here are the beta rhythm seen in active brain with eyes open, the alpha rhythm seen in relaxed brains with eyes closed and the slow wave sleep. Although these EEG rhythms show indisputable periodicities, one of their most salient features is the aperiodic nature of their dynamics. The aim of this section is to show that some of these states could be the manifestation of chaotic dynamics.

Two of the main features of chaotic systems are their fractal geometry and their sensitivity to initial conditions [15, 34, 72, 74]. The "correlation dimension" quantifies the geometric properties of a fractal object [34, 38] whereas the "Lyapunov exponents" measure the exponential divergence of nearby trajectories on the attractor [32, 34]. Moreover, methods have appeared which permit the assessment of these quantities from experimental time series [32, 34, 38]. If applied properly, this kind of analysis provides a powerful mean of investigation of experimental chaotic systems, even if only one variable is accessible to measurements.

1.1 Correlation Dimension from EEG

The correlation dimension D_2 is a generalization of the euclidian dimension and is suitable for the description of fractal objects. $D_2 = 0$ describes a steady state, thus a point in phase space, $D_2 = 1$ for a periodic motion and $D_2 = 2$ indicates quasiperiodicity. Deterministic chaos is characterized by a fractal number greater than two [15, 34, 72, 74]. A purely random process is characterized by an infinite fractal dimension. In general, deterministic chaotic systems are characterized by low fractal dimensions and highly turbulent systems will have larger dimensions.

The correlation dimension can be estimated with the help of the Grassberger-Procaccia algorithm [38] which has been useful in many areas of nonlinear dynamics. As a first step, from the experimental time series $x(t)$, a phase portrait $\{x(t), x(t+\tau), x(t+2\tau), x(t+3\tau), ...\}$ is constructed. This construction involves the single variable x successively delayed in time by a constant value τ. It has been shown that this phase portrait possesses the same correlation dimension, as well as other dynamical properties, as the original phase portrait obtained by considering the complete set of variables of the system [82]. Other constructions of the phase space may be also used and lead to consistent results for relatively low dimensional systems [27].

The correlation dimension D_2 is estimated from the relation [38]

$$\lim_{r \to 0} \frac{Log\ C(r)}{Log\ r} \sim r^{D_2} \tag{1}$$

where r is the distance in the phase space and $C(r)$ is the correlation integral and is the mean number of points laying within a distance r of a reference point, divided by the total number of points. The correlation integral is evaluated by gradually increasing the embedding dimension. A saturation of D_2 towards a constant value is expected for deterministic chaos [38].

In order to obtain reliable correlation dimension estimates, several parameters such as the number of data points, the sampling rate, the time delay τ must be chosen very carefully. A large number of data points is needed such that sufficient pseudo-cycles (> 100) represent the system, otherwise D_2 may be underestimated [6, 10, 21, 76, 83]. The sampling rate may also influence the value of D_2 [6, 10]. We evaluate the time delay by finding a range of τ for which stationarity of D_2 is ensured.

Correlation dimensions can be calculated only if the data set is stationary. The stationarity of a time series can be evaluated with the construction of the recurrence plot from the phase portrait [33]. The recurrence plots have been evaluated for several stages of EEG activity such as alpha rhythm or deep sleep and show that long stretches of EEG have been found which are stationary up to several minutes [8].

In 1985, Babloyantz et al. [7, 13] reported chaotic dynamics for several stages of the human sleep cycle . They could not find saturation for the dimension of the EEG during awake state with eyes open and for REM sleep. However, for sleep stage 2 and 4, they evaluated dimensions of 4.99-5.02 and 4.05-4.37 respectively. The occurrence of chaotic dynamics during the sleep was confirmed later for the stage two [52] and stage four [11, 55]. The Kaplan-Yorke [45] and Mori [59] conjectures applied to the spectrum of Lyapunov exponents also leads to similar values for the deep sleep [36].

The alpha rhythm was also investigated by Albano et al. [3, 68] who found a low value of $D_2 = 2.6$. However we found values of D_2 ranging from 5.95 to 7.4 for three subjects [10, 27]. This latter range of values has been confirmed by other groups [14, 23, 31, 48, 54, 55, 56, 77]. The Kaplan-Yorke and Mori conjectures applied to the spectrum of Lyapunov exponents of alpha rhythm also provides similar values [36]. Moreover, we could show [10] that the surprisingly low value of Albano et al. was due to an insufficient number of data points. They used only 2 s. of alpha EEG (about 20 pseudo-cycles). The dimension D_2 has been found to increase with the data length [10, 52], saturating towards a constant value at about 10 s. for alpha rhythm [10] (about 100 pseudo-cycles). Very similar values of the dimension are found while analyzing data lengths up to 4 min. of alpha rhythm (about 2400 pseudo-cycles) [27].

Graf and Elbert [37] reported large values around 10, for the correlation dimension of alpha rhythm. They also found a significant increase of D_2 as the eyes are opened. On the other hand, for several subjects, low dimensions ranging from 2.8 to 5.9 have also been reported [61, 68, 69] from the analysis of 1 to 2 second epoches of awake and resting EEG.

We also investigated the epileptic petit-mal seizure, for which a very low dimension of 2.05 was found [9]. The burst suppression pattern of the Creutzfeldt-Jakob disease showed correlation dimensions between 3.7 and 5.4 [12, 27]. We believe that the higher values contain physiological noise. Such a decrease of the correlation dimension may be reminiscent of a more general phenomenon of increase of coherence during pathologies [9, 65, 86].

In the case of the Creutzfeldt-Jakob disease, the Kaplan-Yorke and Mori conjectures applied to the spectrum of Lyapunov exponents, confirms the values obtained using the Grassberger-Procaccia algorithm [36]. For the same time series, we could show that interpeak intervals exhibit long-range correlations and cannot be described by a simple random process [28]. This fact constitues a further indication for the presence of deterministic chaos for this EEG. Therefore in this case, evidence from several independent algorithms indicates the presence of deterministic chaos.

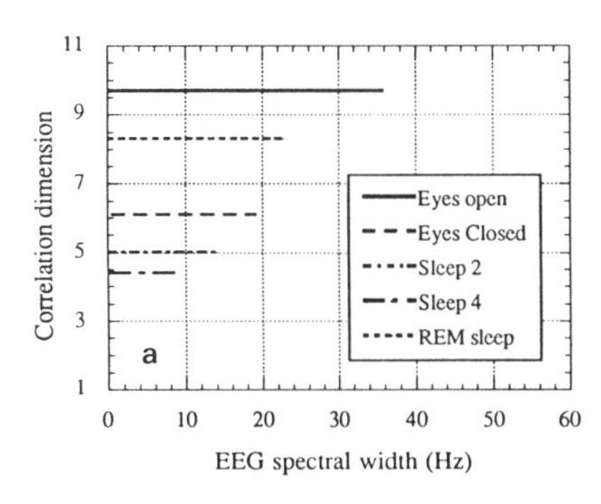

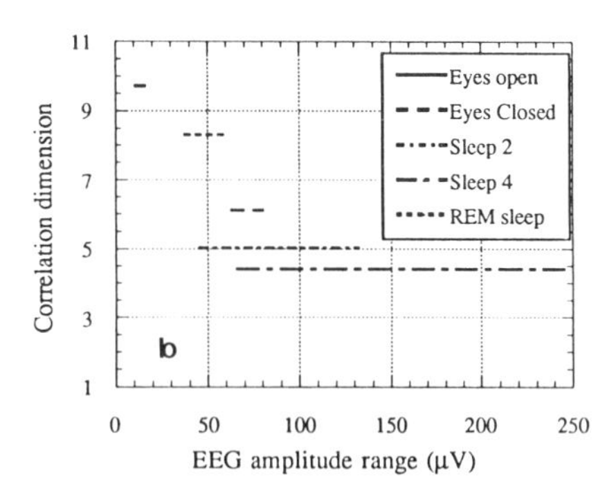

Figure 1. Correlation dimension of the human EEG.
(a) Dimension vs. spectral range. The bars indicate the range of frequency which contains 75% of the spectral energy. (b) Dimension vs. EEG amplitude. Amplitude variations in successive stretches of one second recordings are computed. The bars indicate the range of observed values. Adapted from [12].

Let us note that deterministic chaos has been also reported in the EEG of rabbit olfactory bulb [75], rabbit thalamus [50], and from anesthetized cats [70]. Single neurons can also exhibit chaotic dynamics [40, 53, 60, 67].

The high variability of the results described above leads to reconsidering carefully the meaning of the correlation dimension estimated from the EEG. We feel that the relative values of the dimensions and not their absolute values are meaningful [12, 27, 37, 48].

Such a comparative study has been carried in our group. We could show that there is an intimate relation between correlation dimension, EEG amplitude and EEG frequency [12]. As the brain is active and awake, the EEG is of low amplitude, of high mean frequency and of high correlation dimension. As the eyes are closed, the alpha rhythm displays higher amplitude oscillations with lower frequency and lower correlation dimension. For the normal brain, the deep sleep EEG corresponds to the highest amplitude with the lowest frequency and the lowest correlation dimension. Figure 1 represents the correlation dimension as a function of the spectral width and of the amplitude of these behavioral states. The relation between amplitude, spectra and dimensions reveals a hierarchy of brain attractors which is insensitive to the variability of the measured values of D_2.

1.2 Lyapunov Exponents from EEG

Chaotic dynamics may be studied with the help of another parameter, namely the spectrum of Lyapunov exponents. The Lyapunov exponents are the average exponential divergence of nearby trajectories in the phase space. The presence of at least one positive Lyapunov exponent is the signature of chaotic dynamics [15, 34, 74].

The Lyapunov exponents can be estimated from a time series with help of an algorithm developed by Eckmann et al. [32]. Let us imagine a piece of trajectory contained in a small d_E-dimensional "ball" of radius r (d_E is called the effective dimension and is defined in [32]). The long time evolution of the ball is monitored subsequently. The mean growth rate λ_k of the kth principal axis is:

$$\lambda_k = \lim_{t \to \infty} \frac{1}{t} \frac{P_k(t)}{P(0)} \tag{2}$$

where $P_k(t)$ is the radius of the ball in direction k at time t and $P(0)$ is the radius of the initial ball. λ_k is the Lyapunov exponent corresponding to direction k.

This procedure is repeated for a fixed number of origins and the radius of the ball is adapted at each step such as to keep the number of neighbors constant. $\{\lambda_k\}$ is computed from each origin and the final spectrum of exponents is the average value of these quantities.

Here again, great care must be taken for the choice of the parameters. As discussed in ref. [32], a check for stationarity for these parameters turns out to be the safest way to obtain relevant results.

Positive values for the largest Lyapunov exponent have been reported from the EEG during the sleep cycle [13], the epileptic petit-mal seizure [9] and the alpha rhythm [77]. Recently, the entire spectrum of Lyapunov exponents has been estimated for the Creutzfeldt-Jakob disease, alpha rhythm and deep sleep [36]. It appears that these latter two stages possess respectively three and two positive Lyapunov exponents. Using the Kaplan-Yorke [45] and Mori [59] conjectures, the dimension of the attractor can be estimated from the Lyapunov spectrum. The values obtained by Gallez and Babloyantz from several EEG data [36] are very close the values obtained previously from the Grassberger-Procaccia algorithm (See Section 1.1). Dimensions of 4.44 to 6.12 for the deep sleep, between 5.71 and 7.63 for the alpha rhythm and of 4.0 to 5.22 for Creutzfeldt-Jakob disease have been obtained from this procedure [36]. In this case, we see that two different algorithms provide comparable values.

The evaluation of the correlation dimension and Lyapunov exponents not only quantifies the temporal coherence of the EEG, but it also imposes constraints to model construction. From Figure 1, we see that any model of the cerebral cortex should show the following hierarchy. Increase of EEG amplitude, decrease of frequency and decrease of the correlation dimension as well as of the number of positive Lyapunov exponents.

2 The Model

The emphasis of the model is on the various transitions brought about by the action of periodic inputs to the cerebral cortex. The most salient features of the neurons are incorporated into an electrical analogue of the neuron. However, we try to keep the architecture of the model, as well as the order of magnitude of the parameters, as close as possible to physiological observations.

The simplest equations describing the dynamics of the ionic currents in the neuronal membrane are given by the electrical analogue introduced by Hodgkin and Huxley [41], and described in Fig. 2. According to this model, the equation of the membrane potential is:

$$C_m \frac{dV}{dt} = -g_1 (V - E_1) - g_2 (V - E_2) \tag{3}$$

where C_m is the membrane capacitance, E_1, E_2 and g_1, g_2 are respectively the equilibrium potentials and the membrane conductances associated with each ionic species.

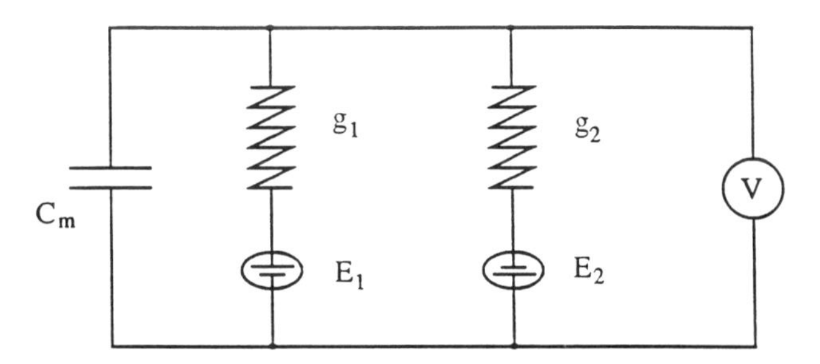

Figure 2. Electrical analogue of the neuronal membrane
Two ionic species with equilibrium potentials (E_1, E_2) and membrane conductances (g_1, g_2) are considered. V is the potential difference between inner and outer sites of the membrane, C_m is the membrane capacitance.

Splitting the conductance $g_i = g_i^0 + g_i^c$ into the "resting" conductance of the membrane (g_i^0; constant in time) and the conductance associated with the opening of specific ion channels (g_j^c; variable in time), one obtains:

$$C_m \frac{dV}{dt} = -g_0 \left(V - V_0\right) - g_1^c \left(V - E_1\right) - g_2^c \left(V - E_2\right) \tag{4}$$

where $g_0 = g_1^0 + g_2^0$ is the resting conductance. In the context of this electrical analogue of the membrane, $V_0 = (g_1^0 E_1 + g_2^0 E_2)/(g_1^0 + g_2^0)$ is the resting potential. g_0 and V_0 are equivalent to the leakage conductance and potential used in electrophysiology.

In the cortical tissue, the population of neurons is divided into approximately 80% of excitatory and 20% of inhibitory neurons [17]. Considering a tissue of N excitatory and M inhibitory neurons, of postsynaptic potential X_i and Y_j respectively, leads to the following set of equations:

$$\begin{aligned} C_m \frac{dX_i}{dt} &= -g_0 \left(X_i - V_0\right) - \left(X_i - E_1\right) \sum_k g_1^{(k->i)} - \left(X_i - E_2\right) \sum_l g_2^{(l->i)} \\ C_m \frac{dY_j}{dt} &= -g_0 \left(Y_j - V_0\right) - \left(Y_j - E_1\right) \sum_k g_1^{(k->j)} - \left(Y_j - E_2\right) \sum_k g_2^{(l->j)} \\ & i, k = 1...N \quad , \quad j, l = 1...M \end{aligned} \tag{5}$$

where $g_i^{(k->i)}$ is the local synaptic conductance associated with the ion j in the synapse that cell k makes to cell i. In the following, only postsynaptic potentials X_i, Y_j constitute the variables of the model.

We assume here that the local conductance $g_j^{(k->i)}$ is proportional to the instantaneous firing rate f_k of the afferent neuron k.

$$g_j^{(k->i)} \sim f_k(t - \tau_{ki}) \tag{6}$$

where τ_{ki} is the propagation time delay between the initiation of the action potential and its arrival at the synaptic junction.

Finally, the firing rate f_k of the neuron k is assumed to be a sigmoidal function of the postsynaptic potential of neuron k:

$$f_k = f_{max}\ F(X_k), \qquad F(X) = \frac{1}{1 + e^{-\alpha(X - V_c)}} \tag{7}$$

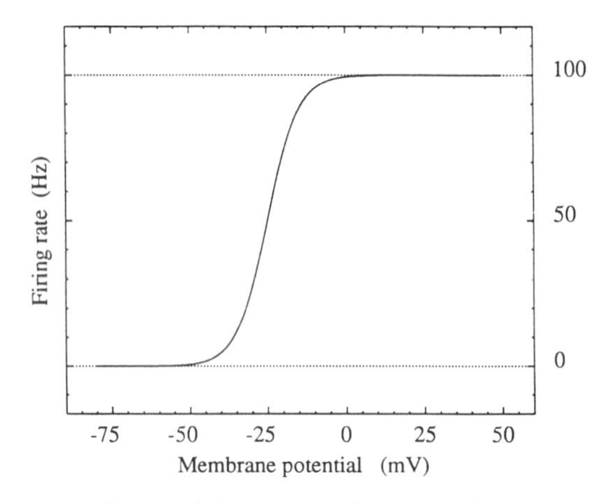

Figure 3. Sigmoidal response function of the neuron.
The firing rate of the neuron as a function of the postsynaptic potential V. $f_{max} = 100\ Hz$, $\alpha = 0.2\ mV^{-1}$, $V_c = -25\ mV$.

Here f_{max}, α and V_c are chosen such as the neuron is silent ($F \simeq 0$) near the resting potential. At a threshold value of the potential, the firing rate increases with the depolarization, and for high values of the potential, it saturates to a maximal activity ($F \simeq 1$; $f_k \simeq f_{max}$). This function is illustrated in Figure 3.

Combining eq. (5) with the relations (6) and (7), the dynamics of the thalamo-cortical tissue is given by [29]:

$$\begin{aligned}
\frac{dX_i}{dt} &= -\gamma(X_i - V_0) - (X_i - E_1)\left[\sum_k \omega_{ki}^{(1)}\ F(X_k(t-\tau_{ki})) + T_i\ g(x)\right] \\
&\quad -(X_i - E_2)\sum_l \omega_{li}^{(2)}\ F(Y_l(t-\tau_{li})) \\
\frac{dY_j}{dt} &= -\gamma(Y_j - V_0) - (Y_j - E_1)\sum_k \omega_{kj}^{(3)}\ F(X_k(t-\tau_{kj})) \\
&\quad -(Y_j - E_2)\sum_l \omega_{lj}^{(4)}\ F(Y_l(t-\tau_{lj})) \\
& \qquad i,k = 1...N \quad , \quad j,l = 1...M
\end{aligned} \tag{8}$$

where γ is the inverse of the time constant of the membrane: $\gamma = g_0\ /\ C_m$. $\omega_{ki}^{(1)}$, $\omega_{li}^{(2)}$, $\omega_{kj}^{(3)}$, $\omega_{lj}^{(4)}$ are respectively the excitatory-to-excitatory, inhibitory-to-excitatory, excitatory-to-inhibitory and inhibitory-to-inhibitory synaptic weights. The τ_{ki} are the propagation time delays from site k to site i of the network. $T_i\ g(x)$ is the thalamic input to excitatory cell i.

The various parameter values used in eq. (8) are based on experimental measurements on pyramidal cells or motoneurons. The following values will be used in the sequel [46]: $\gamma = 0.25\ ms^{-1}$, $V_0 = -60\ mV$, $E_1 = 50\ mV$, $E_2 = -80\ mV$. The parameters of the transfer function $F(x)$ are: $\alpha = 0.2\ mV^{-1}$, $V_C = -25\ mV$ and are estimated assuming the threshold of the neurons at $-40\ mV$ [46] and a saturation of the firing rate to its maximal value at about $-10\ mV$. Such a sigmoidal transfer function was observed experimentally in the frog axon [84].

The two types of neurons are arranged on a two-dimensional regular lattice (see Fig. 4). The propagation time delay τ_{ki} is 2 ms per lattice length. In the hippocampus, time delays of 2.5 ms have been measured between two neighboring pyramidal cells [58].

A given neuron connects to all neurons lying within a fixed neighborhood (see Fig. 4a). Different rules of connectivity, such as first and second neighbor interaction, as well as random connectivity, have been considered. The connections shown in Fig. 4a will be used in the sequel. The boundary neurons have the same connectivity, except that a "mirror" image of the network is repeated outside the boundaries. For first neighbor connections this corresponds to zero flux boundary conditions.

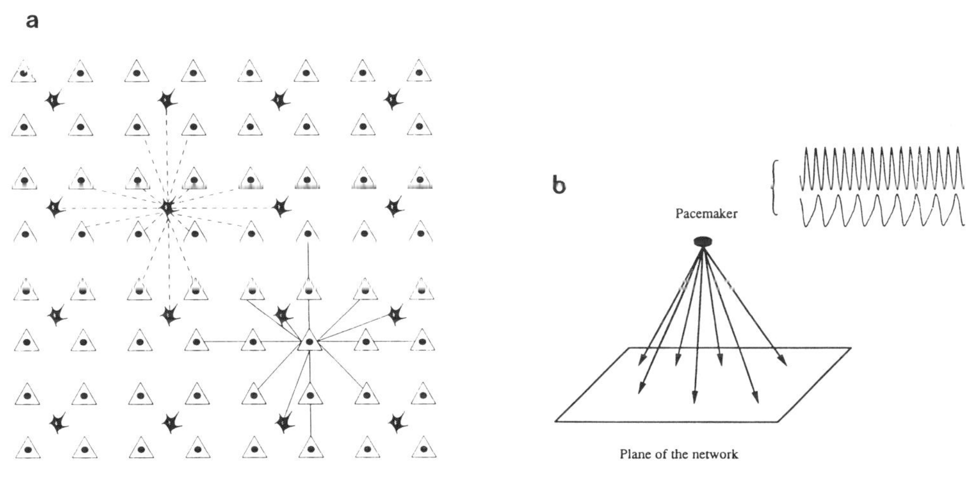

Figure 4. Network architecture.
(a) The excitatory cells (triangular) and inhibitory cells (stellate) form a two dimensional lattice. Excitatory connections (solid lines) and inhibitory connections (dashed lines) are shown for two neurons. (b) schematic representation of the thalamocortical system and the two oscillating states of the pacemaker. Parameters of the oscillator are $I = 2.6$ (above) and $I = 0.6$ (below).

For each type of interaction, the synaptic weights of all neurons are identical and the total sum of the inputs is constant. Therefore, $\sum_k \omega_{ki}^{(1)} = \Omega_1$, $\sum_l \omega_{li}^{(2)} = \Omega_2$, $\sum_k \omega_{kj}^{(3)} = \Omega_3$ and $\sum_l \omega_{lj}^{(4)} = \Omega_4$, $i, k = 1...N$, $l = 1...M$. The dynamical properties of the model are very similar for a wide range of these parameters.

A small fraction of cortical excitatory neurons are submitted to an oscillatory input which mimics thalamic activity. It is well known that the cerebral cortex receives massive inputs from the thalamus. The latter is known as a generator of several types of rhythmical activity characterized by different frequencies.

Llinas and Jahnsen [51] reported that different inputs to the membrane of thalamocortical (TC) neurons *in vitro* elicits various types of oscillatory behavior (6 10 Hz). Recently, pacemaker activities of very low frequency (0.5-2.5 Hz) have also been reported in single TC cells *in vitro* [49]. The thalamus appears therefore as an autonomous oscillator of various frequencies.

We use the phenomenological model of the thalamic oscillations introduced by Rose and Hindmarsh [71]. The two variable version of the model reads:

$$\frac{1}{c}\frac{dx}{dt} = f_1(x) - z + I$$

$$\frac{1}{c}\frac{dz}{dt} = r(h_2(x) - z) \tag{9}$$

where x is the membrane potential of the thalamic neuron, z is an adaptation variable and I is the input to the membrane. $f_1(x) = m_2(x + \frac{1}{2}) + 4(1 - m_2)(x + \frac{1}{2})^3 + \frac{1}{2}$ for $-1 > x > 0$, $f_1(x) = 1 - x^3 - 2x^2$ otherwise, and $f_2(x) = 2(\mu^{x+k+1} - 1)/(\mu^{k+\frac{1}{2}} - 1)$.

This model accounts qualitatively for some of the electrophysiological properties of thalamic neurons [51] such as the coexistence between resting and oscillating activities. I is the control parameter allowing the transition from steady states to oscillatory states. c is a parameter which rescales the time axis. The other parameters are: $\mu = 4.3, k = 0.603, m_2 = -0.1, r = 0.5$. For $I = 0.6$ and $I = 2.6$, two limit cycles of different frequencies are seen (see Fig. 4b).

The thalamic oscillator is connected to 2 % of randomly choosen excitatory neurons via the transfer function $g(x) = 1/\left[1 + \exp(-10\ (x - 0.9))\right]$ with a strength of T_i=15.

The set of eqs. (8) together with eqs. (9) are integrated numerically using a Runge-Kutta algorithm [66] modified for integration of delay differential equations.

Equations describing neuronal tissue without thalamic input and with functional form similar to (8) have been considered previously with a different function F [42, 43], and without propagation delays [16, 39, 64].

3 Spontaneous Dynamics of the Network

In a state of alertness, the EEG is irregular and of low amplitude, which indicates desynchronized behavior. The working hypothesis in this paper is that synchronization of the cortical tissue occurs following periodic inputs from the thalamus. Therefore, in order to model the spontaneous activity of the cortex in the absence of thalamic input, we consider $T_i = 0$ in eqs. (8).

The activity of the network with $T_i = 0$ is a function of the total synaptic weights $\{\Omega_1, \Omega_2, \Omega_3, \Omega_4\}$. For weak values of these parameters, all cells of the network relax to the unique resting potential of -60 mV. However, as the synaptic weights increase, spontaneous sustained activity may appear.

For $\Omega_2 = \Omega_3 = 12.5$ and $\Omega_4 = 0$ and for moderate values of Ω_1, the neurons oscillate periodically in unison and one sees a bulk oscillation. As Ω_1 is increased further, spatiotemporal chaos may appear. In the range of parameters considered, this type of "neuronal turbulence" is not seen in small networks. However, for reasonably large number of neurons ($N \sim 400$), turbulent behavior is always present.

Between the oscillating regime and the turbulent state, there is a range of values of Ω_1 where intermittency is seen. Periods of oscillatory regime alternates with spatiotemporal chaotic activity. For still larger values of Ω_1, spiral waves may be seen (Fig. 5). These type of spatiotemporal dynamics have been described in reaction-diffusion systems [47] and for coupled map lattices [44].

The activity of neurons during the "turbulent" state is seen as an irregular propagation of clusters of activity (Fig. 6). Such clusters spontaneously appear and collide

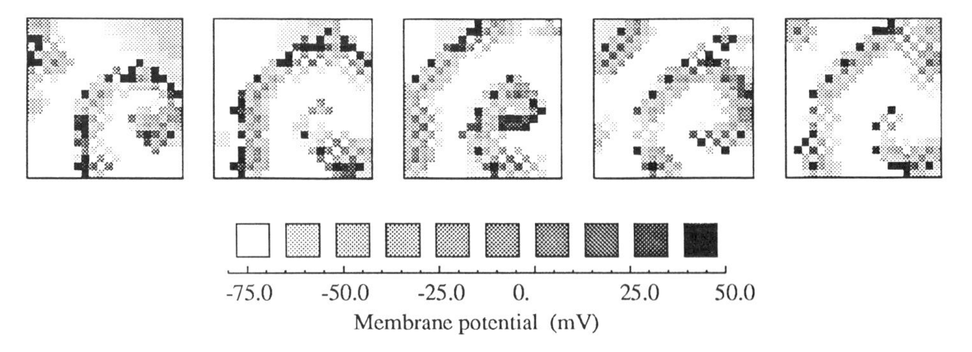

Figure 5. Rotating spiral waves in a N=400 network.
The successive snapshots represent the instantaneous spatial distribution of membrane potentials of excitatory neurons. Each neuron is represented by a shaded square. The higher the potential, the darker the shade (cfr. scale). Snapshots are taken every 10 ms. Parameters are $N = 400$, $M = 100$, $\Omega_1 = 15$, $\Omega_2 = 12.5$, $\Omega_3 = 12.5$ and $\Omega_4 = 0$.

to form very complex patterns. Sometimes, transient spiral waves may also appear. Individually, each neuron behaves as an aperiodic oscillator.

It is assumed that this "desynchronized" state constitues the spontaneous activity of the cortex. The values of $\Omega_1 = 15$, $\Omega_2 = \Omega_3 = 12.5$ and $\Omega_4 = 0$, which give rise to this spatiotemporal chaos, will be used in the sequel.

4 Pacemaker-Induced Synchronization

The onset of oscillatory input with $T_i = 15$ in eqs. (8) generally changes the global activity of the network. Although a small minority of cells receive the input, the vast majority of the neurons may be entrained into coherent behavior (Fig. 7).

We shall investigate the dynamics of the system for two values of the input I and parameter c, $\{I = 0.6, c = 0.06\}$ and $\{I = 2.6, c = 0.125\}$ which correspond to two oscillating states of high and low frequency of different amplitude and wave form (see Fig. 7a,b).

4.1 Spatial Aspects of Synchronization

In Fig. 7a, the system is submitted to the faster rhythm of the pacemaker. From a desynchonized state with $T_i = 0$, the network switches into a partially synchronized dynamics where the activities of the neurons become more phase-locked. As the pacemaker oscillates, the network also oscillates between various spatially synchronized patterns. Fig. 7a is an example of instantaneous activity. Although these oscillations are not strictly regular, the mean frequency of the network is of the same order as that of the pacemaker.

If the parameter I is decreased, the pacemaker switches to a slower oscillating regime. Again the network responds by a partial synchronization of neural activity. However, the synchronization of neurons may attain a remarkable degree of coherence (Fig. 7b). These brief occurrences of almost total synchronization appear irregularly.

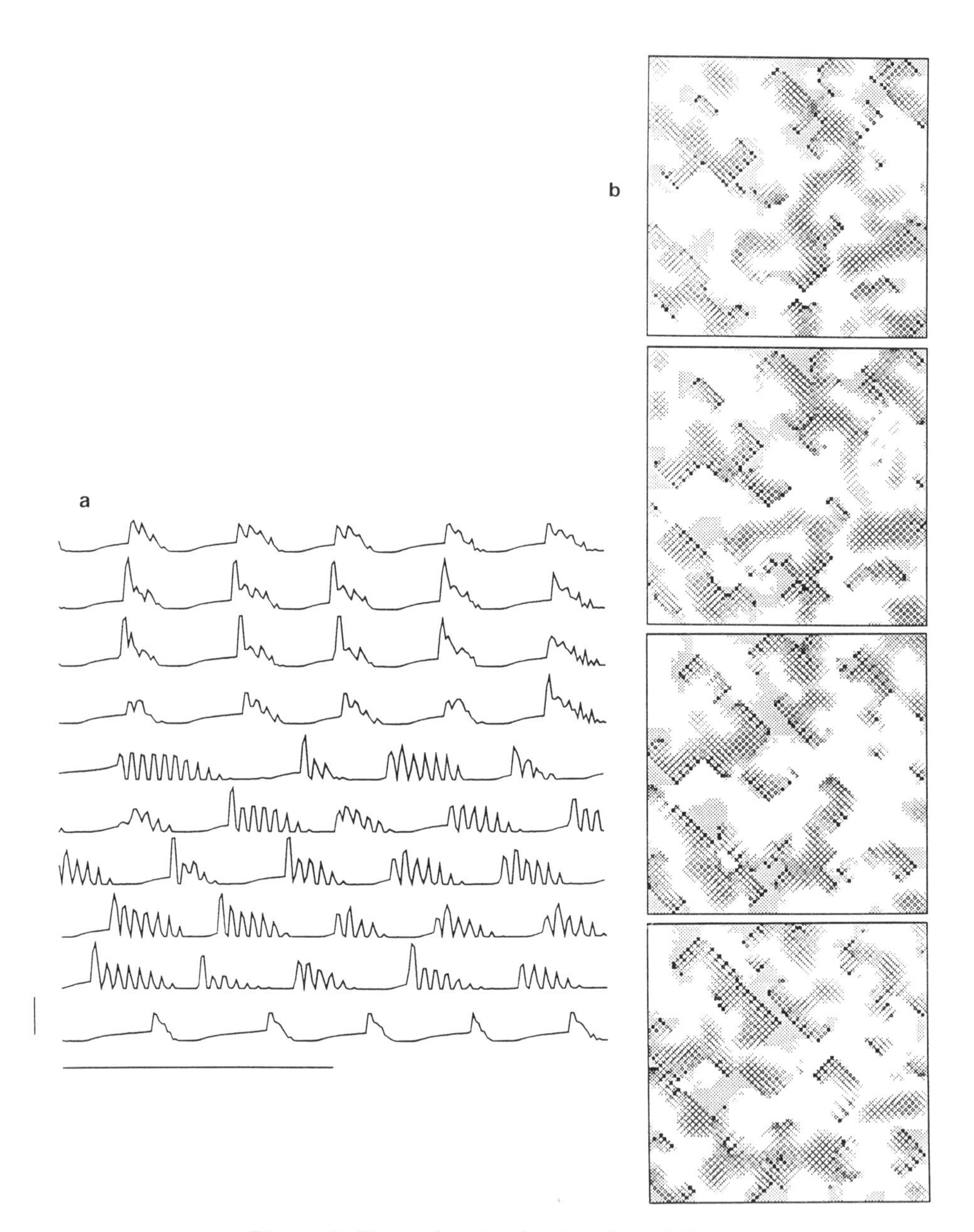

Figure 6. Desynchronized network activity

(a) Electrical potential of 10 excitatory neurons in the network. From top to bottom: four adjacent and six randomly distributed cells. Calibration: vertical bar=100 mV, Horizontal bar=100 ms. (b) Instantaneous spatial distribution of membrane potentials (same coding as in Fig. 5). Snapshots are taken at every 16 ms (from top). $N = 6400$, $M = 1600$, $\Omega_1 = 15$, $\Omega_2 = 12.5$, $\Omega_3 = 12.5$, $\Omega_4 = 0$.

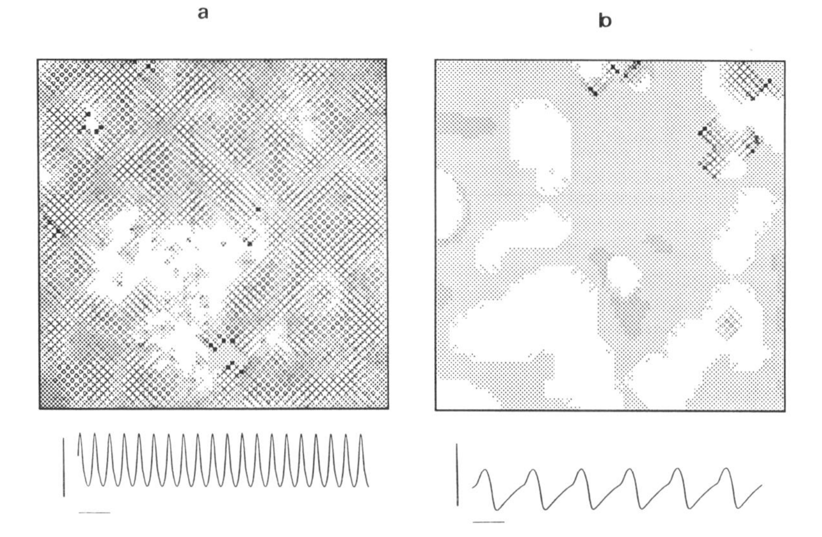

Figure 7. Snapshots of partially synchronized network.
(a) synchronization following the onset of the fast rhythm (I=2.6, c=0.125), (b) synchronization resulting from slow pacemaker input (I=0.6, c=0.06). N=6400, M=1600, Ω_1 = 15, Ω_2 = 12.5, $\Omega_3 = 12.5$ and $\Omega_4 = 0$.

A similar brief and highly synchronized events have been described in the hippocampus [20]. These so-called "sharp waves" are mostly seen during slow wave sleep.

The level of synchronization between neurons can be quantified by evaluating the spatial autocorrelation function:

$$c(r) = \int \sum_{i=1}^{N} \sum_{j \in V_i(r)} X_i(t)\, X_j(t)\, dt \tag{10}$$

where $V_i(r)$ is the set of sites at a distance r of site i. Fig. 8 represents the normalized function $c(r)/c(0)$ evaluated for the system in the absence of input and for slow and fast pacemaker inputs. For the turbulent state, this function vanishes after several lattice lengths, indicating a loss of spatial coherence and the absence of long range spatial correlations. However, as the fast oscillatory input is switched on, the dynamics of the network synchronizes and the autocorrelation function shows a slower spatial decay. During this state, long range spatial correlations appear as a consequence of the more coherent synchronized dynamics. Spatial correlations increase further for the slow periodic input.

Figure 8 shows that there is an obvious oscillation of a period of two lattice lengths superimposed on the correlation function. This fact indicates that the neighboring excitatory cells have a tendency to oscillate out of phase.

4.2 Temporal Aspects of Synchronization

The degree of relative temporal coherence of the network may also be studied by examining the evolution of the spatially averaged membrane potentials in function of

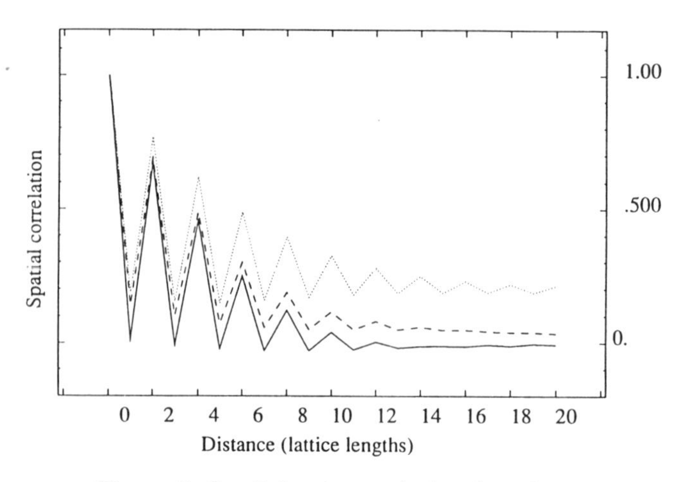

Figure 8. Spatial autocorrelation function.
solid: "turbulent" state in the absence of pacemaker, *dashed:* fast pacemaker input, *dotted:* slow pacemaker input. The parameters are the same as in Fig. 7.

time. Such an average activity constitutes a global variable which may be thought as being the analogue of an EEG.

Figure 9 shows the time evolution of this average value in the absence and in the presence of fast and slow oscillatory inputs. In order to save computer time, a smaller network was considered as qualitatively it shows similar behavior to the larger networks. In the absence of pacemaker, we see low amplitude fast and irregular behavior. However as the periodic input sets in, large amplitude and more regular slower oscillations are seen. The slow pacemaker oscillation organizes the system into a higher amplitude and lower frequency regime.

This increase in coherence may be quantified by considering the spatially averaged activity of the network as a time series. With the help of usual techniques of non-linear time series analysis [38], phase portraits, correlation dimensions and Lyapunov exponents as well as other dynamical parameters can be evaluated.

Figure 9 shows the phase portraits constructed from the three time series. It is seen that in the absence of pacemaker, no obvious attractor exists. However, the oscillatory input organizes the system such as the averaged behavior shows deterministic chaos. The numerical values of the correlation dimension of these attractors are convenient parameters for assessing the degree of synchronization of systems described in Fig. 9.

The correlation dimension is estimated from the Grassberger-Procaccia algorithm [38]. The system is integrated such as a sufficient number of pseudo-cycles covers the attractor. Calculations are made with 10000 data points equidistants of 1.2 ms, which constitutes 80 to 320 pseudo-cycles. The time delay used for phase space reconstruction is estimated from a check for stationarity which leads to 6 ms $< \tau <$12 ms.

In the absence of pacemaker activity, we are in the presence of a very high dimensional dynamical system, and the existing algorithms do not permit the characterization of the system (Fig. 10). This behavior is very similar to the desynchronized EEG, such as beta rhythm or REM sleep, which also show low amplitude high dimensional dynamics [13, 10].

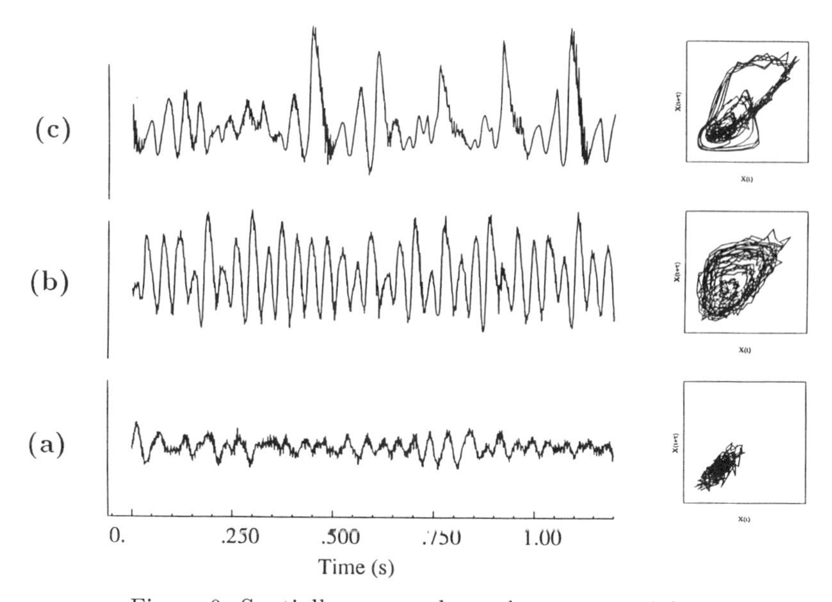

Figure 9. Spatially averaged membrane potentials.
(a) desynchronized "turbulent" state, (b) synchronization during the fast rhythm, (c) synchronization during the slow rhythm. Vertical calibration bars are of 50 mV. The corresponding phase portraits are also shown. The parameters are N=400, M=100, Ω_1 = 15, Ω_2 = 12.5, Ω_3 = 12.5, Ω_4 = 0 and τ = 6 ms.

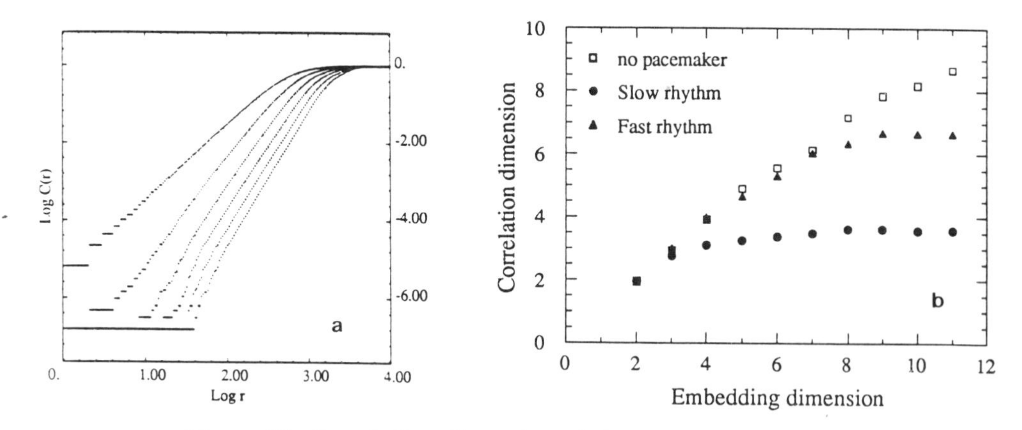

Figure 10. Correlation dimension of the averaged potential.
(a) $\log C(r)$ vs. $\log r$ curves for embedding dimensions from 2 to 7, for the pacemaker with $c = 0.06$, $I = 0.6$ (6 Hz). The slope of the linear region of these curves gives the correlation dimension. (b) Saturation of the slope vs. embedding dimension for the three time series shown in Fig. 9.

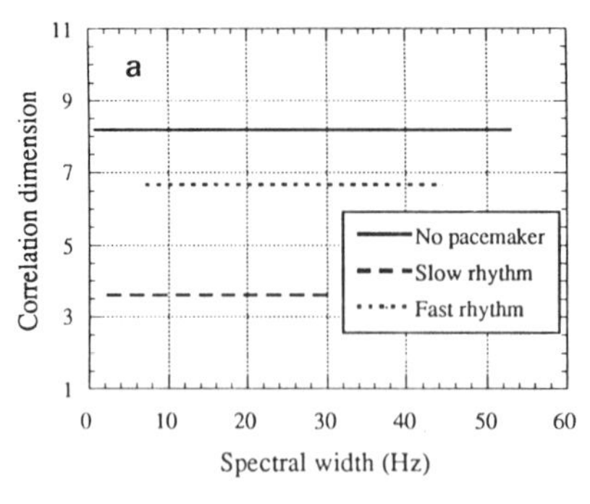

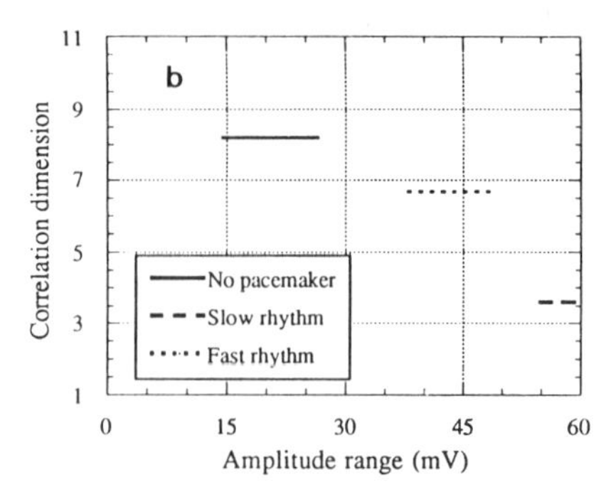

Figure 11. Correlation dimension vs. spectral width and amplitude.
These figures are constructed using the same procedure as in Fig. 1. The dynamics of the chaotic attractors of the model are in qualitative agreement with EEG data, both for dimension-spectral width (a) and dimension-amplitude relations (b). The parameters are the same as in Fig. 9.

As the pacemaker activity sets in, the collective behavior of the network is characterized by low dimensional chaos (Fig. 10). The correlation dimension is $D_2 = 6.6 \pm 0.1$ for the faster rhythm and $D_2 = 3.6 \pm 0.1$ for the slower rhythm. Here again, these values are remarkably similar to the results obtained from the analysis of EEG. The slower rhythms of the EEG, such as deep sleep, correspond to the lower correlation dimensions. The dimension-amplitude and the dimension-spectral range relations obtained from the network (Fig. 11) are qualitatively similar to those of the EEG (Fig. 1).

In order to characterize further the chaotic dynamics seen in the network, the spectrum of Lyapunov exponents has been evaluated from the algorithm of Eckmann et al. [32]. A good convergence has been obtained for the fast rhythm (Fig. 12) as well as for the slow rhythm (not shown). The data length and other parameters for reconstruction of phase portraits were the same as those used for the evaluation of the correlation dimension. The rhythm corresponding to the lowest dimension, therefore a slower thalamic activity, was found to possess two positive Lyapunov exponents (13.6 ± 2.7, $6.6 \pm 1.9\ s.^{-1}$) whereas the dynamics corresponding to the fast rhythm possesses three positive Lyapunov exponents (7.55 ± 0.15 , 4.5 ± 0.6 , $2.1 \pm 0.3\ s.^{-1}$), as illustrated in Fig. 12. The presence of positive Lyapunov exponents constitutes a further indication of chaotic dynamics and the number of positive Lyapunov exponents are in qualitative agreement with EEG data (see section 1.2).

The use of the Kaplan-Yorke conjecture [45] for estimating the dimension from the spectrum of Lyapunov exponents leads to dimensions ranging from 4.6 to 5.16 for the slow rhythm and 5.8 for the fast rhythm. The Mori conjecture [59] furnish values of 3.8 to 3.92 for the slow rhythm and of 4.7 for the fast rhythm. The dimension estimated from this procedure is also lower for the slow rhythm of the thalamic pacemaker. Therefore, the results from the two algorithms show a good consistency as it has been shown for EEG data by Gallez and Babloyantz [36].

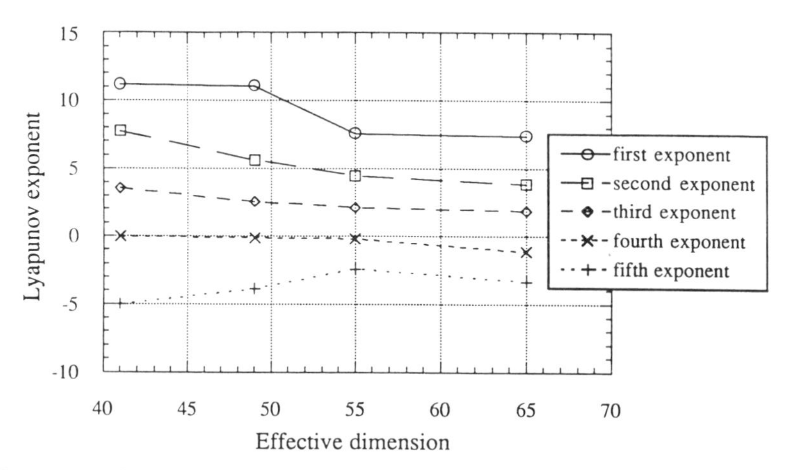

Figure 12. Spectrum of Lyapunov exponents vs. effective dimension d_E. A convergence to 3 positive exponents is seen at d_E=55. $I = 2.6$, $c = 0.125$. The other parameters are the same as in Fig. 9.

4.3 Role of the input frequency

In the preceeding sections, we have shown that a periodic input may organize a neural network into a more synchronizaed and coherent behavior. For slow inputs, both spatial and temporal coherence increase further. It is difficult to see if this increase of coherence is due to the frequency, or to the wave form of the oscillation, as both of these properties change when passing from the fast rhythm to the slow rhythm. In order to see if the frequency is the determinant parameter, we must investigate the network under different conditions of input frequency, with all other parameters unchanged. This can be achieved in eq. (9), if I is kept constant and c is varied. Oscillations of identical wave form and different frequency are produced. The frequency of these oscillations increases with c.

The correlation dimension has been evaluated for the slow rhythm ($I = 0.6$) at four different frequencies. Fig. 13 shows that the correlation dimension increases *linearly* with the pacemaker frequency. Moreover, one of these frequencies is very close to the fast rhythm, however with a different wave form. Nevertheless, the correlation dimension is similar to the dimension of the fast rhythm (Fig. 13). Therefore, we conclude that for the range of frequencies and wave forms considered, the frequency of the pacemaker appears as the most determinant parameter which modulates the coherence of the network.

In order to construct an input/output frequency relationship for the network, several simulations are made under identical conditions, except for the thalamic input pattern which is rescaled over a large range of frequencies. The power spectrum is computed from the average potential, and the prominent peak of this spectrum is evaluated as the main response frequency of the network. The input-output relation is shown in Fig. 14. We see that there exists a large range of frequencies for which the input frequency and the network response are quasi-identical. However, for slow pacemaker inputs, it is seen that the response frequency of the network is different from that of the pacemaker. The network synchronizes at a frequency which seems to depend only on internal parameters.

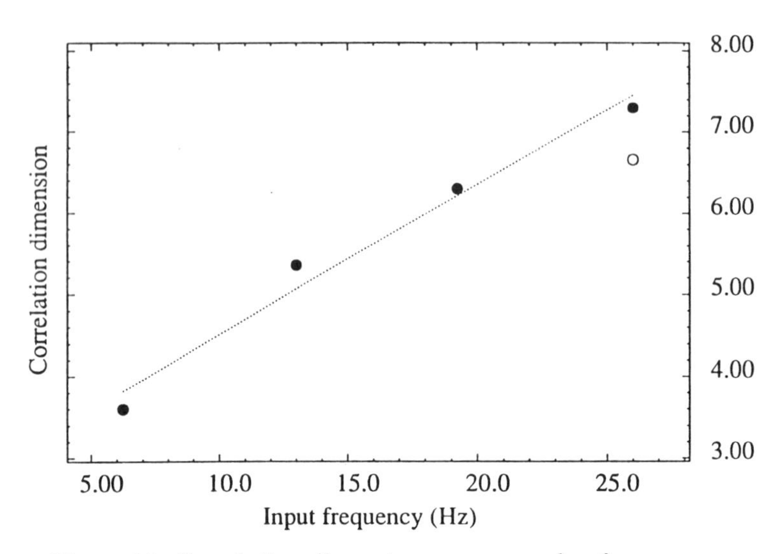

Figure 13. Correlation dimension vs. pacemaker frequency.
(•) pacemaker with $I = 0.6$, $c = 0.06$, $c = 0.125$, $c = 0.185$ and $c = 0.25$. (○) pacemaker oscillation of different wave form ($I = 2.6$, $c = 0.125$).

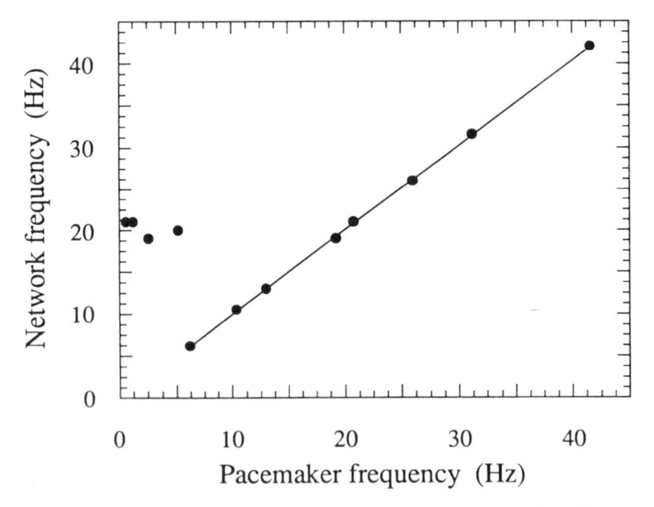

Figure 14. Cortical response frequency vs. pacemaker frequency.
The peak frequency of the averaged membrane potential versus the pacemaker frequency. The parameters are the same as in Fig. 9. 1024 data points.

5 Discussion

Since the pioneering work of Babloyantz et al. who suggested that several stages of human EEG obey deterministic chaotic dynamics, several groups in various countries published similar results (see Section 1). Nevertheless, still there is a lot of debate around the question of the relevance of chaotic attractors for probing the brain dynamics. The controversy stems from two origins. The authors disagree on the applicability of algorithms, data length, techniques for phase space construction and above all the possibility of finding sufficently long stretches of EEG data sets.

However a more serious question arise about the nature of EEG itself. Although EEG stems from the electrical activity of a very large number of spatially distributed and interconnected neurons, the exact relationship between this global activity and the cellular events is not yet completely elucidated. Therefore the nature of the dynamics assessed from the EEG, although related to the cortical events, does not say much about the exact nature of distributed activities at cellular level.

Granted that the underlying dynamics of the EEG is governed by deterministic chaos, it is important to understand the meaning of this finding at the cortical level.

In this paper, we constructed a simple model of the cortex and showed that a periodic input, mimicking thalamic influence into the cortex, partially synchronize the neurons into patterns of collective firing. We also showed that the degree of synchronization, that is the extend of patches where neurons fire in unison, increases as the frequency of the oscillating input decreases. A global network activity was constructed from these networks by a simple averaging of electrical activity of the neurons. This average activity showed an EEG-like etiology.

In the absence of oscillatory input the global activity is of the same type as the human EEG recorded with eyes open. With the onset of fast oscillatory input, the average activity increases in amplitude and the frequency of the waves decreases. This state can be seen as reminiscent of alpha waves. The slower oscillatory input increases the amplitude and decreases the frequency still further, in some way, producing deep sleep type of activity.

Several other frequencies and different wave forms were also introduced into the model cortex. The simulations showed that the frequency is the dominating factor of synchronization.

From these results, it is tempting to suggest that the various behavioral states of the human cortex follows dynamics similar to the one depicted in our model. Thus the temporal chaos seen from time series analysis is the manifestation of spatiotemporal-like chaotic activity at the cortical level. Low dimensional chaotic attractors arise when larger and larger patches of neuronal population synchronize for longer and longer times.

However, a more satisfying model of the thalamo-cortical interaction necessitates a more elaborate model of the thalamus which takes into account the reticular activating system and the incoming sensorial input. Such a model is presently under investigation.

ACKNOWLEDGMENTS We are grateful to D. Gallez, J.A. Sepulchre and X.J. Wang for constructive discussions. Part of numerical calculations were done at IBM ECSEC, Rome, and Cray Research France, Paris. A.D. acknowledges IRSIA fellowships (1986-1988) and grants from the Belgian Government (ARC - 1989-1990).

References

[1] Adrian, E.D. Afferent discharges to the cerebral cortex from peripheral sense organs. *J. Physiol.* **100**: 159-191, 1941.

[2] Adrian, E.D. and Matthews, B.H.C. The interpretation of potential waves in the cortex. *J. Physiol.* **81**: 440-471, 1934.

[3] Albano, A.M., Abraham, N.B., de Guzman, G.C., Tarroja, M.F.H., Bandy, D.K., Gioggia, R.S., Rapp, P.E., Zimmerman, I.D., Greenbaun, N.N. and Bashore, T.R. Lasers and brains: complex systems with low dimensional attractors. in: *Dimensions and Entropies in Chaotic Systems.* Edited by Mayer-Kress, G., *Springer Series in Synergetics* **32**. Berlin: Springer, 1986, pp. 231-240.

[4] Andersen, P. and Andersson, S.A. *Physiological Basis of the Alpha Rhythm* New York: Appelton Century Crofts, 1968.

[5] Andersen, P. and Andersson, S.A. Thalamic origin of cortical rhythmic activity. in: *Handbook of EEG and Clinical Neurophysiology.* Edited by Remond, A. Amsterdam: Elsevier, 1974, pp. 91-118.

[6] Atten, P., Caputo, J.G., Malraison, B. and Gagne, Y. Détermination de dimensions d'attracteurs pour différents écoulements. *Journal de Mécanique Théorique et Appliquée,* special issue, 133-156, 1984.

[7] Babloyantz, A. Evidence of chaotic dynamics of brain activity during the sleep cycle. in: *Dimensions and Entropies in Chaotic Systems.* Edited by Mayer-Kress, G., *Springer Series in Synergetics* **32**. Berlin: Springer, 1986, pp. 241-245.

[8] Babloyantz, A. Some remarks on nonlinear dynamical analysis of physiological time series. in: *Measures of Complexity and Chaos.* Edited by N.B. Abraham, A.L. Albano, A. Passamente and P.E. Rapp, NATO ARW series, Plenum press, New York, 1989, pp. 51-62.

[9] Babloyantz, A. and Destexhe, A., Low dimensional chaos in an instance of epileptic seizure. *Proc. Natl. Acad. Sc. USA* **83**: 3513-3517 , 1986.

[10] Babloyantz, A. and Destexhe, A., Strange attractors in the cerebral cortex. in: *Temporal Disorder in Human Oscillatory Systems.* Edited by Rensing, L., an der Heiden, U. and Mackey, M.C., *Springer Series in Synergetics*, **36**. Berlin: Springer, 1987, pp. 48-56.

[11] Babloyantz, A. and Destexhe, A., Chaos in neural networks. in: *Proceedings of the IEEE First International Conference on Neural Networks.* Edited by Caudill, M. and Butler, C., Vol.4: 31-40 , 1987.

[12] Babloyantz, A. and Destexhe, A., The Creutzfeldt-Jakob disease in the hierarchy of chaotic attractors. in: *From Chemical to Biological Organization.* Edited by Markus, M., Muller, S. and Nicolis, G., *Springer Series in Synergetics*, **39**. Berlin: Springer, 1988, pp. 307-316.

[13] Babloyantz, A., Nicolis, C. and Salazar, M., Evidence for chaotic dynamics of brain activity during the sleep cycle. *Phys. Lett. A* **111**: 152-156 , 1985.

[14] Basar, E., Basar-Eroglu, C., Roschke, J. and Schult, J. Chaos and alpha preparation in brain function. in: *Models of Brain Function* Eds. Cotterill, R.M.J. Cambridge: Cambridge University Press, 1989, pp. 365-395.

[15] Bergé, P., Pomeau, Y. and Vidal, C. *L'ordre dans le Chaos.* Hermann, Paris, 1984. English translation: *Order Within Chaos.* New York: Freeman, 1984.

[16] Bower, J. The simulation of large scale neural networks. in: *Methods in Neuronal Modeling.* Edited by Koch, C. and Segev, I. Cambridge, MA: MIT press, 1988, pp. 295-337.

[17] Braitenberg, V. Two views of the cerebral cortex. in *Brain Theory.* Edited by Palm, G. and Aertsen, A. Berlin: Springer, 1986, pp. 81-96.

[18] Bremer, F. Effets de la déafférentation complète d'une région de l'écorce cérébrale sur son activité électrique. *C.R. Soc. Biol. Paris* **127**: 355-359, 1938.

[19] Burns, B.D. and Webb, A.C. The correlation between discharge times of neighboring neurons in isolated cerebral cortex. *Proc. Roy. Soc. Lond. Ser. B* **203**: 347-360, 1979.

[20] Buzsaki, G. Hippocampal sharp waves: their origin and significance. *Brain Res.* **398**: 242-252, 1986.

[21] Caputo, J.G., Malraison, B. and Atten, P. Determination of attractor dimension and entropy for various flows: an experimentalist's viewpoint. in: *Dimensions and Entropies in Chaotic Systems.* Edited by Mayer-Kress, G., *Springer Series in Synergetics* **32**. Berlin: Springer, 1986, pp. 180-190.

[22] Caton, R. The electric currents of the brain. *Brit. J. Med.* **2**: 278, 1875.

[23] Cerf, R., Farssi, Z., Burgun, B., Trio, J.M. and Kurtz, D. Attracteurs étranges dans les signaux électroencéphalographiques humains. *C.R. Acad. Sci. Paris, Sér. II* **307**: 715-718 , 1988.

[24] Creutzfeldt, O. Neuronal Basis of EEG waves. in: *Handbook of EEG and Clinical Neurophysiology.* Edited by Remond, A. Amsterdam: Elsevier, 1974, pp. 6-55.

[25] Creutzfeldt, O., Watanabe, S. and Lux, H.D. Relation between EEG phenomena and potentials of single cortical cells. I. Evoked responses after thalamic and epicortical stimulation. *EEG Clin. Neurophysiol.* **20**: 1-18, 1966.

[26] Creutzfeldt, O., Watanabe, S. and Lux, H.D. Relation between EEG phenomena and potentials of single cortical cells. II. Spontaneous and convulsoid activity. *EEG Clin. Neurophysiol.* **20**: 19-37, 1966.

[27] Destexhe, A., Sepulchre, J.A. and Babloyantz, A. A comparative study of the experimental quantification of deterministic Chaos. *Phys. Lett. A* **132**: 101-106 , 1988.

[28] Destexhe, A. Symbolic dynamics from biological time series. *Phys. Lett. A* **143**: 373-378, 1990.

[29] Destexhe, A. and Babloyantz, A. Pacemaker-induced coherence in cortical networks. *Neural Computation,* in press, 1990.

[30] Dusser de Barenne, J.G. and Mc Cullogh, W.S. The direct functional interrelation of sensory cortex and optic thalamus. *J. Neurophysiol.* **4**: 304-310, 1941.

[31] Dvorak, I. and Siska, J. One some problems encountered in calculating the correlation dimension of EEG. *Phys. lett. A* **118**: 63-66 , 1986.

[32] Eckmann, J.P., Kamphorst, S.O., Ruelle, D. and Ciliberto, S. Lyapunov exponents from time series. *Phys. Rev. A.* **34**: 4971-4979, 1986.

[33] Eckmann, J.P., Kamphorst, S.O. and Ruelle, D. Recurrence plots of dynamical systems. *Europhys. Lett.* **4**: 973-977, 1987.

[34] Eckmann, J.P. and Ruelle, D. Ergodic theory of chaos and strange attractors. *Rev. Mod. Phys.* **57**: 617-656, 1985.

[35] Elul, R. The genesis of the EEG. *Int. Rev. Neuro.* **15**: 227-272, 1972.

[36] Gallez, D. and Babloyantz, A. Predictability of human EEG: a dynamical approach. Biol. Cybernetics, in press, 1990.

[37] Graf, K.E. and Elbert, T. Dimensional analysis of the waking EEG. Edited by Basar, E. and Bullock, T.H. *Springer Series in Brain Dynamics* **2**. Berlin: Springer, 1989, pp. 174-191.

[38] Grassberger, P. and Procaccia, I. Characterization of strange attractors. *Phys. Rev. Lett.* **50**: 346-349, 1983.

[39] Grossberg, S. Nonlinear difference-differential equations in prediction and learning theory. *Proc. Natl. Acad. Sci. USA* **58**: 1329-1334, 1967.

[40] Hayashi, H., Ishizuka, S., Ohta, M. and Hurakawa, K. Chaotic behavior in the Onchidium giant neuron under sinusoidal stimulation. *Phys. Lett. A* **88**: 435-438, 1982.

[41] Hodgkin, A.L. and Huxley, A.F. A quantitative description of membrane current and its application to conduction and excitation in nerve. *J. Physiol.* **117**: 500-544, 1952.

[42] Kaczmarek, L.K. A model of cell firing patterns in epileptic seizures. *Biol. Cybernetics* **22**: 229-234, 1976.

[43] Kaczmarek, L.K. and Babloyantz, A. Spatiotemporal patterns in epileptic seizures. *Biol. Cybernetics* **26**: 199-208, 1977.

[44] Kaneko, K. Spatiotemporal chaos in one and two-dimensional coupled map lattice. *Physica D* **37**: 60-82, 1989.

[45] Kaplan, J. and Yorke, J. Chaotic behavior of multi-dimensional difference equations. in: *Functional Differential Equations and Approximations of Fixed Points*, Eds. Peitgen, H.O. & Walther, H.O. (Springer, Berlin), *Lectures Notes in Mathematics* **Vol. 330**, 1979, pp. 228-236.

[46] Katz, B. *Nerve, Muscle and Synapse.* New York: Mc Graw-Hill, 1966.

[47] Kuramoto, Y. *Chemical Oscillations, Waves and Turbulence.* Berlin: Springer, 1984.

[48] Layne, S.P., Mayer-Kress, G. and Holzfuss, J. Problems associated with dimensional analysis of electroencephalogram data. in: *Dimensions and Entropies in Chaotic Systems.* Edited by Mayer-Kress, G., *Springer Series in Synergetics* **32**. Berlin: Springer, 1986, pp. 246-256.

[49] Leresche, N., Jassik-Gerschenfeld, D., Haby M., Soltesz, I. and Crunelli, V. Pacemaker-like and other types of spontaneous membrane potential oscillations in thalamocortical cells. *Neurosci. Lett.* **113**: 72-77, 1990.

[50] Leszczynski, K.W. On the dimension of the cortical EEG, *Int. J. Biomed. Computing* **20**: 283-288, 1987.

[51] Llinas, R.R. and Jahnsen, H. Electrophysiology of thalamic neurones in vitro. *Nature* **297**: 406-408, 1982.

[52] Lo, P.C., and Principe, J.C. Dimensionality analysis of EEG segments: experimental considerations. in: *Proceedings of the International Joint Conference on Neural Networks, Washington, june 1989*, IEEE Proceedings, Vol.**I**, pp. 693-698.

[53] Matsumoto, G., Aihara, K., Hanyu, Y., Takahashi, N., Yoshizawa, S. and Nagumo, J. Chaos and phase locking in normal squid axons. *Phys. Lett. A.* **123**: 162-166, 1987.

[54] Mayer-Kress, G. and Holzfuss, J. Analysis of the human electroencephalogram with methods from nonlinear dynamics. in: *Temporal Disorder in Human Oscillatory Systems.* Edited by Rensing, L., an der Heiden, U. and Mackey, M.C., *Springer Series in Synergetics*, **36**. Berlin: Springer, 1987, pp. 57-68.

[55] Mayer-Kress, G. and Layne, S.P. Dimensionality of the human electroencephalogram. in: *Perspectives in Biological Dynamics and Theoretical Medicine*, proceedings, *Ann N.Y. Acad. Sci.* **504**: 62-87, 1987.

[56] Mayer-Kress, G., Yates, F.E., Benton, L., Keidel, M., Tirsh, W., Pppl, S.J. and Geist, K. Dimension analysis of nonlinear oscillations in brain, heart and muscle. *Math. Biosci.* **90**: 155-182, 1988.

[57] Mc Carley, R.W., Benoit, O. and Barrinuevo, G. Lateral geniculate nucleus unitary discharge in sleep and waking: state and rate specific aspects. *J. Neurophysiol.* **50**: 798-818, 1983.

[58] Miles, R., Traub, R.D. and Wong, R.K.S. Spread of synchronous firing in longitudinal slices from the CA3 region in the hippocampus. *J. Neurophysiol.* **60**: 1481-1496, 1988.

[59] Mori, H. Fractal dimensions of chaotic flows of autonomous dissipative systems. *Prog. Theor. Phys.* **63**: 1044-1047, 1980.

[60] Mpistos, G. J., Burton, R.M., Creech, H.C. and Soinila, S.O., Evidence for chaos in spike trains that generate rhythmic motor patterns. *Brain Res. Bull.* **21**: 529-538, 1988.

[61] Nan, X and Jinghua, X. The fractal dimension of EEG as a physical measure of conscious human brain activities. *Bull. Math. Biol.* **50**: 559-565, 1988.

[62] Noda, H. and Adey, W.R. Firing of neuron pairs in cat association cortex during sleep and wakefulness. *J. Neurophysiol.* **33**: 672-684, 1970.

[63] Nunez, P.L. *Electric Fields of the Brain. The Neurophysics of EEG.* New York: Oxford University Press, 1981.

[64] Perkel, D.H., Mulloney, B. and Budelli, R.W. Quantitative methods for predicting neuronal behavior. *Neurosci.* **6**: 823-837, 1982.

[65] Pool, R. Is it healthy to be chaotic ? *Science* **243**: 604-607, 1989.

[66] Press, W.H., Flannery, B.P., Teukolsky, S.A. and Vetterling, W.T. *Numerical Recipes. The Art of Scientific Computing.* Cambridge: Cambridge University Press, 1986

[67] Rapp, P.E., Zimmerman, I.D., Albano, A.M., de Guzman, G.C. and Greenbaun, N.N. Dynamics of spontaneous neural activity in the simian motor cortex: the dimension of chaotic neurons. *Phys. Lett. A* **110**: 335-338 ,1985.

[68] Rapp, P.E., Zimmerman, I.D., Albano, A.M., de Guzman, G.C., Greenbaun, N.N. and Bashore, T.R. Experimental studies of chaotic neural behavior: cellular activity and electroencephalographic signals. in: *Nonlinear Oscillations in Biology and Chemistry.* Edited by Othmer, H.G., *Lectures notes in Biomathematics* **66**. Springer, Berlin, 1986, pp. 366-381.

[69] Rapp, P.E., Bashore, T.R., Martinerie, J.M., Albano, A.M., Zimmerman, I.D. and Mees, A.I. Dynamics of brain electrical activity *Brain Topography* **2**: 99-118, 1989.

[70] Roschke, J. and Basar, E. The EEG is not a simple noise: strange attractors in intracranial structures. in: *Dynamics of Sensory and Cognitive Processing by the Brain.* Edited by Basar, E. *Springer Series in Brain Dynamics* **1**. Berlin: Springer, 1988, pp. 203-216.

[71] Rose, R.M. and Hindmarsh, J.L. A model of a thalamic neuron. *Proc. Roy. Soc. Lond. Ser. B.* **225**: 161-193, 1985.

[72] Ruelle, D. *Chaotic evolution and strange attractors.* Cambridge: Cambridge University Press, 1989.

[73] Sandler, Y.M. Models of neural networks with selective memorization and chaotic behavior. *Phys. Lett. A* **144**: 462-466, 1990.

[74] Schuster, H. *Deterministic Chaos.* Weinheim: Physik Verlag, 1984.

[75] Skarda, C.A. and Freeman, W.J. How brains make chaos in order to make sense of the world. *Behav. Brain. Sci.* **10**: 161-195, 1987.

[76] Smith, L.A. Intrinsic limits on dimension calculations. *Phys. Lett. A.* **133**: 283-288, 1988.

[77] Soong, A.C.K. and Stuart, C.I.J.M. Evidence of chaotic dynamics underlying the human alpha rhythm EEG. *Biol. Cybernetics* **62**: 55-62, 1989.

[78] Steriade, M. and Deschênes, M. The thalamus as a neuronal oscillator. *Brain Res. Rev.* **8**: 1-63, 1984.

[79] Steriade, M., Deschênes, M., Domich, L. and Mulle, C. Abolition of spindle oscillations in thalamic neurons disconnected from nucleus reticularis thalami. *J. Neurophysiol.* **54**: 1473-1497, 1985.

[80] Steriade, M., Domich, L., Oakson, G. and Deschênes, M. The deafferented reticular thalamic nucleus generates spindle rhythmicity. *J. Neurophysiol.* **57**: 260-273, 1987.

[81] Steriade, M. and Llinas, R.R. The functional states of the rhalamus and the associated neuronal interplay. *Physiol. Rev.* **68**: 649-742, 1988.

[82] Takens, F. Detecting strange attractors in turbulence. in: *Dynamical Systems and Turbulence.* Edited by Rand, D.A. and Young, L.S., *Lecture Notes in Mathematics* **898**. Berlin: Springer, 1981, pp. 366-381.

[83] Theiler, J. Spurious dimension from correlation algorithm applied to limited time series data. *Phys. Rev. A* **34**: 2427-2432, 1986.

[84] Verveen, D.A. and Derksen, H.E. Fluctuations in membrane potentials of axons and the problem of coding. *Kybernetik* **2**: 152-160, 1965.

[85] Verzeano, M. Activity of cerebral neurons in the transition from wakefulness to sleep. *Science* **124**: 366-367, 1956.

[86] West, B.J. *Fractals and Chaos in Medicine.* Berlin: Springer, 1989.

CHAOS AND INFORMATION PROCESSING: MULTIFRACTAL ATTRACTORS MEDIATING PERCEPTION AND COGNITION

J.S. Nicolis

Dept. of Electrical Engineering
University of Patras
Greece

Information theory quantifies and optimizes (with respect to given constraints) the efficiency of Designing Performing and Interpreting (the results of) single or multiple measurements.

To this end, information theory devises codes (cascades of one-to-one mapping(s) between sets of symbols) in order to compromise two (conflicting) objectives in signal transmission: speed of transfer and reliability of reception. Specifically, the aim is to ensure that the "receiving site" will be able to detect and correct single or multiple, uncorrelated or correlated errors-in the presence of (quite often overwhalming) noise i.e uncontrolled fluctuations-In short, information theory - as originally devised by C.Shannon deals with symbol manipulations (much like the execution of a program in a digital computer)-at a pure syntactical level-without any reference to meaning of the messages involved. (No"understanding" or "thinking" takes place during this sort of transaction).

After clearing out the received messages from noise, the next step in the cognitive hierarchy is the "semantic"one namely the extraction of meaning: The sine-qua-non condition for that is the unambiguous partitioning of the set of undifferentiated "raw" external stimuli onto coexisting asymptotically stable categories ("minimal-length-algorithms"→chaotic strange attractors) and the establishment of a language of communication (a sequence of jumps) amongst them. The attractors-memories have to be asymptotically stable so that the syndrom of Self-reference-which characterises any linguistic scheme-does not turn into a paradox.

Now,the raw messages that the processor receives may come from 3 different sources (dynamical systems):

a) Dynamical systems with intrinsic relaxation times T_i orders of magnitude above the intrinsic relaxation time(s) τ of the processor itself $T_i >> \tau$

b) D. Systems for which: $T_i << \tau$

c) D. Systems for which: $T_i \sim \tau$

it is not perhaps accidental that the two sciences that Man has developed

Self-Organization, Emerging Properties, and Learning
Edited by A. Babloyantz, Plenum Press, New York, 1991

to the point of perfection are a) Celestial mechanics ($T_i >> \tau$) and b) quantum mechanics ($T_i << \tau$). In case a) one takes a lot of variables and/ or parameters as practically constants and in case b) as boundary conditions.

In case c) on the other hand, the processor has no other choice but to "coevolve" with the system under observation or to wage a game-amounting essentially to a cascade of Decision making under uncertainty and conflict. Such a "game" constitutes the third and last level of the informational edifice we call information hierarchy, namely the pragmatic level.

The algorithmic and dynamical aspects of the on goings at the three distinct hierarchical levels-syntactical, semantic and pragmatic as well as some hints about the way these levels interact are treated in the enclosed references.

Synopsis

We first consider a chaotic dissipative flow (e.g. the Rössler attractor) or a map (e.g the logistic equation) $\dot{\underset{\sim}{x}} = \underset{\sim}{f}(\underset{\sim}{x};\underset{\sim}{\xi}), \nabla \cdot \underset{\sim}{f} < 0$.
We apply an arbitrary (but Markovian) partition of M subintervals on the x-axis thereby imposing an "alphabet" so that transitional points between neighbouring cells-symbols-are mapped on transitional points. (Other wise in case that is, of non-Markovian partitions-the alphabet will get distorted along the flow or along the cascade of iterations).

Taking one subinterval as the pause and iterating the flow or the map we generate-after transients subside-artificial texts of words-strings of symbols separated by the pause-and we investigate the "bulk" statistical properties of such "languages" namely the Shannonian entropy, the Kolmogorov -Sinai entropy the Transinformation and the "Zipf's like"behavior of the text, that is the probability of appearance of a "word"in a long text v.s its ranking order.
we also investigate the Markovian memory of the language (usually of a fifth order) as well as the grammatical and syntactic rules which allow only a small portion,out of all possible words of length k made from M symbols (M^k), to be realise. So we deduce that the chaotic dissipative flow acts as an efficient "symbolic filter"allowing variety and reliability to coexist, and mimicking the evolutionary processes-whatever they are- which have shaped the formation of natural languages.

The statistical properties of texts from contemporary Greek prose are also displayed for comparison. All these deliberations characterise a language at the "syntactical level" as they essentially refer to "word dynamics"-or plain information theory. Moving next to deliberations at the semantic level (e.g. at the hierarchical level where pattern classification and recognition takes place) we deal with correspondances or mappings (not one-to-one) amongst subsets (sub-basins) of external objects and functions (stimuli) and the coexisting categories-attractors of the processor. The empirical study of the ongoings at the semantic level is of course the object of cognitive psychology.
At the sematic level the processor-classifier is simulated by a chaotic dynamics $\dot{\underset{\sim}{x}} = \underset{\sim}{f}(\underset{\sim}{x};\mu), \nabla \cdot \underset{\sim}{f} < 0$ possessing multiple coexisting chaotic strange attractors separated, in general, by fractal basin boundaries. Each attractor is a "cognitive receptor"which far from behaving like a "photographic memory"- as in the case of a steady state-improvises and evolves "from within". The partitioning of a large number of environmental stimuli-time series of length N-onto a set of few chaotic attractors of informational dimensionalities $D_i << N$ amounts to compression of the raw stimuli and the construction of minimal length algorithms (programs)whose length is proportional to the correlation dimensionalities of the attrac-

tors involved. Due to the fractal character of the basin boundaries the above categorisation takes place mostly under conditions of uncertainty and conflict that is it involves an intermittent jumping of a given stimulus from attractor to attractor.

In case where coexisting attractors are stable (not repellers) external "noise" is needed in order to make the different categories to establish a language amongst themselves. In (human) information processing the thalamocortical pacemaker (whose one-dimensional trace is the recorded E.E.G from a lead on the scalp) is responsible for permutating the attractors on a time-division-multiplexing basis.
This is accomplished in a fashion which far from been a "random"process it possesses all the characteristics of a "metalanguage" .Indeed we know that the thalamocortical pacemaker itself possesses a chaotic strange attractor which is more or less uniform during epilepsy, drowsyness or sleep but becomes a multifractal under the demanding regime of execution of mental tasks such as pattern recognition or calculations. Since a multifractal attractor is highly inhomogeneous (it spends practically all the time in the vicinity of few "hot spots"in state-space) we can understand the partimonial reason behind information selection which underlies e.g. recognition of shapes and faces, that is the reason why very few (information rich) details-edges, sharp turns- are enough for the construction of the whole image in the processor, essentially from a contour.

Finally we come to the deliberations taking place at the pragmatic level where games are involved. Here we study the correspondence or mapping amongst the "psychological motivations"or "strategies" of the partners involved and their "objectives"- attractors-under uncertainty and conflict. We show in some carefully chosen examples from paradoxical games (like the game of blackmail or "chicken") that allowing the payoffs of the game to vary with time we come up with an intermittent process between a stable attractor (the "Synergetic" state) where the partners strategies converge asymptotically and stay there, and an inconclusive regime in state space where the exchange of "blows" never ends. In some respects then, non-trivial pattern recognition and paradoxical games share an important feature which makes them isomorphic: They both involve decision making under conditions of uncertainty and conflict.

Finally let us mention that the parameter "attention" is related to the degree of excitation of the ascending branch of the reticular formation of the brain and directly influences the polarization of the specific thalamic nuclei which are the generators of the thalamocortical pacemaking activity. "Attention" then deeply determines (and modifies) the topology of the chaotic strange attractor of the thalamocortical pacemaker thereby dictating the policy of intermittent (multifractal) jumps amongst the coexisting attractors-categories of the cognitive processor.

In Conclusion:

We propose a new selection principle which acting via inhomogeneous intermittency in phase space (e.g. via a multifractal strange chaotic attractor-the thalamocortical pacemaker) is responsible for a highly nonlinear Linguistic filtering process, limiting drastically:

a) at the syntactical level: the grammatically legitimate words.
b) at the semantic level: the "interesting" key features of a pattern.
c) at the pragmatic (Game-playing) level: the set of possible"moves".

This is a apriori selection principle-whereas, "natural selection" is a aposteriori selection principle acting on phenotypes e.g. on attractors via shifting control parameters and resulting in generation/elimination/fusion ("crises") of coexisting attractors- "categories".

Multifractal Chaotic dynamics applied to information processing devices may thus be instrumental:

a) In drastically simplifying Hardware without compromising the breath of the functional repertoire (Software).

b) In reducing or compressing memory capacity;(instead of storing via"pixelization" of a pattern all details, one limits attention to very few key features (associated with small radii of curvature)coding for disproportionally high amounts of stored information.
So,through an inhomogeneous intermittent dynamics in state space we filter out only:

Key words
Key features
Key moves

The principle of natural selection in Linguistics involves further scrutinization on those small, apriori selected subsets.

References

John S. Nicolis "Chaos and Information processing" World Scientific Publishers (1990)

Appendix

In information processing,we design,carry out and interpret measurements. Every measurement includes one "expanding" phase and one "contracting " phase, taking place either in succession or in unison.
During the expanding phase, the processor, via his "motor" activity (by perturbing the universe as it were) makes a subset of alternatives to manifest themselves, both in number and a priori probabilities.
During the contracting phase, the processor, via his "sensors", contracts the basin of attraction onto one (out of many) coexisting attractors-categories.

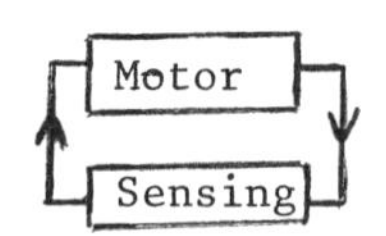

positive Lyapunov exponents

negative Lyapunov exponents

It is imperative then for a "biological "processor to possess coexisting chaotic strange attractors, in order to comply with both above requirements.

SPONTANEOUS ACTIVITY IN A LARGE NEURAL NET: BETWEEN CHAOS AND NOISE

Xiao-Jing Wang

Mathematical Research Branch
National Institutes of Health
Bethesda, Maryland 20892, USA

In what follows I would like to discuss the intrinsic oscillatory dynamics of a large network of neurone-like analog units. Using an appropriate entropic quantity called the ϵ-entropy, the spontaneous activity of the net will be shown to be intermediate between low-dimensional chaos and random noise. This kind of network dynamics seems to reflect some aspects of the ground-state fluctuations of the mammalian brain systems during wakefulness.

Let us consider a model network given by

$$\tau_0 dv_i/dt = -v_i + \sum_{j=1}^{N} J_{ij}\phi(v_j); \qquad i = 1, 2, ..., N \tag{1}$$

where τ_0 is a time constant, J_{ij} is the connection matrix, and $\phi(x)$ is the input-output function of a sigmoid type, e.g. $\phi(x) = tanh(gx)$ with a gain parameter g. Eq.(1) represents a *bona fide* "connectionist" model: each unit alone would merely relax to its rest $v_i \equiv 0$ in a trivial fashion. When such units are connected by strong enough nonlinear interactions, however, collective dynamics and computational abilities can emerge in the network.

In a Hopfield network (Hopfield 1982, 1984) with *symmetric* connection, $J_{ij} = J_{ji}$, we know that the model Eq.(1) has an analogy to a spin glass (a net of magnetic dipoles with random connections). Then, there exists a global Liapunov function, like the Hamiltonian of a spin glass, which ensures that the time evolution of the net would always converge to a fixed point. Such a fixed point attractor can be interpreted as a content-addressable memory, and a main question is how many such memories can be stored in a net of N units, with desired recall properties. With J_{ij} prescribed according to a Hebb-like learning rule, and at the large gain limit ($g \to \infty$), this problem can be treated analytically by statistical mechanics of spin glass (cf. Amit (1989)).

If the symmetry assumption is relinquished, $J_{ij} \neq J_{ji}$, then an attractor need no longer be steady: the net can oscillate. What kind of spacetime pattern(s) the net would have as attractor(s) depend on the network architecture (Amari 1972). One possibility would be to keep the connections in the all-to-all form, hence disregarding the spatial geometry all together. A concrete model system of this kind, studied recently by Sompolinsky *et al* (1988), corresponds to the case where the connection matrix J_{ij} is assumed to be generated by independent random Gaussian variables, with $[J_{ij}] = 0, [J_{ij}^2] = J^2/N$ and $[J_{ij}J_{i'j'}] = 0$ if $i \neq i'$ or $j \neq j'$ (square brackets denote average over the distribution of J_{ij}). I shall henceforth be restricted to this example.

Self-Organization, Emerging Properties, and Learning
Edited by A. Babloyantz, Plenum Press, New York, 1991

Notice that, once J_{ij} is fixed (one says the random interacting links between units are "frozen"), Eq.(1) is a *deterministic* dynamical system. Since the network is not subject to any external stimulation, the dynamics we shall be dealing with will be called its spontaneous activity.

It is easy to see that by a rescaling of v_i and t, Eq.(1) has only one (dimensionless) parameter, gJ. If $gJ = 0$, the solution of Eq.(1) is obviously $v_i \equiv 0$. With increased gJ, this branch of fixed point attractor can become destabilized. Using a time-dependent mean field theory, Sompolinsky *et al* (1988) found that, for an extensively large network, $N \to \infty$, if $gJ < 1$, the only kind of attractor is fixed point. A transition occurs at $gJ = 1$, beyond which the network displays a chaotically oscillatory state (cf. Fig.1). Furthermore, in such a randomly connected network one would not expect any regular spatial pattern or coherence, which would restrict the number of excited degrees of freedom (modes) to a small figure. Consequently, beyond this critical point, the chaotic attractor is immediately "fully developed": it appears to have an infinite dimension, and infinite number of positive Liapunov exponents (Sompolinsky *et al* 1988; Sompolinsky 1990). One observes that this sharp transition is reminiscent of the Manneville-Pomeau intermittency (Pomeau & Manneville 1980) which, as a bifurcation from a *limit cycle* to chaos, describes a universal scenario of "route to chaos" in low dimensional dynamical systems. I intend to discuss this "intermittent transition to spacetime chaos" in a separate communication.

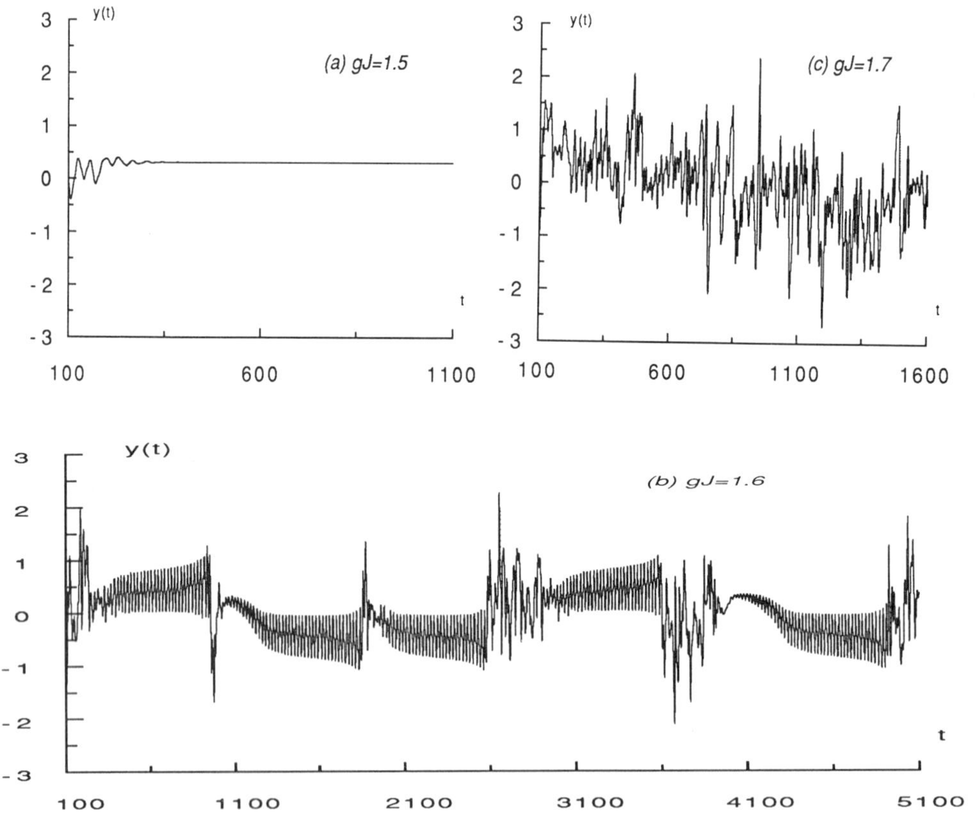

Fig.1. Numerical evidence of intermittency (b) near a transition from a fixed point (a) to a chaotic flow (c). The net Eq.(1) was simulated with $N = 500$ units, J_{ij} were randomly chosen, then fixed. Plotted here are the time evolutions of a single unit for different values of gJ. The intermittent behavior in (b) consists of long regular phases interrupted randomly by chaotic bursts.

Let us focus on this spacetime chaos itself, and attempt to characterize it by its dynamical entropy. Let us recall that, for a finite dynamical system, a chaotic attractor can generate an information flow of finite amount per unit time, quantified by the Kolmogorov-Sinai entropy, h_{KS} (cf. Eckmann & Ruelle 1985; Shaw 1981) This is possible because a *deterministic* chaotic system with *continuous* variables has a correspondence to a *stochastic* process with *discrete* states, via "symbolic dynamics". Then, the Kolmogorov-Sinai entropy is equal to the Shannon entropy per unit time of the associated stochastic process. These discrete states may be viewed as a result of a digitalization of the continuous variables that is *inherent* to the original system. Since the conversion is *exact*, there exits a finite size ϵ_0 of the digitalization accuracy ϵ, beyond which no additional information can be gained. In other words, if one computes the information production rate as a function of ϵ, say $h(\epsilon)$, then it will equal to h_{KS} for all $\epsilon \leq \epsilon_0$.

For a chaotic "flow", with many active degrees of freedom, this Kolmogorov-Sinai entropy is infinite. In such a case one may pretend to associate a deterministic spacetime chaos to a *continuous* variable stochastic process. Indeed, for the network under consideration, Sompolinsky *et al* (1988) showed that, in the limit $N \to \infty$, Eq.(1) is reduced to a dynamical mean-field equation of a single unit, which reads

$$dv_i/dt = -v_i + \eta_i(t), \tag{2a}$$

where $\eta_i(t)$ is a nonMarkovian, Gaussian field, representing the averaged input from other units. Its autocorrelation function

$$< \eta_i(t)\eta_i(t+\tau) >= C(\tau) =< \phi(v_i(t))\phi(v_i(t+\tau)) >, \tag{2b}$$

where $\phi(x) = tanh((gJ)x)$, is evaluated within the mean-field theory (angular brackets denote average with respect to the distribution of $\eta_i(t)$).

The mean-field equation tells us that as $N \to \infty$, the network behaves as if it consisted of *independent* units, embeded in a common field $\eta(t)$. The subscript i becoming unnecessary, will be omitted thereafter.

In case such a connection can be made between a deterministic large system and a probabilistic one, we can introduce (Gaspard & Wang 1990) the notion of ϵ-entropy, initially proposed for continuous stochastic processes (independently by A.N. Kolmogorov and C. Shannon), to deterministic spacetime chaos. The idea of the ϵ-entropy is the following: although the entropy of an analog signal $x(t)$ from a stochastic process is properly speaking infinite, its amount of creating information rate is always bounded, if it is monitored by instruments with finite precision. Suppose the measured signal $y(t)$ differs from the real signal $x(t)$ by a mean quantity, e.g. ϵ is the average of $(x-y)^2$. For instance, $y(t)$ can be a digitalization of $x(t)$ within ϵ. One then can compute the mutual information of x and y, which is well defined. The ϵ-entropy, $h(\epsilon)$, is defined as the minimal mutual information (obtained by optimally choosing the "instrument", or $y(t)$), under the constraint that the error cannot exceed ϵ. For a smaller ϵ, y has to be chosen closer to x, and $h(\epsilon)$ is increased. As $\epsilon \to 0$, $h(\epsilon)$ approaches to the entropy of $x(t)$ itself, i.e. may diverge to infinity. The asymptotic behavior of $h(\epsilon)$ as $\epsilon \to 0$, is a characteristic of the process under consideration.

For a Gaussian stationary process, $h(\epsilon)$ can be written in terms of its power spectrum $f(\omega)$, by the following expressions (due to A.N. Kolmogorov)

$$\begin{aligned} \epsilon^2 &= \frac{1}{2\pi}\int_{-\infty}^{+\infty} min[\theta^2, f(\omega)]d\omega; \\ h(\epsilon) &= \frac{1}{4\pi}\int_{-\infty}^{+\infty} max[0, \log_2 \frac{f(\omega)}{\theta^2}]d\omega. \end{aligned} \tag{3}$$

where $h(\epsilon)$ is parameterized via θ.

This formula implies that the asymptotic form of $h(\epsilon)$ is determined, in the Gaussian cases, by the ultraviolet part of the power spectrum, $f(\omega)$ as $\omega \to \infty$. For instance, an Ornstein-Uhlenbeck process would have a Langevin equation of the same form as Eq.(2), but $\eta(t)$ would be a white noise, $< \eta(t)\eta(t') >= \sigma^2\delta(t-t')$. Then, $f(\omega) = \sigma^2/(2\pi(1+\omega^2))$. From Eq.(3) one can readily show that

$$h(\epsilon) \sim \frac{\sigma^2}{\pi^3}(\frac{1}{\epsilon})^2 \tag{4}$$

That is, the ϵ-entropy of a stationary, Gaussian and Markov process diverges as ϵ^{-2}.

In our case of a large network, the ϵ-entropy is an extensive quantity, proportional to N. In the limit $N \to \infty$, units are essentially decoupled, each being governed by Eq.(2). Thus we can evaluate the ϵ- entropy *per unit*, on the basis of Eq.(2). Since Eq.(2) is linear, and $\eta(t)$ is a Gaussian process, so is $v(t)$. Therefore Eq.(3) is applicable, provided that we know the power spectrum of $v(t)$. A self-consistent equation for the correlation function of $v(t)$ has been given in Sompolinsky *et al* (1988), and solved for $0 < \vartheta \equiv gJ - 1 \ll 1$, yielding

$$\Delta(\tau) \equiv < v(t)v(t+\tau) > \simeq \vartheta cosh^{-2}(\vartheta\tau/\sqrt{3}). \tag{5}$$

The power spectrum $f_v(\omega)$ can be obtained by the Fourier transform of $\Delta(\tau)$,

$$\begin{aligned} f_v(\omega) &= \sqrt{\frac{2}{\pi}}\int_0^\infty cos(\omega\tau)\Delta(\tau)d\tau \\ &= \sqrt{\frac{6}{\pi}}(\omega/\omega_0)sinh^{-1}(\omega/\omega_0) \sim_{\omega\to\infty} \omega/\omega_0 \exp(-\omega/\omega_0). \end{aligned} \tag{6}$$

where $\omega_0 = \frac{2\vartheta}{\sqrt{3}\pi}$.

The fact that $f_v(\omega)$ decreases *exponentially* as $\omega \to \infty$ implies that $v(t)$ is smooth in time. A way to see this is to rewrite $v(t)$ in a "spectral representation" (Yaglom 1962):

$$v(t) = \int_{-\infty}^{+\infty} e^{i\omega t} z(\omega)d\omega, \tag{7a}$$

where $z(\omega)$ are independent Gaussian random variables, with

$$< z(\omega)z^*(\omega') > = f_v(\omega)\delta(\omega - \omega'). \tag{7b}$$

Then, the n-th derivative of $v(t)$ with respect to time t is clearly well defined, for all n, and is also a Gaussian process. Its power spectrum given by $\omega^{2n} \cdot f_v(\omega)$ behaves well because $f_v(\omega)$ decays fast enough for large ω.

Now, applying the Kolmogorov formula Eq.(3) to $f_v(\omega)$, one obtains

$$h(\epsilon) \sim \frac{2\vartheta(\ln 2)^2}{\sqrt{3}\pi^2}(\log_2 \frac{1}{\epsilon})^2. \tag{8}$$

One can draw several interesting conclusions from this brief calculation. On one hand, $h(\epsilon)$ is unbounded. This implies that the spacetime chaos in the neural net Eq.(1) with $gJ > 1$ has an unbounded dynamic entropy, and *a fortiori* of infinite dimension. On the other hand, recalling that ϵ is essentially the accuracy limit by which one monitors the output of a continuous stochastic process, one remarks that for a given, small ϵ, the information production rate per unit time for this system of formal neurones, is qualitatively different from that of a "random noise", such as the Ornstein-Uhlenbeck one (compare Eq.(4) with Eq.(8)). To illustrate our point, let $\vartheta = 0.1$, $\sigma^2 = 0.1$. If $\epsilon = 10^{-3}$, then $h(\epsilon) \sim 0.5$ bits per unit time for the Eq.(2), and $h(\epsilon) \sim 3 \times 10^3$ bits per unit time for the Ornstein-Uhlenbeck process. And if $\epsilon = 10^{-7}$, we have $h(\epsilon) \sim 3$ bits per unit time for the former, and $h(\epsilon) \sim 3 \times 10^{11}$ bits per unit time for the latter case!

The behavior of this network is intermediate between a low dimensional chaos and noise in the following sense: As mentioned above, for a determinstic chaos one would expect

$$h(\epsilon) \sim_{(1/\epsilon)\to\infty} h_{KS}; \quad \text{and} \quad \frac{d}{d(1/\epsilon)}h(\epsilon) \sim_{(1/\epsilon)\to\infty} 0; \tag{9a}$$

while for a Ornstein-Uhlenbeck process one has

$$h(\epsilon) \sim_{(1/\epsilon)\to\infty} \infty; \quad \text{and} \quad \frac{d}{d(1/\epsilon)} h(\epsilon) \sim_{(1/\epsilon)\to\infty} \infty; \tag{9b}$$

and for the network model

$$h(\epsilon) \sim_{(1/\epsilon)\to\infty} \infty; \quad \text{and} \quad \frac{d}{d(1/\epsilon)} h(\epsilon) \sim_{(1/\epsilon)\to\infty} 0. \tag{9c}$$

In other words, even so $h(\epsilon)$ diverges to infinity as $\epsilon \to 0$, its growth rate tends to zero. If it is plotted versus $1/\epsilon$, one would observe an apparent plateau as $1/\epsilon \to \infty$, where the information production rate $h(\epsilon)$ appears seldom sensitive to the change of ϵ.

Of course, the chaotic state we are describing here being the spontaneous activity of the network Eq.(1), one may prefer not to speak of "information flow", if the word is reserved to information that the network can process about its *external world*. Nevertheless, $h(\epsilon)$ is a characteristic of the intrinsic dynamics of the net and, if a random noise is present, it can be used to distinguish the signal (the output of Eq.(2), $v(t)$) from noise. Let us add a white noise to Eq.(2), with a variance $\sigma^2 \ll \vartheta$. Then, the power spectrum of $v(t)$ is merely the sum of its previous part (Eq.(6)) and that of an Ornstein-Uhlenbeck process. When we again compute $h(\epsilon)$, it is straightfoward to see that there exits a cross-over value of ϵ, say ϵ^*, marking the accuracy limit beyond which the noise starts to manifest. ϵ^* can be estimated as

$$\epsilon^{*2} \sim \frac{\sigma^2}{\pi^2 \omega_0} (\ln(1/\sigma^2))^{-1}. \tag{10}$$

Thus, ϵ^{*2} differs from the "noise/signal ratio", σ^2/ω_0, by a logarithmic correction. For $\epsilon \gg \epsilon^*$, the signal is dominant, and $h(\epsilon)$ is given by Eq.(8) as before. On the other hand, for $\epsilon \ll \epsilon^*$, the noise prevails, and $h(\epsilon)$ eventually becomes the same as for a geniune random noise (Eq.(4)). Presumably, when the net is subject to external inputs, a similar approach can be useful to describe the interplay between "real signals" relevant to the external world and the spontaneous fluctuations, using the notion of ϵ-entropy.

To summarize, the spontaneous activity of a network with random connections (Eq.(1)) can be a dynamic state where individual units are effectively uncorrelated, and fluctuate temporally with no apparent regularities. It is distinguished from both a low-dimensional chaos and "noise", as characterized by its own $h(\epsilon)$ form.

If one wishes to ask whether the present discussion could be of any suggestive value to biological nervous systems, one needs to know which conclusions based on Eq.(1) are dependent on the details of the model, and which are not. Some features of the network are likely vulnerable to such changes. For instance, if one adds a cubic *damping* term to the equation of each unit, Eq.(1), the dynamical mean-field theory would lead to an equation similar to Eq.(2),

$$dv/dt = -v - v^3 + \eta(t), \tag{11}$$

where $\eta(t)$ is again a Gaussian field. Now, because Eq.(11) is not linear, $v(t)$ will no longer be Gaussian.

On the other hand, the condition under which the asymptotic behavior of $h(\epsilon)$ as Eq.(8) is expected, seems quite general, namely that the network dynamics must guarantee the smoothness of its solutions: the trajectory of each unit can be chaotic enough to mimic a continuous stochastic process, and yet remains infinitely differentiable, perhaps even analytic.

Acknowledgements

It is my pleasure to thank Prof. John Rinzel for his invaluable support, Prof. Ilya Prigogine for the invitation to this NATO workshop, and Prof. H. Sompolinsky for a conversation. Numerical simulations were carried out at the National Cancer Institute Advanced Scientific Computing Laboratory. The present work is closely related to an ongoing research in collaboration with Dr. Pierre Gaspard.

References

Amari, S.-I., 1972, Characteristics of random nets of analog neuron-like elements, *IEEE Trans. Syst. Man & Cybern.*, SMC-2:643.

Amit, D.J., 1989, *Modeling Brain Function: the world of attractor neural networks*, Cambridge University Press, New York.

Eckmann,J.-P. & Ruelle,D., 1985, Ergodic theory of chaos and strange attractors, *Rev. Mod. Phys.*, 57:617.

Gaspard, P. & Wang, X.-J., 1990, in preparation.

Hopfield,J., 1982, Neural networks and physical systems with emergent collective computational abilities, *Proc. Natl. Acad. Sci. USA*, 79:2554.

Hopfield,J., 1984, Neurons with graded response have collective computational properties like those of two-state neurons, *Proc. Natl. Acad. Sci. USA*, 81:3088.

Pomeau,Y. & Manneville,P., 1980, Intermittent transition to turbulence in dissipative dynamical systems, *Commun. Math. Phys.*, 74:189.

Shaw,R.S., 1981, Strange attractors, chaotic behavior and information flow, *Zeitschrift für Naturforschung*, A36:80.

Sompolinsky, H., Crisanti, A. & Sommers, H., 1988, Chaos in random neural networks, *Phys. Rev. Lett.*, 61:259.

Sompolinsky, H., 1990, personal communication.

Yaglom, A.M., 1962, *An Introduction to the Theory of Stationary Random Functions*, Dover, New York.

NEURAL LEARNING ALGORITHMS AND THEIR APPLICATIONS IN ROBOTICS

Carme Torras i Genís

Institut de Cibernètica (CSIC-UPC)
Diagonal 647, 08028-Barcelona. SPAIN

1. INTRODUCTION

Computationally expensive processes that must be executed in real time —such as those typically arising in Robotics— need to exploit any feasible parallelism they can admit. Three types of parallelism can be distinguished according to the kind of information processing they are suitable for: numerical, symbolic or pattern processing (Recce and Treleaven 1988).

The simplest model of *numerical parallel computation* is that of SIMD machines (e.g., vector processors, systolic arrays). They typically consist of many simple processors with mesh connections and operate under a synchronous local control scheme. Another model is that of MIMD machines, which typically include few complex processors and few connections, their control strategy being that of message-passing under the supervision of a central scheduler.

The architectural model for *symbolic parallel processing* is provided by distributed processing in open systems which have no global controls (Huberman 1988). This model with asynchronous communication between processors has recently been investigated within the field of Artificial Intelligence to implement cooperative computation among several agents (Durfee et al. 1987).

Finally, the architectural model for *pattern processing* —which is intrinsically parallel— is that of neural computers (Eckmiller and von der Malsburg 1988) that implement a connectionist style of computation (Feldman and Ballard 1982). This model relies on a massive number of highly-interconnected processing elements that operate under either a synchronous or an asynchronous local control scheme.

Within the Robotics field, there are instances of processes belonging to each of the three types of information processing above.

Low-level vision and computational geometry algorithms are examples of numerical information processing tasks for which parallel solutions have already been devised (Davis and Rosenfeld 1981).

Task decomposition, subprocess scheduling and integration of partial results are symbolic information processing tasks that have begin to be dealt with under the cooperative computation paradigm (Durfee et al. 1987; Hayes-Roth et al. 1986).

Finally, there are many tasks requiring not only *massive parallelism* to carry out pattern processing in real time, but also *learning capabilities* to adapt to different environments. In general, these tasks are tied to the peripheral elements (sensors and actuators) and involve the learning of reflexes linking perception to action.

The latter kind of tasks constitutes the subject of this paper, which is structured as follows. Section 2 provides an overview of the different classes of *learning tasks* that have been tackled using neural networks. Then, in Section 3, the *learning rules* proposed to carry out these tasks are described and classified into three groups: correlational, error-minimization and reinforcement rules. In addition, the *network learning models* that incorporate each such rule are mentioned. The triple distinction between learning tasks, learning rules and learning models is important, because the same learning rule can be incorporated into different network models in order to accomplish different tasks. Section 4 is devoted to describing the *robotics applications* where neural networks have been used. The learning task underlying each application is identified and the network models developed for each application are settled in the unified framework for learning rules set forth above. Finally, some conclusions are sketched in Section 5.

2. LEARNING TASKS

The learning tasks that have been undertaken by using neural networks fall into four general categories: pattern reproduction, pattern association or classification, feature discovery or clustering, and reward maximization. Task categories differ in the type of problem information used. In one extreme of the spectrum, one finds feature discovery or clustering, which uses no problem information at all and is thus totally *unsupervised.* Pattern reproduction, association and classification are placed in the opposite extreme of the spectrum, since they require complete target information —in the form of input/output pairs— and are thus totally *supervised.* Somewhere in between both extremes lies reward maximization, which makes use only of reinforcement information —i.e. an assessment of how good is an output to a given input.

2.1. Pattern Reproduction

The task of pattern reproduction consists in retrieving one of a set of stimulus patterns that had been repeatedly shown to the system during training, upon presentation of a portion or a distorted version of the original pattern. When a portion of the original pattern is used as retrieval cue, the task is also referred to as *pattern completion.* Furthermore, Rumelhart and Zipser (1985) call this task *auto-association*, interpreting that each pattern becomes associated with itself.

2.2. Pattern Association or Classification

The most classical and widely studied learning task is that of pattern association or classification, which constitutes the main objective of the Pattern Recognition discipline. It consists in training the system with pairs of patterns —of which the second can be interpreted as the class to which the first belongs— so that when the first pattern of a pair is presented, the system produces the second. Although some authors consider association and classification to be two different tasks, we envisage the first as a limit case of the second in which each class consists of only one pattern.

2.3. Feature Discovery or Clustering

This learning task has also been called unsupervised learning within the Pattern Recognition literature. Its goal is to find out statistically salient features of the population of stimulus patterns that permit establishing clusters of patterns with similar features.

2.4. Reward Maximization

The task of reward maximization requires that the system is given reward or punishment depending on the action it takes in response to each stimulus pattern. The system has to configure itself in a way that maximizes the amount of reward it receives. The open loop version of this task, in which the actions of the system do not have any influence on the choice of the stimulus patterns subsequently presented, has been called *associative reinforcement learning* by Barto and Anandan (1985). The closed-loop version amounts to solving complex control problems (Barto et al. 1983).

3. NEURAL LEARNING ALGORITHMS

Most neural learning algorithms proposed to date rely on the uniform application of a unineuronal learning rule throughout a network of formal neurons, each being in a state x_j governed by the following generic equation:

$$x_j(t + \delta t) = f[\sum_{i \in I_j} w_{ij}(t) x_i(t)] \qquad (1)$$

where I_j is the set of inputs to neuron j —which can come from the environment or be the outputs of other neurons—, w_{ij} is the synaptic weight of the connection from neuron i to neuron j, t is the time instant, and f takes usually the form of either a deterministic or a stochastic threshold function.

The majority of the unineuronal learning rules proposed work by modifying the synaptic weights. Since they are intended to constitute biologically plausible hypotheses, most such rules postulate parameter changes on the basis of only local information available to a single neuron.

It must be noted that the time scale for short-term neural functioning is assumed to be very fine-grained in comparison to that ruling the learning process and thus δt is taken to be smaller enough to allow for the propagation of activity through the network before the weight update is carried out.

The remainder of this section is devoted to the description of the most widely known learning rules, grouped into three categories: correlational, error-minimization and reinforcement rules. Although at a first glance it might seem that this corresponds neatly to the distinction between unsupervised, supervised and reinforcement tasks made in the preceding section, this is not exactly the case, since one can devise neural architectures (models) that accomplish supervised tasks by using only correlational rules —e.g. the associative memories mentioned in the following subsection. For this reason, in describing the rules we mention some of the network models in which they have been incorporated.

3.1. Correlational Rules

These rules adjust a connection weight according to the correlation between the states of the two neurons connected. They can all be considered variants of the classical *Hebbian learning rule* (Hebb 1949), whose expression in terms of the generic neuron model defined by (1) is:

$$w_{ij}(t+1) = w_{ij}(t) + cx_i(t)x_j(t) \tag{2}$$

where c is a positive constant that determines the speed of learning. To prevent the unbounded rise of weights, different normalization procedures have been used, the most common one being that based on the euclidean metric; notice that in this case the threshold must be included as a component in the vector to be normalized.

The most extensively studied application of the Hebbian rule is the implementation of "associative memories" (Spinelli 1970; Nakano 1972; Amari 1977 a & b; Anderson et al. 1977; Kohonen 1977; Hinton and Anderson 1981), which are network learning models able to carry out pattern association tasks as described in the preceding section. The three most interesting aspects of this kind of network: resistance to noise, addressing by content, and generalization capability, derive from the distributed way in which information is stored.

This rule has also been incorporated into "competitive learning models" (von der Malsburg 1973; Rumelhart and Zipser 1985) able to accomplish feature discovery tasks. These models consist of a set of hierarchically layered neurons, each neuron receiving excitatory input from the layer immediately below. Futhermore, the neurons in each layer are grouped into disjoint clusters, each neuron in a cluster inhibiting all other neurons within the cluster. The name "competitive learning" comes from the fact that the neurons within a cluster "compete" with one another to respond to the pattern appearing on the layer below; the more strongly any particular neuron responds, the more it shuts down the other members of its cluster (Didday 1970), which therefore becomes a winner-take-all network as defined by Feldman and Ballard (1982). A cluster containing n neurons can be considered an n-ary feature, every stimulus pattern being classified as having exactly one of the n possible values of this feature. It has been proved that, if the stimulus patterns naturally fall into classes, the system will find exactly these classes and the attained classification will be very stable.

However, when presented with arbitrary input environments, competitive learning models can become very unstable (Grossberg 1987) and the need appears of stabilizing their response through the use of specialized mechanisms, leading to "adaptive resonance models" (Grossberg 1976, 1982). These are recurrent modular networks able to form a new cluster whenever they are presented with an input pattern that is very different from the patterns previously seen.

In the same line of competitive learning, Kohonen (1988) proposes to use "self-organizing feature maps" to construct mappings that preserve topography (i.e. neurons that are spatially close in the network are maximally activated by input vectors close according to the euclidian metrics). Essentially, this is realized through 2-layer networks with intralayer lateral inhibition and interlayer plastic excitatory connections.

3.2. Error-Minimization Rules

These rules work by comparing the response to a given input pattern with the desired response and then modifying the weights in the direction of decreasing error. Depending on whether the desired response is specified at the single neuron or overall network levels, the gradient of the error with respect to each synaptic weight will necessarily be or may only be locally computable; moreover, the repertoire of learning tasks that can be carried out in the latter case is wider than that accomplishable in the former case.

The most classic error-correction rule that requires specification of the desired response for each single neuron is the *perceptron learning rule* (Rosenblatt 1962). Its expression in terms of the generic neuron model that we use as reference (equation 1) is:

$$w_{ij}(t+1) = w_{ij}(t) + c(x_j^*(t) - x_j(t))x_i(t) \tag{3}$$

where both the desired response $x_j^*(t)$ and the actual response $x_j(t)$ of the j-neuron are binary, since the f in (1) is taken to be a threshold function. The same consideration about normalization made for the Hebbian rule applies also here.

The networks of neurons using the perceptron learning rule, called "perceptrons", are especially well-suited to carry out pattern classification tasks. The simplest such network proposed (Rosenblatt 1962) has three layers: sensory, preprocessor and actuator; only the weights of the connections between the second and the third layer are modifiable through learning. If $x_j^*(t)$ supplies the desired classification for each input pattern $X(t) = (x_1(t), \ldots, x_{d_j}(t))$ to the actuator neuron j, and the classes are linearly separable, then the synaptic weights will converge to a configuration that classifies all patterns correctly. The rule provides thus an iterative algorithm to find the solution of a system of linear inequalities.

Rosenblatt (1962) generalized this simple version of perceptron in different ways: incorporating more actuators, considering a variable number of intermediate layers, and introducing both intra- and inter-layer feedback. Nilsson (1965) proved convergence theorems for different types of perceptron, always presupposing the existence of a solution, or in other words, of a configuration of weights that would give rise to the correct classification. Minsky and Papert (1969) approached the complementary question, characterizing the limits of the discriminative ability for various types of perceptron. Essentially, they proved that topological features such as connectedness and symmetry cannot be discriminated by simple perceptrons —the work must be done in the preprocessing layer.

The *Widrow-Hoff learning rule* (1960) is expressed by the same equation (3), but considering $x_j^*(t)$ and $x_j(t)$ to take real values and the f in (1) to be the identity.

Through this rule, the repeated presentation of pairs (X_l, z_l), $l = 1, \ldots k$, with $X_1, \ldots X_k$ linearly independent, to the simplest perceptron described before, causes the convergence of the synaptic weights toward the proper configuration for the response to each stimulus X_l to be the desired real number z_l. The rule thus provides an iterative procedure to find the solution of a system of linear equations. If the patterns $X_1, \ldots X_k$ are not linearly independent, the rule can be slightly modified (converting the parameter c into a variable that goes to zero with time) so that the convergence of the weights minimizes the quadratic error between the actual output and the desired one. This slightly modified version of the rule computes a linear regression iteratively (Duda and Hart 1973).

Kohonen and Oja (1976), Amari (1977 a & b) and Albus (1979) have studied the implementation of "associative memories" through networks of neurons that incorporate the Widrow-Hoff rule. While, as we saw previously, the Hebbian rule only provides a perfect association if the stimulation patterns are orthogonal, the Widrow-Hoff rule relaxes the restriction to their being linearly independent. As could be foreseen, if c approaches zero with time, optimal associations are formed according to the criterion of quadratic minimization, even when the stimulation patterns are linearly dependent.

An extension of the error-minimization rules so far described to the case where the desired response is specified only for a subset of neurons (those whose outputs constitute the output of the network) is *back-propagation*. As its name indicates, it proceeds by propagating error signals from the output neurons back to the sensory neurons, through all intermediate layers, so that appropriate corrections can be applied to all connection weights. Note that this procedure works only for layered networks with unidirectional interlayer connections and no intralayer connections.

A back-propagating error-correction procedure based on the perceptron learning rule was initially proposed by Rosenblatt (1962). Recently, this procedure has been generalized (Le Cun 1985; Rumelhart et al. 1986) to minimizing the mean square error E between the actual and the desired responses to all input patterns, by repeatedly applying the rule:

$$w_{ij}(t+1) = w_{ij}(t) - c\frac{\partial E}{\partial w_{ij}} \tag{4}$$

To prevent weights from getting very large, a fixed percentage weight-decay per update has usually been implemented in the networks using this rule, instead of the normalization procedure employed by the previously described learning rules. Observe that, except for this point, the Widrow-Hoff learning rule is a particular instance of this generic rule, since for the former the f in (1) is the identity and thus $\partial x_j / \partial w_{ij} = x_i$, leading to:

$$\frac{\partial E}{\partial w_{ij}} = \frac{\partial E}{\partial x_j} \cdot \frac{\partial x_j}{\partial w_{ij}} = (x_j^* - x_j)x_i \tag{5}$$

By incorporating this result into (4), equation (3) is obtained.

Rumelhart et al. (1986) and Hinton (1987) have applied the generic rule (4) to multilayered networks of neurons having a sigmoidal input-output function (i.e., the f in (1) takes the form $f(x) = 1/(1 + e^{-x})$). In this case, $\partial x_j / \partial w_{ij} = x_i x_j (1 - x_j)$ and the factor $\partial E/\partial x_j$ has to be calculated from the activity levels x_k of the neurons in the next layer. Hence, starting with the neurons in the output layer, for which $\partial E/\partial x_j = (x_j^* - x_j)$, the computation proceeds backwards:

$$\frac{\partial E}{\partial x_j} = \sum_k \frac{\partial E}{\partial x_k} \cdot \frac{\partial x_k}{\partial x_j} = \sum_k \frac{\partial E}{\partial x_k} w_{jk} x_k (1 - x_k) \tag{6}$$

and therefore:

$$\frac{\partial E}{\partial w_{ij}} = \frac{\partial E}{\partial x_j} \cdot \frac{\partial x_j}{\partial w_{ij}} = x_i x_j (1 - x_j) \sum_k \frac{\partial E}{\partial x_k} w_{jk} x_k (1 - x_k) \tag{7}$$

Back-propagation is especially well-suited for solving feature discovery tasks; features relevant for discrimination between input patterns get progressively encoded in the activity of the neurons belonging to intermediate layers. This learning procedure has the drawbacks of all gradient descent techniques, namely the possibility of getting stuck in local minima and a slow convergence rate.

The last error-minimization learning rule that we are going to describe here is that implemented in "Boltzmann machines (Hinton et al. 1984). Adhering to the neural interpretation of these machines, according to which they consist of a network of binary neurons symmetrically influencing one another, and assuming that there are two types of neurons (visible and hidden), we can ask them to perform a learning task as follows. Suppose that several patterns are successively clamped on the visible neurons, each pattern recurring with a given probability and being held until thermal equilibrium is reached, the task of the network is then to reproduce when running freely the patterns previously clamped with the probabilities with which they have been presented. Phrased in this way, this is a pattern reproduction task, in which not only the patterns themselves but also the probability distribution over these patterns are reproduced. If the set of visible neurons is further subdivided into input and output subsets, then it is possible to carry out pattern association tasks with these machines. Notice that the aforementioned tasks are accomplished by building compressed internal representations of the patterns in the hidden neurons, a task which can be conceptualized as feature discovery.

In order to define the learning rule, we have first to introduce an error measure between the probabiltity distribution for the clamped patterns and that for the states of the visible neurons when running freely:

$$E = \sum_{(x_1,\ldots x_n)\in\{0,1\}^n} P(x_1,\ldots x_n)\; ln \frac{P(x_1,\ldots,x_n)}{P'(x_1,\ldots,x_n)} \tag{8}$$

where n is the number of visible neurons, x_j is the state of the jth neuron, $P(x_1,\ldots x_n)$ is the probability of appearance of the clamped pattern $(x_1,\ldots x_n)$, and $P'(x_1,\ldots x_n)$ is the probability of reproduction of the same pattern when the network is running freely, measured at equilibrium.

The learning rule applied amounts to carrying out gradient descent in E by adequately changing the synaptic weights. Hinton et al. (1984) proved that:

$$\frac{\partial E}{\partial w_{ij}} = -\frac{1}{T} \sum_{(x_1,\ldots x_n)\in\{0,1\}^n} x_i x_j [P(x_1,\ldots x_n) - P'(x_1,\ldots x_n)] \tag{9}$$

where T is a temperature parameter, according to the thermodynamic interpretation. Thus, by substituting (9) into (4), the *Boltzmann machine learning rule* is obtained:

$$w_{ij}(t+1) = w_{ij}(t) + c \sum_{(x_1,\ldots x_n)\in\{0,1\}^n} x_i x_j [P(x_1,\ldots x_n) - P'(x_1,\ldots x_n)] \qquad (10)$$

The strategy commonly used to keep the weights small when using this rule is to include the squared euclidean norm of the weight vector as a summation term in the quantity E to be minimized.

Note the similarity of this rule with the Hebbian one (equation 2). The length of time over which the probabilities are estimated before changing the weights is an important factor that can have a significant impact on the learning process.

The same two drawbacks of gradient descent techniques mentioned in relation to back-propagation apply also here, the slowness of convergence being aggravated by the need to wait first for equilibrium and second for statistics collection. As a counterpart, randomness is already built-in because probabilities are necessarily estimated and, therefore, it is relatively straightforward to implement an annealing search for the global minimum of E.

3.3. Reinforcement Rules

These rules do not require being supplied with the desired responses, either at the single neuron or at the overall network levels, but instead a measure of the adequacy of the emitted responses suffices. This measure is reinforcement, which is used to guide a random search process to maximize reward. Hence, reinforcement rules can be considered neuronal analogs of instrumental conditioning in that the neuronal spontaneous responses are favored or weakened through the application of certain reinforcement schemes. Depending on whether the reinforcement signal is provided at the overall network level or is particularized for each single neuron, the credit-assignment problem (Minsky 1961) does or does not arise. This is the problem of correctly assigning credit or blame to the action of each neuron that contributed to the overall evaluation received (refer to Barto (1985) for details).

The two reinforcement learning rules that we are going to describe in this section incorporate the required source of randomness in the input-output function of the neuron model they use, i.e. the f in (1) is assumed to be a random threshold function with a sigmoidal distribution for the threshold noise.

The *associative search learning rule* (Barto et al. 1981) is the reinforcement-based counterpart of the correlational rules (equation 2). Its simplest expression is:

$$w_{ij}(t+1) = w_{ij}(t) + c x_i(t) x_j(t) r(t) \qquad (11)$$

where $r(t)$ is the reinforcement signal.

A neuron model equipped with this rule learns to maximize $r(t)$ for each stimulus situation. If $r(t)$ is a random variable, its mathematical expectation is instead maximized. Two-layered neural networks that incorporate the above rule are called Associative Search Networks (ASN) and, if certain conditions are satisfied, they learn to respond to each stimulus situation $X_l = (x_{l1}, \ldots x_{ln})$ of a set $\{X_1, \ldots X_k\}$ repeatedly presented, with the vector $Y = (y_1, \ldots y_m)$ that maximizes the reinforcement function r. The conditions that have to be satisfied are: (a) the function r has to be unimodal, and (b) for each neuron, the subset of stimulus situations in which the optimum response is 0 has to be linearly separable from the corresponding subset in which the optimum response is 1. To the advantages already described for classical associative memories —resistance to noise, generalization capacity, and addressing by content— the above network adds two more: not needing the explicit presentation of the desired response for each stimulus situation, and using a single reinforcement signal common to all neurons.

The rule just described has quite limited pattern discrimination abilities and is not able to accomplish reward maximization tasks that require the discovery of complex features in the stimulus patterns and in the reinforcement contingencies. Since hidden neurons are needed to represent such features, this amounts to saying that this rule is not well-suited for networks incorporating neurons that do not contribute directly to the response that is being reinforced. Thus, the associative search learning rule suffers from the credit-assignment problem.

To overcome this problem, Barto and Anandan (1985) proposed the *associative reward-penalty learning rule* which, being the reinforcement-based counterpart of the error-minimization rules (equation 3), has the following expression:

$$w_{ij}(t+1) = w_{ij}(t) + c_{r(t)}[r(t)x_j(t) - E\{x_j(t) \mid \sum_{i=1}^{d_j} w_{ij}(t)x_i(t)\}]x_i(t) \tag{12}$$

where the parameter $c_{r(t)}$ makes the rule asymmetric by adopting different values for reward ($r(t) = +1$) and penalty ($r(t) = -1$), and $E\{\cdot\}$ is the expected value for the output given the weighted sum of inputs.

Convergence of this rule under broad conditions has been proved for a single neuron (Barto and Anandan 1985) and several network configurations including hidden neurons behaving according to this rule have been shown through simulation to be able to accomplish difficult nonlinear associative reinforcement learning tasks (Barto 1985). Provided each input pattern is presented with a given frequency, and since neurons are here stochastic components, the reward maximization attained is in fact a maximization of the probability of reward.

The models described so far have addressed a structural credit-assignment issue, which is sufficient for dealing with static pattern classification tasks. However, when dynamical situations need to be considered, because what is of interest is a temporal sequence of events, then a temporal credit-assignment problem also arises. Instead of assigning credit or blame to the action of each neuron in the network, credit or blame must here be assigned to each action in a sequence. Sutton (1984, 1988) has proposed to use *temporal-difference methods* for this purpose. Methods of this type have been embodied, for example, in "critic modules" used in conjunction with reinforcement learning procedures. The goal of these modules in this setting is to produce an heuristic reinforcement signal which, by predicting future outcomes, is more informed than that directly supplied by the environment.

4. ROBOTIC APPLICATIONS

Within the research field of robotics we can distinguish four general subfields: perception, control, programming and planning.

Many problems have been addressed within the *perception subfield* by using neural learning algorithms. Among them we mention object recognition, object tracking, and sensor fusion (both using equal sensors, such as in stereo matching, and sensors from different modalities, like vision and touch, for example). This is no doubt the largest subfield of application of neural networks and a survey of the results attained in this area goes beyond the objectives of this paper. In order to tackle a manageable task and in view of the fact that *action* is the key ingredient of robotics, we will confine ourselves to describing the applications in the subfields related to action, some of which may of course involve some perception.

In the *control subfield*, neural learning algorithms have been used to solve inverse kinematics, inverse dynamics, and to carry out sensorimotor integration in grasping, navigation and insertion operations. As we will see in the following subsections, these three tasks amount to *constructing nonlinear mappings*.

As regards the *programming subfield*, the main application is the encoding of trajectories that entails the capacity to *generate sequences*.

Finally, the main area of application in the *planning subfield* is path finding, although there have been some efforts to tackle symbolic task planning by using neural networks (Torras 1990). The final aim of the developments in this subfield is *optimization*.

Summarizing, the applications of neural learning algorithms in robotics deal mainly with the construction of nonlinear mappings, the generation of sequences and optimization.

4.1. Inverse Kinematics

This is probably the robotic application related to action that has received most attention from the neural networks community. The goal is to build a map from the world coordinates of a workspace onto the robot joint coordinates. The name "inverse kinematics" comes from the fact that the natural (direct) map is that which relates the values of the joint coordinates defining a robot configuration to the position and orientation of its end-effector in the workspace.

Essentially, *error-minimization learning rules, under a supervised scheme*, have been applied to solving this problem. The two rules most widely used have been the Widrow-Hoff one and back-propagation. Among those labs that have applied the former, we find the University of New Hampshire (Miller et al. 1987; Miller 1989) and Osaka University (Kawato et al. 1987 a & b). Back-propagation has been used at Carnegie-Mellon University (Goldberg and Pearlmutter 1988), Case Western University (Pao and Sobajic 1987; Sobajic et al. 1988) and Drexel University (Guez and Selinsky 1988), among others.

4.2. Inverse Dynamics

Inverse dynamics refers to the problem of determining a map from the gripper velocities to the forces and torques exerted at the different joints. Again, the name comes from the fact that the natural (direct) map is that which goes the other way around.

Both supervised and unsupervised approaches have been used to tackle this problem. In the supervised case, the same error-minimization rules described for inverse kinematics have also been applied to determining inverse dynamics, by essentially the same labs. Concerning unsupervised learning, self-organizing feature maps have been used at the University of Illinois (Ritter and Schulten 1988) to construct the inverse dynamics mapping, while the ART system has been used at Boston University (Grossberg and Kuperstein 1986) for the same purpose.

4.3. Sensorimotor Integration

Sensorimotor integration requires the construction of a map relating the stimulation patterns to the appropriate motor commands. Maps of this kind are needed for carrying out operations such as grasping, navigation and insertion. Work in this area using neural learning approaches has concentrated on the first two types of operations.

Both supervised and unsupervised approaches have also been used in this context. The supervised ones have been applied in the same labs mentioned above by making use of the Widrow-Hoff and the back-propagation learning rules. The most widely known unsupervised approaches are those of the systems INFANT (Kuperstein 1987, 1988), DARWIN-III (Edelman 1987; Reeke and Edelman 1987) and MURPHY (Mel 1989), as well as the controller developed at Carleton University (Graf and LaLonde 1988, 1989), which makes use of Kohonen's self-organizing feature maps.

4.4. Robot Programming

Only supervised approaches have been applied to robot programming. Essentially, the robot is led along the trajectory that it has to learn and, by using a back-propagation algorithm on a recurrent network, it encodes the trajectory to be subsequently followed in a repetitive way. This approach has been pursued at Caltech (Sideris et al. 1987) and at the University of Düsseldorf (Eckmiller 1988).

4.5. Path Finding

This entails generating a path from an initial to a target robot configuration that does not lead to collision with any obstacle. Usually, there is an optimization criterion guiding the generation of the path, such as minimizing path length while maintaining a safety distance from obstacles.

Only reinforcement approaches have been used in this context, which is easily understandable since the task to be carried out can be formulated as reward maximization. The seminal works in this area were carried out at the University of Massacchusetts (Barto and Sutton 1981), and later on they have also been pursued at GTE (Sutton 1990) and the Institut de Cibernètica (Millán and Torras 1990 a & b).

5. CONCLUSION

In this paper, the most representative neural learning models have been surveyed, together with their applications in the field of robotics. It is worth stressing that only models and applications entailing *iterative synaptic learning* have been considered, and thus others relying on attractor dynamics (Babloyantz et al. 1990; Barhen et al. 1988, 1989) or concentrating on modelling neurophysiological phenomena (Arbib 1981; Ewert and Arbib 1989) have been left out. A similar survey concentrating on neural architectures for robotics was carried out by Kung and Hwang (1989) and a wider one encompassing all neural network applications was compiled by DARPA (1988).

The main conclusion arising from the survey is that the robotic tasks to which neural learning approaches have been applied are those that entail either the construction of nonlinear mappings, the generation of sequences, or optimization. To construct nonlinear mappings (for inverse kinematics, inverse dynamics and sensorimotor integration) both supervised and unsupervised learning approaches have been used, while only supervised approaches have been applied to the generation of sequences (underlying robot programming) and reinforcement-based approaches have been applied to tasks involving optimization (such as path finding).

References

Albus J.S., 1979, Mechanisms of planning and problem solving in the brain, Mathematical Bioscience 45, 247-293.

Amari S., 1977a, A mathematical approach to neural systems, in "Systems Neuroscience", Metzler J., ed., Academic Press, New-York.

Amari S., 1977b, Neural theory of association and concept-formation, Biological Cybernetics 26, 175-185.

Anderson J.A., Silverman J.W., Ritz S.A., Jones R.S., 1977, Distinctive features, categorical perception and probability learning: some applications of a neural model, Psychology Review 85, 413-451.

Arbib M.A., 1981, Perceptual structures and distributed motor control, in "Handbook of Physiology - The Nervous System II. Motor Control", Brooks V.B., ed., 1449-1480, American Physiological Society, Bethesda, MD.

Babloyantz A., Sepulchre J.A., Steels L., 1990, A network of oscillators can perform tasks without prior training, Technical Report, Université Libre de Bruxelles.

Barhen J., Dress, W.B., Jorgensen C.C., 1988, Applications of concurrent neuromorphic algorithms for autonomous robots, in "Neural Computers", Eckmiller R., von der Malsburg, C., eds., ASI NATO Series F: Computer and Systems Sciences 41, Springer-Verlag.

Barhen J., Gulati S., Zak M., 1989, Neural learning of constrained nonlinear transformations, IEEE Computer, June, 67-76.

Barto A.G., 1985, Learning by statistical cooperation of self-interested neuron-like computing elements, Human Neurobiology 4, 229-256.

Barto A.G., Anandan P., 1985, Pattern-recognizing stochastic learning automata, IEEE Trans. on Syst., Man and Cybern. 15(3), 360-375.

Barto A.G., Sutton R.S., 1981, Landmark learning: an illustration of associative search, Biological Cybernetics 42, 1-8.

Barto A.G., Sutton R.S., Anderson Ch.W., 1983, Neuron-like adaptive elements that can solve difficult learning control problems, IEEE Trans. on Syst., Man and Cybern. 13(5), 834-846.

Barto A.G., Sutton R.S., Brouwer P.S., 1981, Associative search network: a reinforcement learning associative memory, Biological Cybernetics 40, 201-211.

"DARPA Neural Network Study", 1988, AFCEA International Press, Fairfax, Virginia.

Davis L., Rosenfeld A., 1981, Cooperating processes for low-level vision: a survey, Artificial Intelligence 17:412.

Didday R.L., 1970, The simulation and modelling of distributed information processing in the frog visual system, Ph.D. Thesis, Stanford University.

Duda R.O., Hart P.E., 1973, "Pattern Classification and Scene Analysis", Wiley, New-York.

Durfee E.H., Lesser V.R., Corkill D.D., 1987, Coherent cooperation among communicating problem solvers, IEEE Trans. on Computers 36(11), 1275-1291.

Eckmiller R., 1988, Neural networks for motor program generation, in "Neural Computers", Eckmiller R., von der Malsburg C., eds., ASI NATO Series F: Computer and Systems Sciences 41, Springer-Verlag, Berlin Heidelberg New-York Tokyo.

Eckmiller R., von der Malsburg C., eds., 1988, "Neural Computers", NATO ASI Series F: Computer and Systems Sciences 41, Springer-Verlag, Berlin Heidelberg New-York Tokyo.

Edelman G.M., 1987, "Neural Darwinism", Basic Books, New-York.

Ewert J-P., Arbib M.A., eds., 1989, "Visuomotor Coordination: Amphibians, Comparisons, Models, and Robots", Plenum Press, New-York London.

Feldman J.A., Ballard D.H., 1982, Connectionist models and their properties, Cognitive Science 6, 205-254.

Goldberg K., Pearlmutter B., 1988, Using a neural network to learn the dynamics of the CMU direct-drive arm II, Technical Report CMU-CS-88-160, Computer Science Department, Carnegie-Mellon University.

Graf D.H., LaLonde W.R., 1988, A neural controller for collision-free movement of general robot manipulators, Proc. 2nd IEEE Intl. Conf. on Neural Networks, Vol. I, 77-84.

Graf D.H., LaLonde W.R., 1989, Neuroplanners for hand/eye coordination, Proc. Intl. Joint Conf. on Neural Networks, Vol. II, 543-548.

Grossberg S., 1976, Adaptive pattern classification and universal recoding. II - Feedback, expectation, olfaction, and illusions, Biological Cybernetics 23, 187-202.

Grossberg S., 1982, "Studies of Mind and Brain: Neural Principles of Learning, Perception, Development, Cognition, and Motor Control", Reidel Press, Boston.

Grossberg S., 1987, Competitive learning: from interactive activation to adaptive resonance, Cognitive Science 11, 23-63.

Grossberg S., Kuperstein M., 1986, "Neural Dynamics of Adaptive Sensory-motor Control: Ballistic Eye Movements", Elsevier, Amsterdam.

Guez A., Selinsky J., 1988, A trainable neuromorphic controller, Journal of Robotic Systems 5(4), 363-388.

Hayes-Roth B., Johnson M.V., Garvey A., Hewett M., 1986, Application of the BB1 blackboard control architecture to arrangement-assembly tasks, Artificial Intelligence, Computational Mechanics Publications 1(2), 85-94.

Hebb D.O., 1949, "The Organization of Behavior", Wiley, New-York.

Hinton G.E., 1987, Learning translation invariant recognition in massive parallel networks, Proc. of PARLE (Parallel Achitectures and Languages Europe), de Bakker J.W., Nijman A.J., Treleaven P.C., eds., Lecture Notes in Computer Science 258, 1-13.

Hinton G.E., Anderson J.A., 1981, "Parallel Models of Associative Memory", Erlbaum, Hillsdale, NJ.

Hinton G.F., Sejnowski T.J., Ackley D.H., 1984, Boltzmann machines: constraint satisfaction networks that learn, Technical Report CMU-CS-84-119, Carnegie-Mellon University.

Huberman B.A., 1988, Asynchrony and concurrency, in "Neural Computers", Eckmiller R., von der Malsburg C., eds., ASI NATO Series F: Computer and Systems Sciences 41, Springer-Verlag, Berlin Heidelberg New-York Tokyo.

Kawato M., Furukawa K., Suzuki R., 1987a, A hierarchical neural network model for control and learning of voluntary movement, Biological Cybernetics 57, 169-185.

Kawato M., Uno Y., Isobe M., Suzuki R., 1987b, A hierarchical model of voluntary movement and its application to robotics, Proc. IEEE 1st Intl. Conf. on Neural Networks, San Diego.

Kohonen T., 1977, "Associative Memory: A System Theoretic Approach", Springer-Verlag, Berlin.

Kohonen T., 1987, "Content-Addressable Memories" (second edition), Springer-Verlag, Berlin Heidelberg New-York Tokyo.

Kohonen T., 1988, "Self-Organization and Associative Memory" (second edition), Springer-Verlag, Berlin Heidelberg New-York Tokyo.

Kohonen T., Oja E., 1976, Fast adaptative formation of orthogonalizing filters and associative memory in recurrent networks of neuron-like elements, Biological Cybernetics 21, 85-95.

Kung S-Y., Hwang J-N., 1989, Neural network architectures for robotic applications, IEEE Trans. on Robotics and Automation 5(5), 641-657.

Kuperstein M., 1987, Adaptive visual-motor coordination in multijoint robots using a parallel architecture, Proc. of the IEEE Intl. Conf. on Robotics and Automation, 1595-1602.

Kuperstein M., 1988, An adaptive neural model for mapping invariant target position, Behavioral Neuroscience, 148-162.

Le Cun Y., 1985, Une procedure d'aprentissage pour reseau au seuil assymetrique, Proc. of COGNITIVA, 599-604.

von der Malsburg C., 1973, Self-organization of orientation sensitive cells in the striate cortex, Kybernetik 14, 80-100.

Mel B.W., 1989, Murphy: a neurally-inspired connectionist approach to learning and performance in vision-based robot motion planning, Ph.D. Thesis, Technical Report CCSR-89-17, Center for Complex Systems Research, University of Illinois at Urbana-Champaign.

Millán J. del R., Torras C., 1990a, Reinforcement learning: discovering stable solutions in the robot path finding domain, Proc. 9th European Conference on Artificial Intelligence (ECAI), Stockholm, 219-221.

Millán J. del R., Torras C., 1990b, Reinforcement learning in robot path finding: a comparative study, Proc. 3rd. COGNITIVA, Madrid.

Miller W.T., 1989, Real-time application of neural networks for sensor-based control of robots with vision, IEEE Trans. on Syst., Man and Cybern. 19(4), 825-831.

Miller W.T., Glanz F.H., Kraft L.G., 1987, Application of a general learning algorithm to the control of robotic manipulators, Intl. Journal of Robotics Research 6(2), 84-98.

Minsky M.L., 1961, Steps toward artificial intelligence, Proc. of the Institute of Radio Engineers 49, 8-30. (Reprinted in "Computers and Thought", 1963, Feigenbaum E.A., Feldman J., eds., McGraw-Hill, New-York, 406-450.

Minsky M.L., Papert S., 1969, "Perceptrons: An Introduction to Computational Geometry", MIT Press, Cambridge.

Nakano K., 1972, Associatron –a model of associative memory, IEEE Trans. on Syst., Man and Cybern. 2, 380-388.

Nilsson N.J., 1965, "Learning machines", McGraw-Hill.

Pao Y-H., Sobajic D.J., 1987, Artificial neural-net based intelligent robotics control, Proc. 6th SPIE Conf. on Intelligent Robots and Computer Vision, 542-549.

Recce M., Treleaven P.C., 1988, Parallel architectures for neural computers, in "Neural Computers", Eckmiller R., von der Malsburg C., eds., ASI NATO Series F: Computer and Systems Sciences 41, Springer-Verlag, Berlin Heidelberg New-York Tokyo.

Reeke G.N., Edelman G.M., 1987, Selective neural networks and their implications for recognition automata, Intl. Journal of Supercomputing Applications 1, 44-69.

Ritter H., Schulten K., 1988, Extending Kohonen's self-organizing mapping algorithm to learn balistic movements, in "Neural Computers", Eckmiller R., von der Malsburg C., eds., ASI NATO Series F: Computer and Systems Sciences 41, Springer-Verlag, Berlin Heidelberg New-York Tokyo.

Rosenblatt F., 1962, "Principles of Neurodynamics", Spartan Books.

Rumelhart D.E., Zipser D., 1985, Feature discovery by competitive learning, Cognitive Science 9, 75-112.

Rumelhart D.E., Hinton G.E., Williams R.J., 1986, Learning representations by back-propagating errors, Letters to Nature 323, 533-535.

Sideris A., Yamamura A., Psaltis D., 1987, Dynamic neural networks and their application to robot control, Proc. IEEE Conf. on Neural Information Processing Systems – Natural and Synthetic.

Sobajic D.J., Lu J-J., Pao Y-H., 1988, Intelligent control of the Intelledex 605T robot manipulator, Technical Report TR 88-106, Center for Automation and Intelligent Systems Research, Case Western Reserve University.

Spinelli D.N., 1970, Occam: a computer model for a content addressable memory in the central nervous system, in "The Biology of Memory", Pribram K., Broadbent D., eds., Academic Press.

Sutton R.S., 1984, Temporal credit assignment in reinforcement learning, Ph.D. Thesis, Dept. of Computer and Information Science, University of Massachusetts, Amherst.

Sutton R.S., 1988, Learning to predict by the methods of temporal differences, Machine Learning 3, 9:44.

Sutton R.S., 1990, Integrated architectures for learning, planning and reacting based on approximating dynamic programming, Proc. 7th Annual Conf. of the Cognitive Science Society.

Torras C., 1989, Relaxation and neural learning: points of convergence and divergence, Journal of Parallel and Distributed Computing 6, 217-244.

Torras C., 1990, Report of the group discussion about "neural networks in robotics", in "Sensor-Based Robots: Algorithms and Architectures", NATO ASI Series, Springer-Verlag, Berlin Heidelberg New-York London Paris Tokyo.

SELF-ORGANIZATION OF SPATIAL REPRESENTATIONS AND ARM TRAJECTORY CONTROLLERS BY VECTOR ASSOCIATIVE MAPS ENERGIZED BY CYCLIC RANDOM GENERATORS

Paolo Gaudiano† and Stephen Grossberg‡

Center for Adaptive Systems
and
Graduate Program in Cognitive and Neural Systems
Boston University
111 Cummington Street
Boston, MA 02215 USA

1. Introduction: A New Model of Unsupervised, Real-Time, Error-Based Learning

This chapter describes some recent results about biological models of unsupervised, real-time, error-based learning. In particular, we describe a new model called a Vector Associative Map, or VAM, and illustrate it with examples drawn from the learning of multidimensional associative maps and adaptive sensory-motor control.

Prior to the recent discovery of the VAM model in Gaudiano and Grossberg (1990a, 1990b) and Grossberg (1990), models of error-based learning, such as perceptrons and back propagation, have typically been supervised, off-line models whose learning rate is slow in order to avoid unstable learning properties. The VAM model suggests a possible approach o overcoming some of these limitations.

The chapter begins with a description of a model for self-organization of a neural network hat controls arm movement trajectories during visually guided reaching. The discussion hen indicates how this model suggests a more general framework for unsupervised, real-ime error-based learning.

The motor control model clarifies how a child, or untrained robot, can learn to reach for bjects that it sees. Piaget (1963) has provided basic insights with his concept of a *circular eaction*: As an infant makes internally generated movements of his hand, the eyes automatically follow this motion. A transformation is learned between the visual representation of and position and the motor representation of hand position. Learning of this transformaion eventually enables the child to accurately reach for visually detected targets. Grossberg

† Supported in part by the National Science Foundation (NSF IRI-87-16960).

‡ Supported in part by the Air Force Office of Scientific Research (AFOSR 90-0175), DARPA (AFOSR 90-0083), and the National Science Foundation (NSF IRI-87-6960).

Acknowledgements: The authors wish to thank Cynthia E. Bradford for her valuable ssistance in the preparation of the manuscript.

lf-Organization, Emerging Properties, and Learning
dited by A. Babloyantz, Plenum Press, New York, 1991

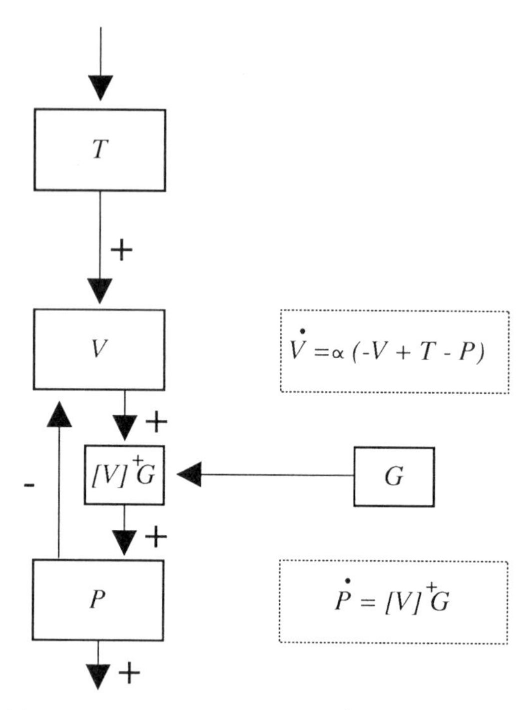

Figure 1. Main variables of the VITE circuit: T = target position command, V = difference vector, G = GO signal, P = present position command. The circuit does not include the opponent interactions that exist between the VG and P stages of agonist and antagonist muscle commands.

and Kuperstein have shown how the eye movement system can use visual error signals to correct movement parameters via cerebellar learning. Here it is shown how endogenously generated arm movements lead to adaptive tuning of arm control parameters. These movements also activate the target position representations that are used to learn the visuo-motor transformation that controls visually guided reaching.

The Adaptive Vector Integration to Endpoint (AVITE) model presented here is an adaptive neural circuit based on the VITE model for arm and speech trajectory generation of Bullock and Grossberg (1988a, 1988b, 1990). In the VITE model, a Target Position Command (TPC) represents the location of the desired target. The Present Position Command (PPC) encodes the present hand-arm configuration (Figure 1). The Difference Vector (DV) population continuously computes the difference between the PPC and the TPC. A speed-controlling GO signal multiplies DV output. The PPC integrates the (DV)·(GO) product and generates an outflow command to the arm. Integration at the PPC continues at a rate dependent on GO signal size until the DV reaches zero, at which time the PPC equals the TPC.

Because of its location within the VITE model, the GO signal affects the rate at which the PPC is continuously moved toward the TPC, without altering the resulting trajectory. For example, as long as the GO signal is zero, instatement of a TPC generates a non-zero DV, but the PPC remains unaltered. This "primed" DV codes the difference between the arm's present position and desired position. If the arm is passively moved through space by external forces while the GO signal is zero, the PPC is updated through sensory feedback from the muscles via a Passive Update of Position, or PUP, circuit that was described by Bullock and Grossberg (1988a). See Figure 2. The DV also changes to reflect the change in arm position, so that onset of the GO signal during a subsequent voluntary movement will still result in formation of a correct trajectory.

When the GO signal is nonzero, any activation in the DV is integrated by the PPC at a rate proportional to the product (DV)·(GO). Integration ceases when the PPC equals the TPC and the DV equals zero, even if the GO signal remains positive. Other things being equal, a larger GO signal causes the PPC to integrate at a faster rate, so the same target is reached in a shorter time.

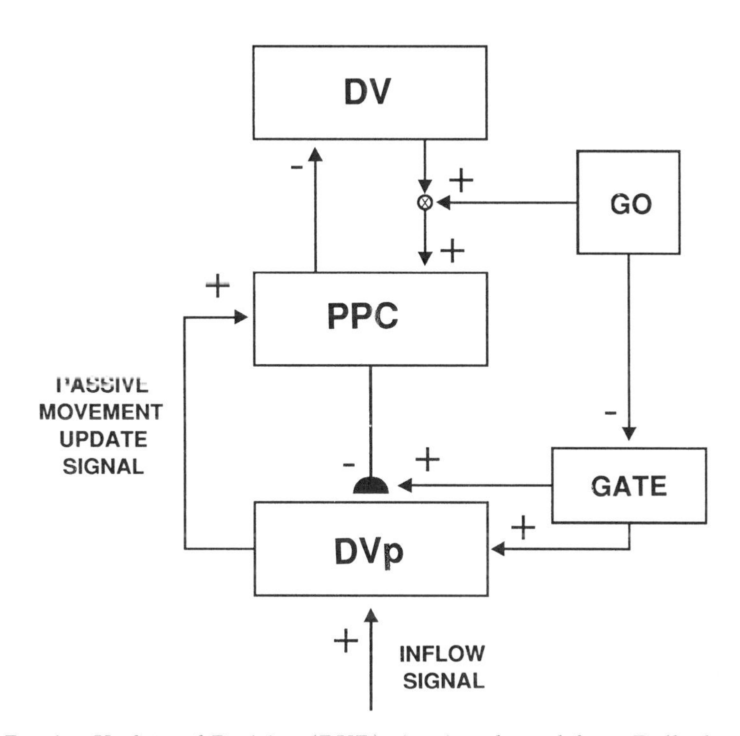

Figure 2. The Passive Update of Position (PUP) circuit, adapted from Bullock and Grossberg (1988a). DV and PPC are same as in Figure 1. The adaptive pathway PPC→DV_p calibrates PPC outflow signals to match inflow signals during intervals of posture. DV output equals zero during passive arm movements, so that DV_p can update the PPC until it equals the new position. GO signal activation disables passive update to allow discrimination between voluntary movements and movements caused by external forces.

Bullock and Grossberg (1988a, 1988b, 1990) and Grossberg and Kuperstein (1989) have ummarized experimental evidence suggesting that the TPC is computed in parietal cortex, he DV in motor cortex, and the GO signal in globus pallidus. The VITE model then predicts hat the motor cortex and globus pallidus give rise to output pathways that converge upon a rocessing stage where DV and GO signals are multiplied to compute a measure of movement peed and direction. This processing stage, in turn, is predicted to generate excitatory inputs o a neural (leaky) integrator which computes PPC outflow command signals.

The AVITE model (Figure 3) explains how self-consistent TPC and PPC coordinates are utonomously generated and learned. Learning of AVITE parameters is initiated by activaion of a self-regulating Endogenous Random Generator (ERG) of random training vectors.

Each vector is integrated at the PPC, giving rise to a movement. The generation of each vector induces a complementary postural phase during which ERG output stops and learning occurs. Then a new vector is generated and the cycle is repeated. This cyclic, biphasic behavior is controlled by a specialized gated dipole circuit (Grossberg, 1972, 1982). ERG output autonomously stops in such a way that, across trials, a broad sample of workspace target positions is generated. When the ERG shuts off, a modulator gate opens, copying the PPC into the TPC. Learning of a transformation from TPC to PPC occurs using the DV as an error signal that is zeroed due to learning.

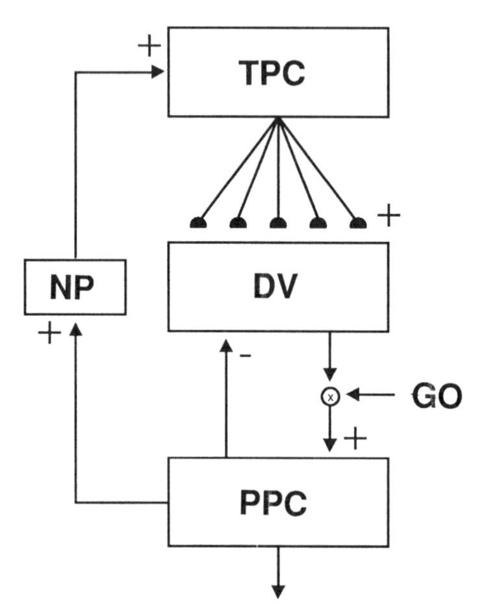

Figure 3. A schematic diagram of the Adaptive VITE (AVITE) circuit. See Section 3 for details of TPC, DV, PPC, and GO populations. The Now Print (NP) gate copies the PPC into the TPC when the arm is stationary, and the plastic synapses (semicircles in the TPC→DV pathways) learn to transform target commands into correctly calibrated outflow signals at the PPC.

This learning scheme contains the essence of the Vector Associative Map, or VAM model. The VAM model is a general-purpose device for autonomous real-time error-based learning and performance of associative maps. The DV stage serves the dual function of reading out new TPCs during performance and reading in new adaptive weights during learning, without a disruption of real-time operation. VAMs thus provide an on-line unsupervised alternative to the off-line properties of supervised error-correction learning algorithms. VAMs and VAM Cascades for learning motor-to-motor and spatial-to-motor maps are described. VAM models and Adaptive Resonance Theory (ART) models (Carpenter and Grossberg, 1987a, 1987b, 1988, 1990) exhibit complementary matching, learning, and performance properties that together provide a foundation for designing a total sensory-cognitive and cognitive-motor autonomous system.

2. Autonomous Learning of AVITE Coordinates

In order for the AVITE model to generate correct arm trajectories, the TPC and PPC must be able to activate dimensionally consistent signals TPC→DV and PPC→DV for com

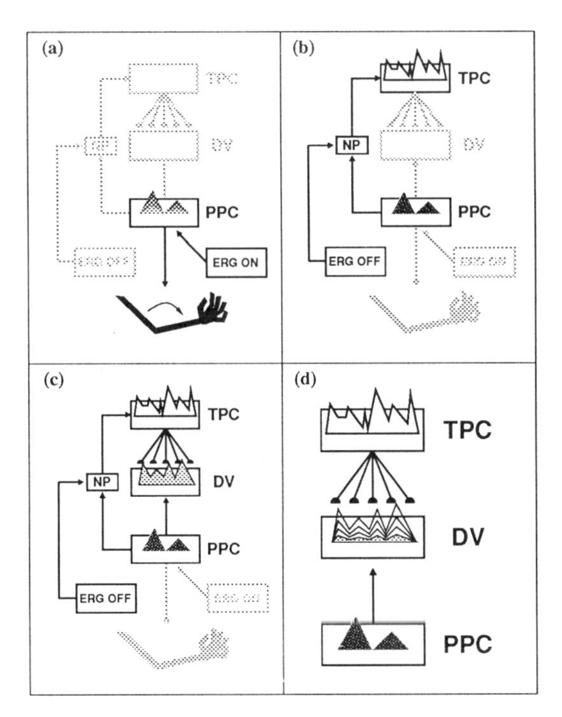

Figure 4. A diagrammatic illustration of a single babbling cycle in the AVITE. (a) The Endogenous Random Generator ON channel output (ERG ON) is integrated at the PPC, giving rise to random outflow signals that move the arm. (b) When the arm stops moving at ERG ON offset, a complementary ERG OFF signal opens the Now Print (NP) gate, copying the current PPC into the TPC through an arbitrary transformation. (c) The filtered TPC activation is compared to the PPC at the DV stage. DV activation would be zero in a properly calibrated AVITE. (d) The learning law changes TPC→DV synapses to eliminate any nonzero DV activation, thus learning the reverse of the PPC→NP→TPC transformations.

parison at the DV. There is no reason to assume that the gains, or even the coordinates of these signals are initially correctly matched. Learning of an adaptive coordinate transformation is needed to achieve self-consistent matching of TPC- and PPC-generated signals at the DV.

To learn such a transformation, TPCs and PPCs that represent the same target positions must be simultaneously activated. This cannot be accomplished by activating a TPC and then letting the AVITE circuit integrate the corresponding PPC. Such a scheme would beg the problem being posed, which is to discover how TPC→DV and PPC→DV signals are calibrated so that a TPC *can* generate the corresponding PPC. An analysis of all the possibilities that are consistent with VITE constraints suggests that PPCs are generated by internal, or endogenous, activation sources during a motor babbling phase. After such a babbled PPC is generated and a corresponding action taken, the PPC itself is used to activate a TPC representation which *a fortiori* represents the same target position (Figure). Thus motor babbling samples the work space and, in so doing, generates a representative set of pairs (TPC, PPC) for learning the AVITE coordinate transformation.

3. Associative Learning during the Motor Babbling Phase

Further analysis suggests that the only site where an adaptive coordinate change can take place is at the synaptic junctions that connect the TPC to the DV. These junctions are represented as semi-circular synapses in Figures 3 and 4. Moreover, DV activation can be used as an internal measure of error, in the sense that miscalibrated signals TPC→DV and PPC→DV from TPCs and PPCs corresponding to the same target position will generate a nonzero DV. Learning is designed to change the synaptic weights in the pathways TPC→DV in a way that drives the DV to zero. After learning is complete, the DV can only equal zero if the TPC and PPC represent the same target position. If we accept the neural interpretation of the TPC as being computed in the parietal cortex (Anderson, Essick, and Siegel, 1985; Grossberg and Kuperstein, 1986, 1989) and the DV as being computed in the precentral motor cortex (Bullock and Grossberg, 1988a; Georgopoulos *et al.*, 1982, 1984, 1986), then associative learning is predicted to occur from parietal cortex to motor cortex during the motor babbling stage, and drives activation of the difference vector cells in the motor cortex to zero, or to their tonic resting level, during postural intervals.

4. Vector Associative Map: On-Line DV-Mediated Learning and Performance

When such a learning law is embedded within a complete AVITE circuit, the DV can be used for on-line regulation of both learning and performance. During a performance phase, a new TPC is read into the AVITE circuit from elsewhere in the network, such as when a reaching movement is initiated by a visual representation of a target. The new DV is used to integrate a PPC that represents the same target position as the TPC. Zeroing the DV here creates a new PPC while the TPC is held constant. In contrast, during the learning phase, the DV is used to drive a coordinate change in the TPC→DV synapses. Zeroing the DV here creates new adaptive weights while both the PPC and TPC are held fixed.

Both the learning and the performance phases use the same AVITE circuitry, notably the same DV, for their respective functions. Thus learning and performance can be carried out on-line in a real-time setting, unlike most traditional off-line supervised error correction schemes. The operation whereby an endogenously generated PPC activates a corresponding TPC, as in Figure 4b, "back propagates" information for use in learning, but does so using local operations without the intervention of an external teacher or a break in on-line processing.

We call the class of models that use this on-line learning and performance scheme a Vector Associative Map (VAM) because it uses a difference vector to both learn and perform an associative mapping between internal representations.

Autonomous control, or gating, of the learning and performance phases is needed to achieve effective on-line dynamics, at least when learning is fast. For example, the network needs to distinguish whether $DV \neq 0$ because the TPC and PPC represent different target positions, or because the TPC→DV synapses are improperly calibrated. In the former case learning should not occur; in the latter case, it should occur. Thus some type of learning gate may be needed to prevent spurious associations from forming between TPCs and PPCs that represent different target positions. The design of the total AVITE network shows how such distinctions are computed and used for real-time control of the learning and performance phases. We now summarize how this is accomplished.

5. The Motor Babbling Cycle

During the motor babbling stage, an Endogenous Random Generator (ERG) of random vectors is activated. These vectors are input to the PPC stage, which integrates them thereby giving rise to outflow signals that move the arm through the workspace (Figure 3a). After each interval of ERG activation and PPC integration, the ERG *automatically* shut off, so that the arm stops at a random target position in space.

Offset of the ERG opens a Now Print (NP) gate that copies the PPC into the TPC through some fixed, arbitrary transformation (Figure 4b). The top-down adaptive filter from TPC to DV learns the correct reverse transformation by driving the DV toward zero while the NP gate is open (Figures 4c–4d). Then the cycle is automatically repeated. When the ERG becomes active again, it shuts off the NP gate and thus inhibits learning. A new PPC vector is integrated and another arm movement is elicited.

The ERG is designed so that, across the set of all movement trials, its output vectors generate a set of PPCs that form an unbiased sample of the workspace. This sample of PPCs generates the set of (TPC, PPC) pairs that is used to learn the adaptive coordinate change TPC→DV via the VAM.

6. The Endogenous Random Generator of Workspace Sampling Bursts

The ERG design embodies a specialized opponent processing circuit. Opponent processing is needed to control two complementary phases in the motor babbling cycle, during which the ERG is an *active* phase or a *quiet* phase, respectively. During the active phase, the ERG generates random vectors to the PPC, thereby moving the arm randomly through the workspace. During the quiet phase, input to the PPC from the ERG is zero, thereby providing the opportunity to learn a stable (TPC, PPC) relationship. In addition, there must be a way for the ERG to signal onset of the quiet phase, so that the NP gate can open and copy the PPC into the TPC. The NP gate must not be open at other times: If it were always open, any incoming commands to the TPC could be distorted by contradictory inputs from the PPC. Therefore, offset of the active ERG phase must be accompanied by onset of a complementary mechanism whose output energizes opening of the NP gate.

The signal that opens the NP gate can also be used to modulate learning in the adaptive filter. In general, no learning should occur except when the PPC and TPC encode the same position.

Figure 5 provides a schematic diagram of the ERG circuit. The design is a specialized gated dipole (Grossberg, 1972, 1982, 1984). A gated dipole is a neural network model for the type of opponent processing during which a sudden input offset within one channel can trigger activation, or antagonistic rebound, within the opponent channel. Habituating transmitter gates in each opponent channel regulate the rebound property by modulating the signal in their respective channel. In applications to biological rhythms, each channel's offset can trigger an antagonistic rebound in the other channel, leading to a rhythmic temporal succession of rebound events. An example of such an endogenously active rhythm generator was developed by Carpenter and Grossberg (1981; reprinted in Grossberg, 1987a) to explain parametric data about control of circadian rhythms by the suprachiasmatic nuclei of the hypothalamus.

In the present application, note the complementary time intervals during which the ON and OFF channels of the ERG are active: The ON channel output must be different during each active phase so that integrated PPCs result in random movements that sample the workspace. In contrast, OFF channel activation must be fairly uniform across trials, thereby providing intervals during which learning can stably occur.

Figure 6 illustrates the main characteristic of the simplest type of feedforward gated dipole: When a phasic input J^+ is applied to the ON channel, the corresponding ON channel output O^+ exhibits a transient overshoot that decays, or habituates, to a new, lower resting level. Offset of the phasic input causes the ON output to quickly drop to zero, while the OFF channel output O^- exhibits a transient antagonistic rebound followed by a decay to zero. Hence the gated dipole embodies a mechanism for generating a transient antagonistic rebound to offset of a phasic cue.

The OFF rebound is due to opponent interactions between two channels whose signals are multiplicatively gated by chemical transmitters. The chemical gates (rectangular synapses in Figures 5 and 6) are presumed to act on a time scale slower than the time scale of neuronal activation, so that sudden shifts in input are followed by slower changes in the amount of available transmitter substance.

The basic gated dipole circuit needs to be specialized to design an effective ERG circuit. Such an ERG circuit needs to convert a continuous stream of random inputs to the ON channel (J^+ in Figure 5) into cyclic output bursts from the ON channel, interspersed with OFF intervals whose duration is relatively stable across trials.

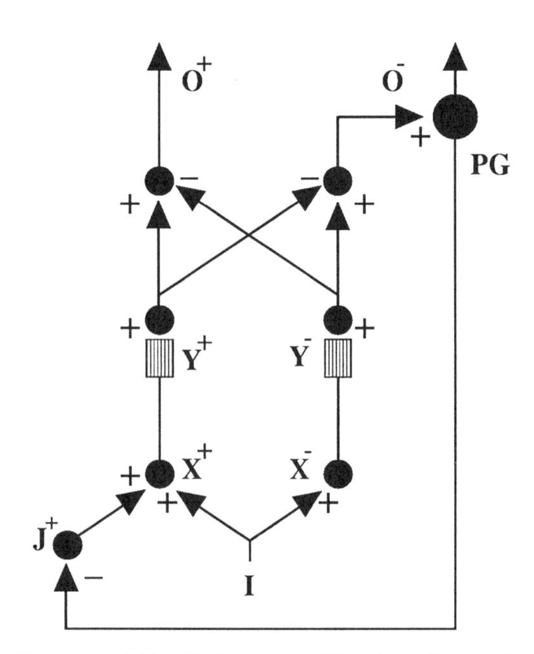

Figure 5. Schematic diagram of the Endogenous Random Generator (ERG). PG = Pauser Gate; J^+ = phasic input to the ON channel; I = tonic input to both channels; X^+,X^- = input layer activation; Y^+,Y^- = available chemical transmitter (chemical gates represented by rectangular striped synapses); O^+ = ERG ON output to the PPC; O^- = ERG OFF output controls PG activation. See text for description of ERG dynamics.

In order to convert a stream of random inputs into a series of output bursts, activation of the ON channel must initiate a process that spontaneously terminates ON channel output even while the random inputs remain on. This can be achieved if the net signal through the transmitter gate is an inverted-U function of input size. Then the gated ON output can "crash" on the time scale of transmitter habituation. The usual transmitter law of a gated dipole needs to be modified to achieve this property, because the net signal through the transmitter gate in the simplest gated dipole is an increasing function, not an inverted-U function, of input size.

In order to achieve cyclic output bursts from the ON channel, the ON chemical transmitter gate must be allowed to recover from habituation after crashing. To this end, the random input stream to the ON channel must be blocked after the ON gate crashes. Our solution is to let OFF channel activation (which becomes positive when the ON channel crashes) gate shut the source of phasic input J^+, which will cause a transient increase of activity in the OFF channel while the ON transmitter gate recovers from habituation. This process is represented in Figure 5 as a feedback pathway from the OFF channel of the ERG to the input source (J^+) through a Pauser Gate (PG) whose output is constant above its firing threshold. Figure 7 illustrates the dynamics of the ERG as it goes through a complete processing cycle.

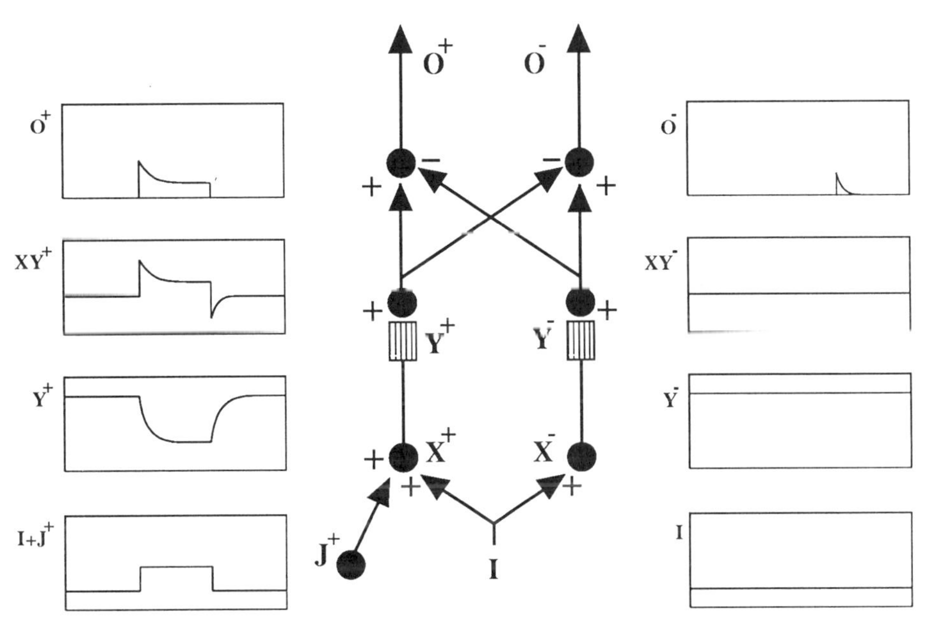

Figure 6. Schematic representation of linear, feed-forward gated dipole. Plots indicate response of each dipole stage over time. A sudden increase in phasic input (J^+, bottom left of figure) leads to a transient overshoot in the ON channel (top left), followed by habituation to a positive plateau. Removal of the differential input leads to a transient rebound in the OFF channel (top right) due to the depleted chemical transmitter in the ON channel gate.

7. Computer Simulations of Parameter Learning and Trajectory Formation through Motor Babbling

This section provides a qualitative overview of the major results obtained through simulation of the ERG–AVITE system during the babbling phase of adaptive tuning. Figure 8 is a schematic diagram of the complete system used in the simulations described below to control a two-jointed arm. Each AVITE module consists of one agonist channel and one antagonist channel, coupled in a push-pull fashion. Each channel receives inputs from its own ERG circuit. All ERG OFF channels cooperate to activate a single PG, denoted by $\sum$ in Figure 8, to insure synchronous activation of all muscle pairs.

Figure 9 shows the graphical output of the simulation program during babbling. Each grid shows a different configuration of the two-joint arm, with each joint regulated by one

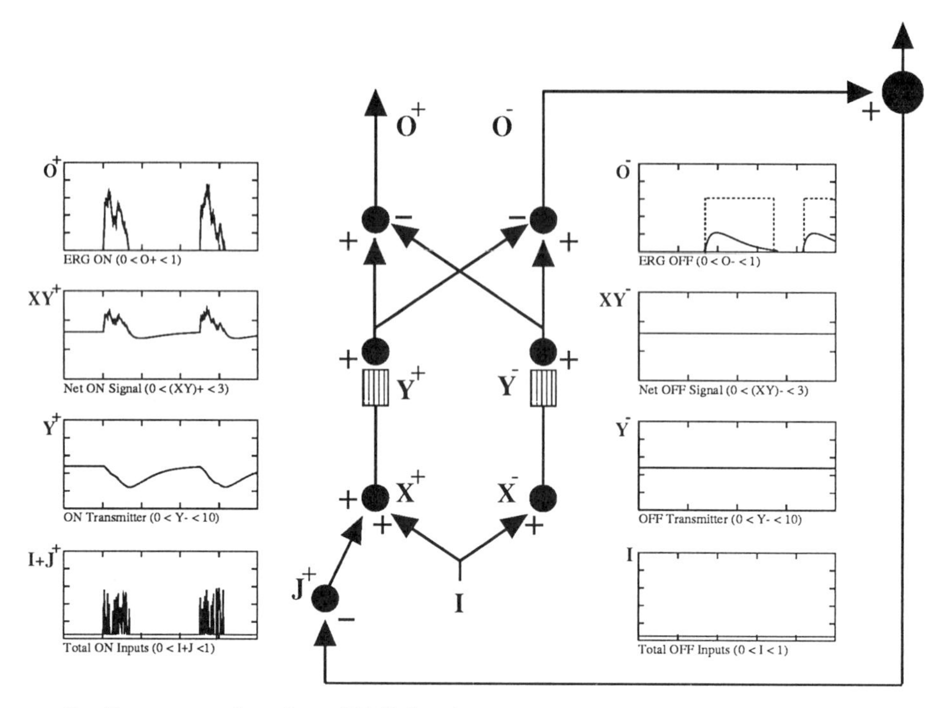

Figure 7. Response of various ERG levels to a continuous differential input J^+. The inverted-U transfer function through the chemical gates (rectangular synapses) leads to a transient ON response (top left), followed by activation of the OFF channel. Sufficient OFF channel activation energizes the Pauser Gate (PG), which shuts off phasic input to the ON channel, causing a larger rebound in the OFF channel. Removal of the phasic input allows ON channel transmitter to replenish, eventually shutting off the PG and starting a new cycle. The ON channel output is choppy due to noisiness of phasic input J^+. Dashed lines in upper right-hand plot represent PG activation (not drawn to scale).

AVITE module. The figure illustrates some of the positions attained during the quiet phases of motor babbling. Note the broad sample of arm positions. A more quantitative demonstration of the relatively uniform distribution of endogenously-generated arm positions is given in Figures 10–12.

Figure 13 illustrates the convergence of the learning process as motor babbling progresses. The plot shows the DV values of agonist and antagonist, as well as the total DV, at the end of successive quiet phases, when the PPC equals the TPC. Learning successfully drives the DV to zero at an approximately exponential rate. The agonist and antagonist DVs are defined below in our summary of the mathematical equations of the AVITE and ERG models. Further details may be obtained in Gaudiano and Grossberg (1990b).

8. AVITE Equations

In the AVITE equations, the subscript i refers to the ith module in the simulation. Each module consists of an agonist-antagonist pair of channels, and a single module controls a single joint. Unless otherwise indicated, each equation below describes the behavior of variables for the agonist channel, labeled by the (+) superscript in Figures 5 and 14. The corresponding equations for the antagonist variables in the same module—omitted for clarity—can be obtained by exchanging every (+) superscript with a (−) superscript, and vice versa.

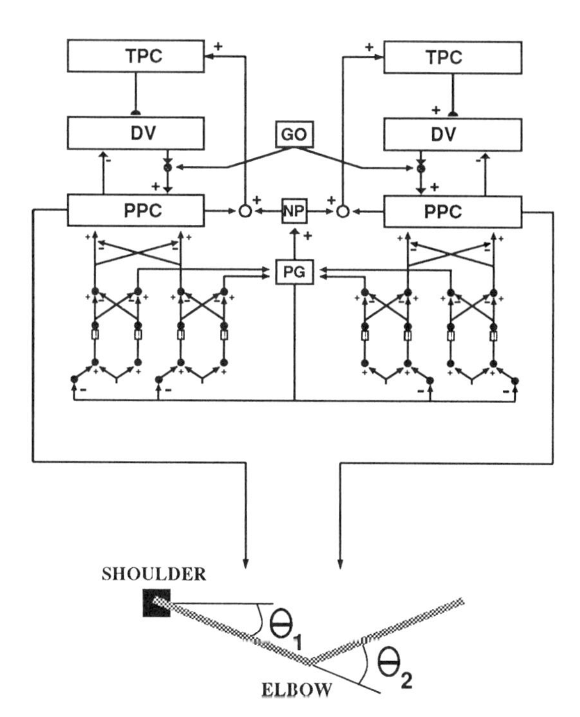

Figure 8. Diagram of the complete ERG AVITE system used for the two-joint simulations. Each AVITE module is driven by two ERG modules. AVITE outflow commands control movement of a simulated two-joint arm. The GO signal, NP gate, and PG of all modules are coupled together to ensure synchronous movement and learning for both synergies.

Each AVITE module requires the input of two ERG ON channels coupled in a push-pull fashion to insure that contraction of the agonist muscle is accompanied by relaxation of the antagonist. Because ON and OFF channel variables for each ERG circuit are also distinguished with (+) and (−) superscripts, we will use the following notation: Variable O^{+}_{2i-1} indicates the output of an ERG ON channel to the ith *agonist* PPC, whose activity is denoted by P^{+}_{i}. Variable O^{+}_{2i} indicates the output of a different ERG ON channel to the ith *antagonist* PPC, whose activity is denoted by P^{-}_{i} (see bottom of Figure 14).

Present Position Command

Let PPC variable P^{+}_{i} obey the equation

$$\frac{dP^{+}_{i}}{dt} = (1 - P^{+}_{i})\left(G\left[V^{+}_{i}\right]^{R} + O^{+}_{2i-1}\right) - P^{+}_{i}\left(G\left[V^{-}_{i}\right]^{R} + O^{+}_{2i}\right), \tag{1}$$

where $[w]^{R} = \max(w, 0)$ represents rectification. This is the rate-determining equation for the entire system: We assume an integration rate of 1 and adjust the time constant of all other equations relative to this one.

Difference Vector

The DV variable V^{+}_{i} obeys the additive equation

$$\frac{dV^{+}_{i}}{dt} = \alpha\left(-V^{+}_{i} + T^{+}_{i}Z^{+}_{i} - P^{+}_{i}\right). \tag{2}$$

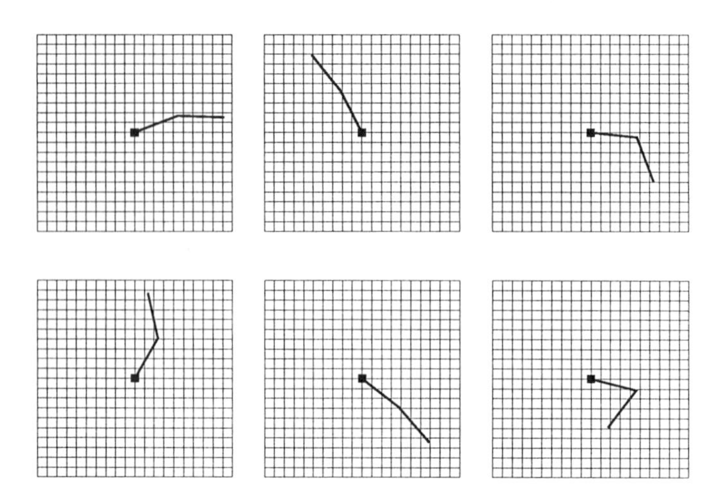

Figure 9. Pictorial representation of ERG-AVITE simulations. Each grid represents the configuration of the simulated arm during a quiet phase (ERG OFF is active). Note the diversity of attained positions.

The DV tracks the difference between a filtered copy of the TPC, namely $T_i^+ Z_i^+$, and the PPC variable P_i^+ at rate α.

Adaptive Filter LTM Traces

The LTM trace, or adaptive weight, Z_{ji}^+ from the TPC component T_i^+ to the DV component V_i^+ obeys the learning equation

$$\frac{dZ_i^+}{dt} = g_n f(T_i^+)(-\beta Z_i^+ - \gamma V_i^+), \tag{3}$$

where

$$f(T_i) = \begin{cases} 1 & \text{if } T_i > 0 \\ 0 & \text{if } T_i = 0. \end{cases} \tag{4}$$

Equations (3) and (4) define a *gated vector learning law* whereby changes in adaptive weights are driven by deviations of the DV from zero when the learning gate g_n is opened and the presynaptic node T_i is active. Other types of $f(T_i)$ would work as long as learning is prevented when $T_i = 0$. As the correct scaling factors from PPC to TPC channels are learned, the DV values converge to zero. Term g_n in (4) represents the Now Print Gate. The Now Print Gate enables the PPC to activate a TPC that represents the same target position of the arm. This gate can be coupled to the Pauser Gate g_p of equation (17) below, or to activation of the GO signal.

Target Position Command

The TPC variable T_i^+ obeys the equation

$$\frac{dT_i^+}{dt} = \delta\left[-\epsilon T_i^+ + (1 - T_i^+)(E_i^+ + F_i^+ + T_i^+) - T_i^+(E_i^- + F_i^- + T_i^-)\right]. \tag{5}$$

The input terms to each TPC are of three types:

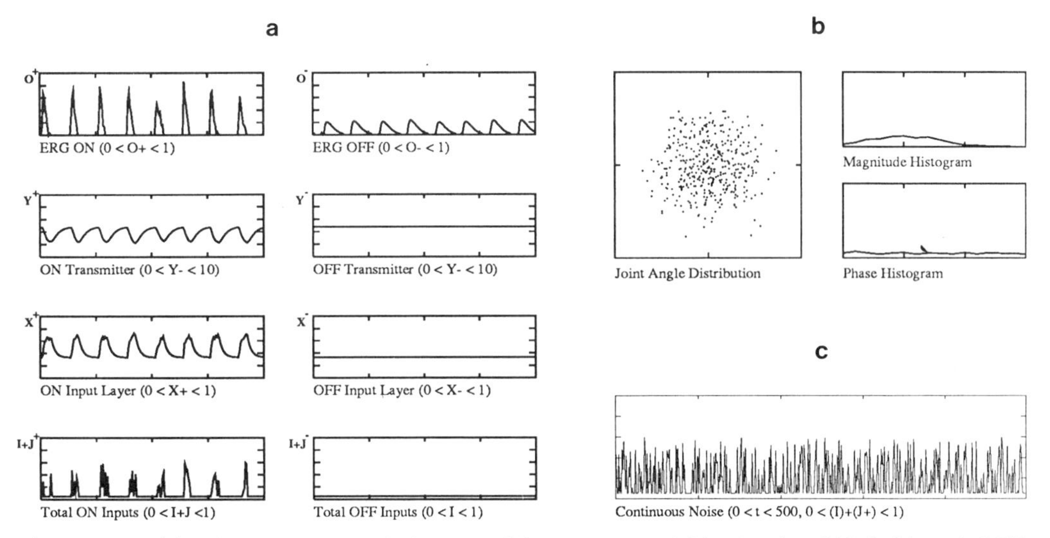

Figure 10. (a) The time course behavior of four state variables in the ON (left) and OFF (right) ERG channels during 2000 steps of the simulation. The range of each plot is indicated in parentheses. From top to bottom: ERG outputs (O^+,O^-), available transmitter (Y^+,Y^-), input layer activation (X^+,X^-), and total input signal ($I+J^+$,I). Note that bottom three plots for OFF channel (right) are always constant. This is due to lack of phasic input to OFF channel. However, baseline activation is necessary to energize OFF channel rebounds (top right plot). (b) Distribution of joint angles (between $-\pi$ and π radians) attained during 100,000 steps of the simulation. Left: each dot in the scatterplot represents the angle of the two joints (θ_1, θ_2) during each quiet phase. Center point represents resting position. Right: histograms of the distribution of joint angles around resting position. Magnitude histogram represents the number of dots falling within each of 16 evenly spaced concentric rings about the center; the unimodal distribution on the left side of the histogram indicates a tendency for less extreme joint angles. Phase histogram represents the number of dots at each of sixteen evenly spaced quadrants about the resting position; a flat phase histogram indicates a uniform distribution of joint angle combinations. (c) Representative sample of uninterrupted total input $I+J^+$ for 500 steps. The PG threshold Γ_P was raised to disable phasic input gating. The tonic input I causes the shift above zero.

(i) Intermodal Target Commands: These are feedforward external inputs (E_i^+, E_i^-) that instate new TPCs from other modalities, say from visual inspection of a moving target;

(ii) PPC-to-TPC Conversions: These are feedback inputs (F_i^+, F_i^-) from the PPC to the TPC. These inputs instate the TPC value corresponding to the PPC value attained during an active phase of ERG input integration. Terms F_i^+ and F_i^- turn on when the Now Print gate g_n turns on; that is,

$$F_i^+ = l(P_i^+, g_n) \tag{6}$$

and

$$F_i^- = l(P_i^-, g_n), \tag{7}$$

where function l represents a fixed mapping.

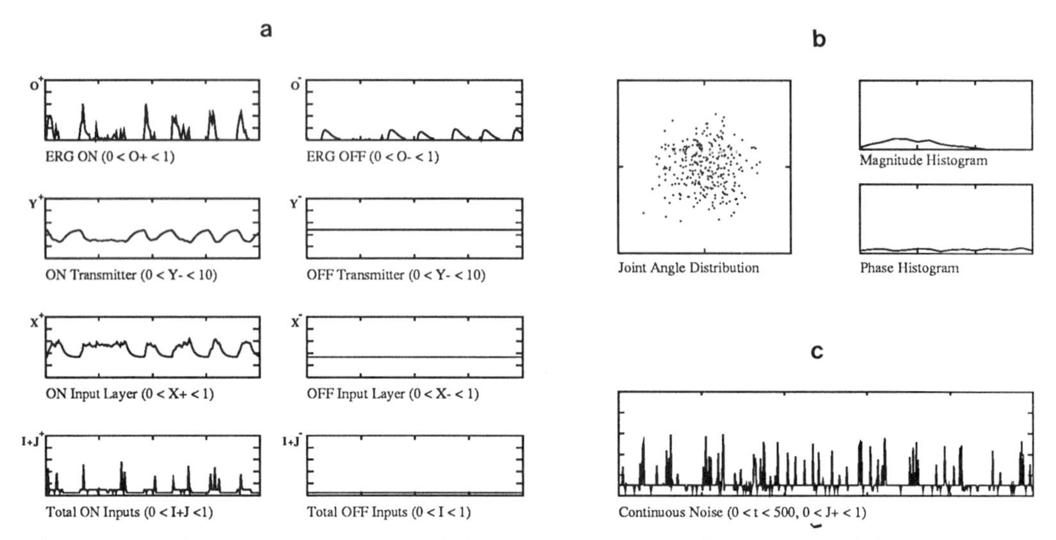

Figure 11. The average period π_J of the noise was raised. See equation (9). As a result, the phasic input J^+ is much more sparse, and the overall ERG dynamics are significantly affected. In spite of this, the overall distribution of joint angles is similar to that obtained with the standard parameters.

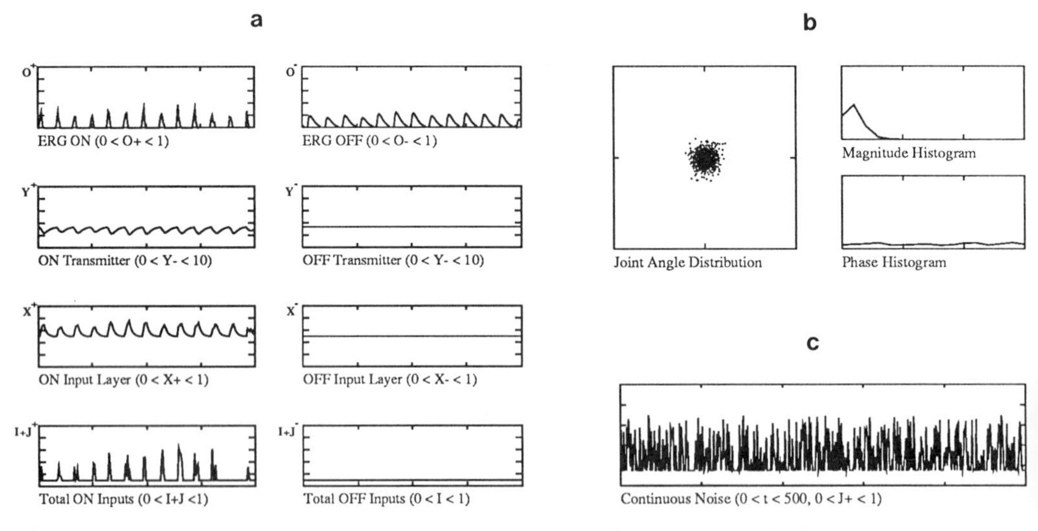

Figure 12. The tonic arousal level I was raised. See equation (10). As a result, the dynamics of the input layer X are dominated by the tonic input I, and the differential input J^+ becomes less effective. The ERG ON output bursts (O^+) are much smaller, resulting in smaller joint angles during babbling.

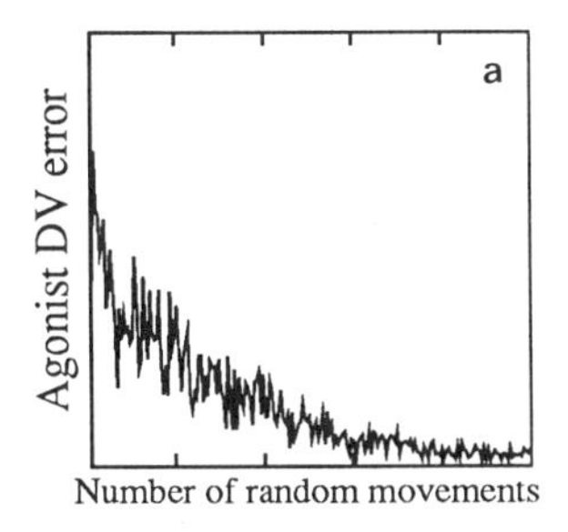

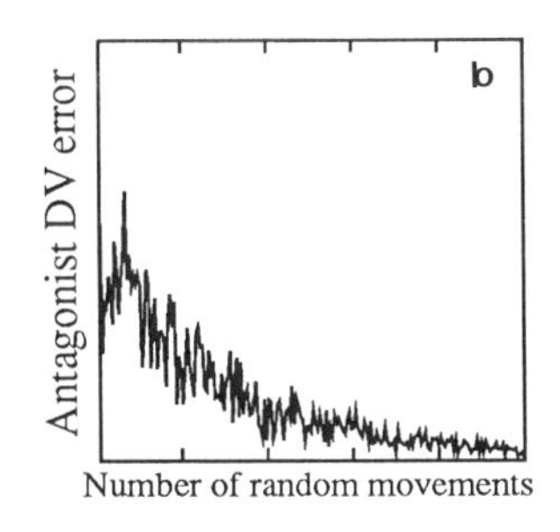

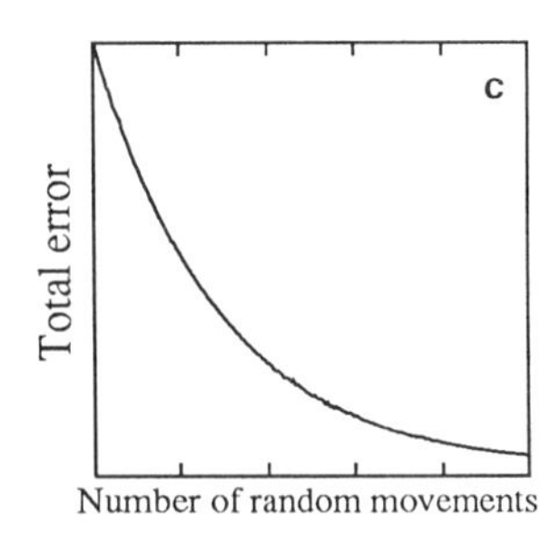

Figure 13. (a) Error in the agonist and antagonist DV shortly after onset of quiet phase. (b) Absolute value of the sum of errors in (a). Note approximately exponential decay. Learning rate was slowed down to illustrate smooth, slow error decay over 50,000 steps (approximately 200 babbled movements).

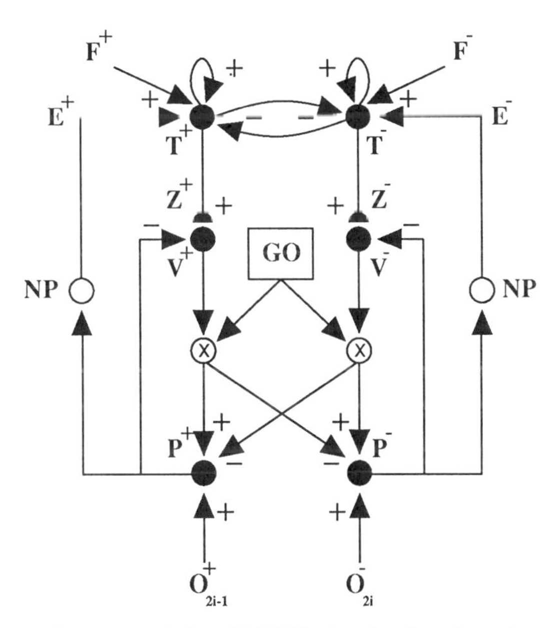

Figure 14. Schematic diagram of the AVITE circuit showing the existence of opponent channels for control of agonist-antagonist muscle pairs, indicated by (+) and (−) superscripts, respectively. Push-pull interactions at the TPC layer and in the (DV·GO)→PPC pathways insure that contraction in one channel will result in relaxation of the opponent channel, and vice versa.

(iii) Short-Term Memory Storage: These are feedback signals (T_i^+, T_i^-) from TPCs to themselves such that each agonist excites itself and inhibits its antagonist via a linear function of its activity.

9. ERG Equations

Tonic Input to the ERG

Let the tonic inputs I_k^+ and I_k^- to the kth ERG ON channel and ERG OFF channel obey the equations

$$I_k^+ = I_k^- = I \text{ (constant)}. \tag{8}$$

The tonic input I provides a constant baseline of activation in both ERG channels (Figure 5). This provides the energy for the transient rebound in the kth OFF channel after the random input J_k^+ to the kth ON channel is gated off by the Pauser Gate g_p. Without tonic input, OFF channel activation could never exceed zero.

Random Input to the EG

Let input J_k^+ to the kth ON channel of the ERG obey the equation

$$J_k^+ = \begin{cases} \max\left(0, J \in [\mu_J - \frac{\sigma_J}{2}, \mu_J + \frac{\sigma_J}{2}]\right) & \text{with probability } 1/\pi_J \\ \mu_J & \text{with probability } (1 - 1/\pi_J). \end{cases} \tag{9}$$

Random noise values J_k^+ are chosen from an interval of size σ_J centered around the average level μ_J. The term π_J represent the average time that elapses between activation "spikes."

ON and OFF Channel Input Layer Activations

The kth ERG ON channel input layer activity X_k^+ obeys the equation

$$\frac{dX_k^+}{dt} = -\zeta X_k^+ + (\eta - X_k^+)[I + J_k^+(1 - g_p)]. \tag{10}$$

Habituating Transmitter Gates

Let the transmitter gate Y_k^+ in the kth ON channel obey the equation

$$\frac{dY_k^+}{dt} = \kappa(\lambda - Y_k^+) - h(X_k^+)Y_k^+. \tag{11}$$

In (11), transmitter Y_k^+ accumulates to a maximal level λ at the constant rate κ and is inactivated, or habituates, at the activity-dependent rate $h(X_k^+)$, where

$$h(X) = \nu X^2 + \xi X. \tag{12}$$

The net ON channel signal through the gate is

$$X_k^+ Y_k^+, \tag{13}$$

which is proportional to the rate of transmitter release. When solved at equilibrium, the system (11), (12) and (13) give rise to an inverted-U function of X_k^+; namely,

$$X_k^+ Y_k^+ = \frac{\kappa \lambda X_k^+}{\kappa + \nu (X_k^+)^2 + \xi X_k^+}. \tag{14}$$

Opponent Output Signals

The net output O_k^+ of the kth ERG ON channel, after opponent processing, obeys the equation

$$O_k^+ = [X_k^+ Y_k^+ - X_k^- Y_k^-]^R. \tag{15}$$

The outputs O_k^+ are the inputs to the PPC populations of the AVITE model, as in equation (1). The ERG OFF outputs O_k^- obey the analogous equation

$$O_k^- = [X_k^- Y_k^- - X_k^+ Y_k^+]^R. \tag{16}$$

These signals activate the Pauser Gate in the manner described below.

Pauser Gate

The Pauser Gate g_p obeys the equation

$$g_p = \begin{cases} 1 & \text{if } \sum_k O_k^- > \Gamma_P \\ 0 & \text{otherwise} \end{cases} \tag{17}$$

where θ_P is a fixed threshold. When multiple ERG modules are simulated, all of the OFF channel outputs O_k^- are summed at the Pauser Gate via term $\sum_k O_k^-$ in (17). This insures that all AVITE modules are in their quiet phase at the same time, and that learning is synchronous across all movement-controlling joints.

10. A VAM Cascade for Learning a Multimodal Invariant Spatial-to-Motor Map

The results above illustrate the ability of the simplest VAM, the AVITE model, to control autonomous real-time error-based learning. We conclude this chapter with an example of a VAM Cascade that learns a spatial-to-motor map followed by a motor-to-motor map to handle invariant, multimodal adaptive control of arm movements. This example may be generalized in several directions.

The act of reaching for visually-detected targets in space is known to involve a number of different modalities: For instance, the position of the target on the retina, the position of the eyes in the head, the position of the head in the body, and the position of the arm with respect to the body must all be taken into account for correct execution of a reaching movement (e.g., Soechting and Flanders, 1989). In particular, the position of a target with respect to the body can be represented by many combinations of eye positions in the head and target positions on the retinas. We now show that the VAM is able to learn an invariant multimodal mapping; that is, it can learn to generate a correct muscle-coordinate command for all combinations of retinal and eye positions corresponding to a single target in space.

In Figure 15, the two top spatial maps represent the horizontal position of the target on the retina, and the horizontal position of the eyes within the head. For simplicity, we consider one-dimensional spatial maps, and we assume a linear relationship between the change in arm position and the total change in retinal position and eye position. That is,

$$i_E + j_R = H, \tag{18}$$

where i_E represents activation of the ith node of the eye position map; j_R represents activation of the jth node in the retinal map; and H is linearly related to arm position. In particular, if there are N nodes in the eye-position map and M nodes in the retinal map, we let

$$H = (N + M)P^+. \tag{19}$$

By (18), each fixed target position H can be represented by many combinations of eye position and retinal position. In particular, equations (18) and (19) indicate that for a fixed AVITE outflow command (P^+, P^-), a rightward shift in eye position (i_E increases) is canceled by a leftward shift in retinal position (j_R decreases), and vice versa. This set-up is similar that used to learn the Invariant Target Position Map of Grossberg and Kuperstein (1986, 1989, Chapter 10). Our results herein show how to learn such a map using a VAM Cascade.

For the simulations, the arm position H during each quiet phase of babbling is mapped into a random (i_E, j_R) pair that satisfies equations (18) and (19). In other words, the eyes track the arm. Then the active node i_E in the eye position map and i_R in the retinal

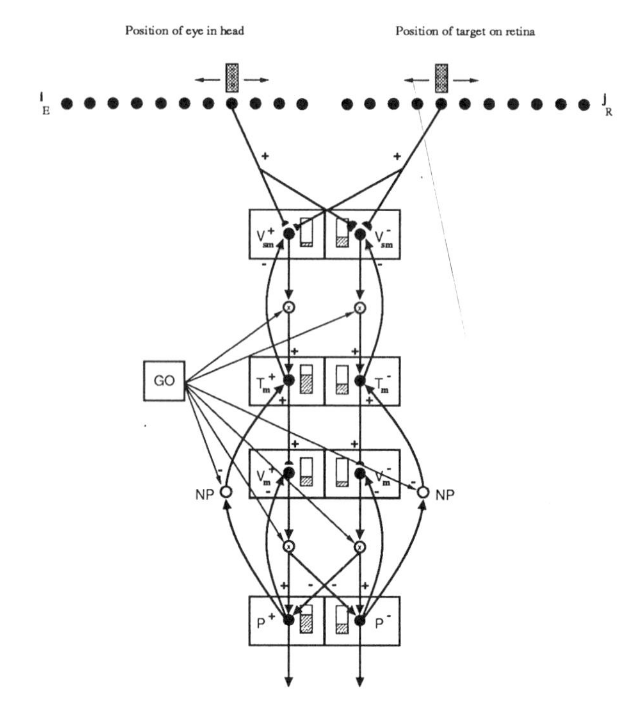

Figure 15. The Multimodal VAM. Activation of the upper left map represents eye position, and that of the upper right map represents target position on the retina. Activation from these two maps contribute at the Multimodal VAM. A given shift in eye position can be cancelled by an equal and opposite shift in retinal target position.

position map can sample the current arm position registered at the AVITE TPC. However, the VAM activation is affected by activity in both populations, so that the filtered signal from each population only needs to be half as strong as it would be if only one population were present. After training, instatement of a target (i_E, j_R) moves the arm to the correct location according to equations (18) and (19). Changes in i_E and j_R such that $i_E + j_R$ remains unchanged do not change the position of the arm.

Similar results hold if the two intermodal populations are not in the same coordinate system. For example, the horizontal eye position could be coded by a pair of nodes that represent the muscle lengths for an agonist-antagonist pair of oculomotor muscles.

11. Towards a System-Level Synthesis of Complementary ART and VAM Designs

In learning and performance by a VAM, the matching event is *inhibitory*. For example, in an AVITE model, matching a TPC with a PPC zeroes the DV. This is the basis for saying that VAM learning is *mismatch* learning: Learning occurs only when DV$\neq 0$ and drives the DV mismatch to zero. In contrast, a complementary type of learning occurs in ART models, which are also capable of autonomous real-time learning. In ART, learning is *approximate match* learning; that is, learning occurs only if the match between the learned top-down expectation (cf., the AVITE TPC$\rightarrow$DV signals) and the bottom-up input pattern (cf., the AVITE PPC$\rightarrow$DV signals) are sufficiently close that the orienting, or novelty, subsystem is inhibited, and matching by the 2/3 Rule (Carpenter and Grossberg, 1987a, 1987b, 1988) uses a fusion event, or attentional focus, or resonant state to develop which drives the rning process.

These complementary learning rules coexist with complementary rules for top-down iming. In a VAM model such as AVITE, the top-down TPC→DV signals prime a *mo- expectation.* When this expectation is matched by a PPC, the limb has already moved to target. No further movement is needed, and the DV is zeroed, or inhibited. In ART, by ntrast, a top-down *sensory expectation* prepares the network for an anticipated bottom-up ent that may or may not occur. If the event does occur, then matching causes resonant citation, not inhibition. Thus the two complementary learning rules coexist with two mplementary rules for top-down priming, or intentionality.

Sensory-cognitive circuits seem to be designed according to ART style processing where- cognitive-motor circuits seem to be designed according to VAM-style processing. In rticular, Carpenter and Grossberg (1988) and Grossberg (1988) have noted that ART-style rning is stable in response to an arbitrary sequence of sensory input patterns, for purposes recognition learning and reinforcement learning. These authors contrasted the stability of RT learning with instabilities of mismatch learning, notably learning by perceptrons and ck propagation, when they are used for recognition learning and reinforcement learning. rpenter and Grossberg also analysed why back propagation is not a real-time model; rather needs to be run off-line under carefully controlled conditions, including a slow learning e.

VAM models, in contrast to back propagation, are capable of real-time processing. They designed to carry out both learning and performance within the same processing channel using self-controlled real-time gating of complementary learning and performance modes. ch gating also enables VAM learning to be fast. Within an intramodal VAM, such as ITE, fast learning is always stable because the "back propagation" from PPC to TPC tomatically assures that correct (TPC, PPC) correlations are learned. In this sense, al- ough AVITE learning is *mismatch* learning, it is based upon self-controlled *matches* of C and TPC. Within an intermodal VAM that takes its data partly from prior stages of sory processing, fast learning is stable because ART mechanisms assure the stability of sensory representations themselves.

Taken together, the ART and VAM models provide a framework for designing stable l-time fast-learning systems that exploit both approximate match learning and mismatch rning. ART networks can achieve stable real-time fast-learning of recognition and rein- cement codes. VAM networks, fed by outputs from the stable ART networks, can be used achieve stable real-time fast learning of sensory-motor maps. Thus, all the benefits of proximate match learning and mismatch learning, and the corresponding benefits of exci- ory matching and inhibitory matching, can be achieved, without their limitations. This y be accomplished by incorporating them both into an appropriately cascaded neural hitecture wherein they coexist as complementary aspects of a larger system design.

EFERENCES

derson, R.A., Essick, G.K., and Siegel, R.M. (1985). Enclosing of spatial location by poste- rior parietal neurons. *Science*, **230**, 456–458.

llock, D. and Grossberg, S. (1988a). Neural dynamics of planned arm movements: Emergent nvariants and speed-accuracy properties during trajectory formation. *Psychological Review*, **95**, 49–90.

llock, D. and Grossberg, S. (1988b). The VITE model: A neural command circuit for gener- ating arm and articulatory trajectories. In J.A.S. Kelso, A.J. Mandell, and M.F. Shlesinger (Eds.), **Dynamic patterns in complex systems**. Singapore: World Scientific Publishers.

Bullock, D. and Grossberg, S. (1989). VITE and FLETE: neural modules for trajectory formation and postural control. In W.A. Hershberger (Ed.), **Volitional action**. Amsterdam: North-Holland-Elsevier.

Bullock, D. and Grossberg, S. (1990). Adaptive neural networks for control of movement trajectories invariant under speed and force rescaling. *Human Movement Science*, **9**, in press.

Carpenter, G.A. and Grossberg, S. (1981). Adaptation and transmitter gating in vertebrate photoreceptors. *Journal of Theoretical Neurobiology*, **1**, 1–42.

Carpenter, G.A. and Grossberg, S. (1987a). A massively parallel architecture for a self-organizing neural pattern recognition machine. *Computer Vision, Graphics, and Image Processing*, **37**, 54–115.

Carpenter, G.A. and Grossberg, S. (1987b). ART 2: Stable self-organization of pattern recognition codes for analog input patterns. *Applied Optics*, **26**, 4919–4930.

Carpenter, G.A. and Grossberg, S. (1988). The ART of adaptive pattern recognition by a self-organizing neural network. *Computer*, **21**, 77–88.

Carpenter, G.A. and Grossberg, S. (1990). ART 3: hierarchical search using chemical transmitters in self-organizing pattern recognition architectures. *Neural Networks*, **3**, 129–152.

Gaudiano, P. and Grossberg, S. (1990a). A self-regulating generator of sample-and-hold random training vectors. In M. Caudill (Ed.), **Proceedings of the international joint conference on neural networks, 2**, 213–216. Hillsdale, NJ: Erlbaum Associates.

Gaudiano, P. and Grossberg, S. (1990b). Vector associative maps: Unsupervised real-time error-based learning and control of movement trajectories. *Neural Networks*, in press.

Georgopoulos, A.P., Kalaska, J.F., Caminiti, R., and Massey, J.T. (1982). On the relations between the direction of two-dimensional arm movements and cell discharge in primate motor cortex. *Journal of Neuroscience*, **2**, 1527–1537.

Georgopoulos, A.P., Kalaska, J.F., Crutcher, M.D., Caminiti, R., and Massey, J.T. (1984). The representation of movement direction in the motor cortex: Single cell and population studies. In G.M. Edelman, W.E. Goll, W.M. Cowan (Eds.), **Dynamic aspects of neurocortical function**. Neurosciences Research Foundation, pp. 501–524.

Georgopoulos, A.P., Schwartz, A.B., and Kettner, R.E. (1986). Neural population coding of movement direction. *Science*, **233**, 1416–1419.

Grossberg, S. (1972). A neural theory of punishment and avoidance, II. Quantitative theory. *Mathematical Biosciences*, **15**, 39–67.

Grossberg, S. (1982). **Studies of mind and brain: Neural principles of learning, perception, development, cognition, and motor control.** Boston: Reidel Press.

Grossberg, S. (1984). Some normal and abnormal behavioral syndromes due to transmitter gating of opponent processes. *Biological Psychiatry*, **19**, 1075–1118.

Grossberg, S. (Ed.) (1987a). **The adaptive brain I: Cognition, learning, reinforcement, and rhythm**. Amsterdam: Elsevier/North-Holland.

Grossberg, S. (Ed.) (1987b). **The adaptive brain II: Vision, speech, language, and motor control**. Amsterdam: Elsevier/North-Holland.

Grossberg, S. (1988). Nonlinear neural networks: Principles, mechanisms, and architectures. *Neural Networks*, **1**, 17–61.

Grossberg, S. (1990). Self-organizing neural architectures for motion perception, adaptive sensory-motor control, and associative mapping. In M. Caudill (Ed.), **Proceedings of the international joint conference on neural networks, 2**, 26–29. Hillsdale, NJ: Erlbaum Associates.

Grossberg S. and Kuperstein, M. (1986). **Neural dynamics of adaptive sensory-motor control: Ballistic eye movements**. Amsterdam: Elsevier/North-Holland.

Grossberg S. and Kuperstein, M. (1989). **Neural dynamics of adaptive sensory-motor control: Expanded edition**. Elmsford, NY: Pergamon Press.

Piaget, J. (1963). **The origins of intelligence in children**. New York: Norton.

Soechting, J.F. and Flanders, M. (1989). Errors in pointing are due to approximations in sensorimotor transformations. *Journal of Neurophysiology*, **62**, 595–608.

NEURAL DARWINISM AND SELECTIVE RECOGNITION AUTOMATA:

HOW SELECTION SHAPES PERCEPTUAL CATEGORIES

George N. Reeke, Jr.
Olaf Sporns

The Neurosciences Institute and
The Rockefeller University
New York, New York

SUMMARY

It is becoming increasingly clear that animals generally do not use classical algorithmic strategies to classify objects and events. In this chapter, we summarize an alternative approach, namely, the theory of neuronal group selection (TNGS). The TNGS provides a framework related to Darwinian selection for understanding perceptual systems without a priori *assumptions about properties of the stimulus world. We present two models for aspects of visual perception based on synthetic neural modelling, a paradigm we have developed for testing theories about the nervous system by computer simulation. One of these models explores the implications of recent experimental evidence for coherent oscillations in visual cortex. The second is Darwin III, a neurally based simulated automaton capable of autonomous behavior involving categorization and motor acts with an eye and an arm. Automata of this type not only provide new insights into biological mechanisms of categorization, but also indicate how to construct a new class of perception machines with interesting properties.*

INTRODUCTION--THE CLASSIFICATION PROBLEM

The dual abilities to recognize and categorize objects and events in the environment are among the most basic functions of human and animal cognition. In fact, perception and adaptive behavior depend critically on these functions. Without categorization, perceptual systems would quickly be overwhelmed by the infinite combinatorial variety of sensory experience. For this reason, perceptual categorization has been and remains a central issue in artificial intelligence and neural modelling. Occasionally impressive results have been achieved, but some vital aspects of biological categorization have been ignored in the classical theories. It has proved to be very difficult to construct a pattern recognition device that combines adaptiveness with reliability and a wide range of applicability.

The partitioning of world objects or events into categories by an organism depends critically upon the requirements of that organisms's econiche and must ultimately support its survival. The criteria that the organism uses may be very different from those of another organism living in a different environment, and may also differ from scientifically "objective"

Self-Organization, Emerging Properties, and Learning
Edited by A. Babloyantz, Plenum Press, New York, 1991

or "logical" criteria. Categories are not fixed, objective, and immutable entities that only need to be learned or discovered by the organism; they are constructed by the perceptual system as it interacts with the environment.[1,2] Categorization depends on a variety of secondary factors such as context, affective and intentional states of the organism, and on memory.

As these facts have become more widely recognized, psychologists and philosophers have begun to change their view of categories and concepts. The classical view is that categories are defined by lists of necessary and sufficient features.[3] Class membership is unambiguous and ties every instance of a class to its definition with essentially the same strength. This view, if taken literally, implies that words refer to classes of items in the world in a unique and unequivocal way. However, analyses by Wittgenstein[4] and more recently by Lakoff[5] show that this view is untenable. Nonclassical theories of category are needed to fit these linguistic data. Such theories hold that membership in a category is not the result of applying a set of rules, but rather the result of assessing similarity between the item in question and some stored representation of previous encounters with members of the class.[3]

Experiments suggest that human categories indeed have a nonclassical internal structure; membership of an item in a category is graded rather than all-or-none.[6,7] Human subjects are able reliably and consistently to rate the typicality of a given example with respect to a category name. For example, *robin* serves as a consistently more typical member of the object class *birds* than does *penguin* or *stork*. Additional evidence for graded category membership comes from differences in learning and verification speed and from the probability of spontaneous recall upon presentation of the category name.[8] Although the interpretation of typicality ratings as evidence for the nonclassical internal structure of categories has been challenged,[9,10] the evidence for the complexity of everyday categories seems to be overwhelming, and is corroborated by findings from animal studies. Despite the fact that the classical view of categorization cannot serve as a general basis for the explanation of human or animal perceptual categorization, most extant pattern recognition and categorization models are based on some version of it. We now briefly describe some of these models to provide a framework for what follows.

NEURAL MODELS OF PATTERN RECOGNITION

Abstract neural modelling began with the introduction of networks of neuron-like logical units by McCulloch and Pitts.[11] The design principles of such networks are closely related to those of digital computers, which were developed at around the same time. In 1947 Pitts and McCulloch[12] suggested a method for carrying out pattern recognition using their networks. Later, Rosenblatt tried to incorporate perceptual learning into a model for pattern recognition.[13] He studied a class of devices known as perceptrons, which consist of multiple layers of neuronal elements connected by feedforward connections. Perceptrons use a learning rule in which the weights of connections are incrementally adjusted to minimize errors in the output. They are trained by presenting them with suitable input patterns together with the desired outputs for those patterns. Rosenblatt proved that two-layer perceptrons will converge to optimal behavior if the input patterns are linearly separable.[14]

Minsky and Papert pointed out that the original two-layer perceptron is incapable of handling certain large classes of categorization problems.[15] While newer models have largely overcome these limitations, perceptrons have other critical shortcomings which are still found in most of the more recent pattern recognition models.

Most importantly, these models depend on the prior existence of an external teacher with the ability to pair stimuli with correct responses. Obviously, no model that requires such a teacher can provide a satisfactory explanation for primary human or animal categorization because it implies an infinite regress of teachers. Put another way, it is the categorization of the teacher, not that of the pupil, which requires explanation.

A second critical problem is that perceptrons and related systems[16] depend on the assumption that input patterns can be classified in terms of weighted sums of independent features. In such systems, boundaries between object classes must not only be definable as hyperplanes in a multi-dimensional feature space, but they also must be "objectively" accessible to the machine without contextual or attentional information. This notion has been trenchantly criticized by Bongard,[17] who writes that "the search for a plane dividing classes ... for many problems is hopeless. This is not because the search method is unsatisfactory, but because there is no such plane" (p. 21).

A third shortcoming was pointed out by Gyr *et al.*,[18] who remarked that sensorimotor activity plays an active and important role in both pattern recognition and perceptual learning which has not been considered sufficiently in the design of perceptual models. These authors present compelling evidence that perception is an active process involving motor activity and attention. This last point has also been stressed by Gibson[19] and his school of ecological psychology. The incorporation of motor activity in exploration and sampling of sensory signals as well as the influence of attentional states has remained a challenge to pattern recognition models.

Many of the more recent models of pattern recognition and categorization are based on extensions and refinements of classical architectures and algorithms, and as such, share these shortcomings. "Parallel distributed processing" (PDP) or "connectionist" models, for example, are multi-layer perceptrons equipped with more sophisticated learning rules.[20,21] Their capacities exceed those of simple perceptrons by virtue of added layers of "hidden units," which allow the formation of higher-order predicates. However, in spite of their success in numerous demonstrations,[22-26] they are unsatisfactory as models of biological processes.[27,28] Connectionist models incorporate many features that contradict our present knowledge of brain anatomy and physiology.[29]

A perceived analogy between neural networks and statistical systems has led to the construction of "spin glass" models of content-addressable memory.[30,31] In such models, memories are identified with stable states in the dynamics of the system and memory retrieval with the approach of an initial state vector to equilibrium at one of these low-energy states. Such networks can retrieve prototype patterns from noisy inputs and correct false details.[32,33] Their storage capacity however, is limited and the retrieval of patterns becomes more and more error-prone as the number of stored patterns increases.[34,35] The analogy between spin glasses and human memory[36] is almost certainly flawed, since everything we know about the brain as a dynamical system indicates that convergence to fixed points does not occur. In addition, the architecture of the cerebral cortex is strikingly different from the very dense connectivity pattern required for these models.

An interesting model for the formation of stable category recognition codes by clustering of similar input patterns has been formulated by Grossberg and Carpenter.[37] Their adaptive resonance theory (ART) proposes that categorization consists of two parallel processes. An input pattern is first filtered and processed in a competitive learning module. This results in the activation of category-specific units in a higher-level network. These units then signal back to the lower level their expectancy about what kind of pattern should be present for a particular output. ART avoids the use of non-local learning algorithms such as back-propagation and does not rely on instructive procedures to transmit category-specific information directly to the system. It has been successfully applied to simple classification tasks involving discrete and analog patterns.

Despite these recent advances, the modelling of pattern recognition by animals has made very little conceptual progress over the last forty years. The thinking of most neural network designers is firmly wedded to the classical notion of a world full of information that only needs to be transmitted to and analyzed by animals or automata. Virtually all modern models are more or less subtle variants of the template matching or feature analysis strategies.

Although analytical and computational techniques have been much refined in recent years the underlying concepts have not changed much. It seems that our knowledge of human and animal pattern recognition and categorization has increased far beyond the horizon of current work in modelling and engineering.

CATEGORIZATION AND THE THEORY OF NEURONAL GROUP SELECTION

Animals of many species are capable of grouping sensory stimuli in meaningful ways. Categorization and pattern recognition, therefore, are not exclusively human cognitive faculties. Pigeons, for example, can be trained to respond to the presence of *trees*, *fish*, or *humans* in photographs in a great variety of contexts.[38] Herrnstein introduced the term "natural concept" for categories of this kind that are characterized by open-ended variability rather than reproducibility or definability.[39] His results seem to rule out the possibility that pigeons are pattern recognition machines like the ones discussed above. Many primates can also form concepts that are isomorphic to natural language categories like *humans*[40] or *dogs*, *flowers*, and *shoes*.[41] One must conclude that perceptual categorization does not depend on language since it can be performed by non-verbal animals.

These experimental findings argue strongly for the generality of perceptual categorization. Any theory of brain function must offer an explanation for this phenomenon that is at once based on the known anatomy and physiology of the nervous system and that takes into account the many nonclassical properties of categorization. We believe that a theory based on selectional principles, the theory of neuronal group selection (TNGS),[1,2,42] offers a plausible solution to these problems.

At the heart of the TNGS is the notion that the nervous system operates by the selection of adaptively responding recognizing elements (neuronal groups) arranged in preestablished diverse repertoires. Like other biological theories based on selectional principles (e.g. the theory of natural selection in evolution and the theory of clonal selection in immunology) the TNGS has to fulfill certain requirements:[42-44] (a) Variance must exist before selection can occur--in the TNGS, embryonic development produces repertoires of variant circuits. (b) These variant entities must encounter the environment--in the TNGS, repertoires of neuronal groups are stimulated by and respond to patterns of sensory signals. (c) Differential amplification of selected elements must occur--in the TNGS, neuronal groups contributing to behaviors having adaptive value for the organism are amplified by differential modification of the synaptic weights connecting them with other circuits. These modifications may be depend on local conditions only or on global signals related to the adaptive value of previous behavioral acts.

According to the TNGS, selection in the nervous system occurs in three partially overlapping processes (Fig. 1). First, neuronal circuitry is generated during embryonic development (Fig. 1, top). Genetic and epigenetic control mechanisms are insufficient to specify the position and strength of each individual synaptic junction, as evidenced by the large degree of variability in the morphology of individual neurons, even in invertebrate species in which specific neurons can be identified in multiple individuals. There is mounting evidence that the fine structure of neuronal dendrites and axons is determined largely by neuronal activity and resulting competitive processes.[45] Such processes also contribute to the second stage of selectional events.

Within structures containing an abundance of variant circuitry, competition between neurons and consequent selection leads to the generation of local populations of neurons, called "neuronal groups" (Fig. 1, middle). These groups form the basic unit of response in the nervous system. They consist of strongly or densely interconnected collectives of neurons that tend to fire in a correlated fashion to afferent inputs. Neuronal groups are arranged in maps, which can be topographic (as in the primary sensory cortices) or nontopographic (as in frontal or prefrontal cortices). Evidence for the existence of neuronal groups has been found in various regions of the cerebral cortex.[46,47]

Communication between neuronal groups occurs along anatomically ordered reciprocal connections linking different neuronal maps (Fig. 1, bottom). This ongoing bidirectional exchange of signals is called *reentry*. Reentry is distinguished by a number of properties: (a) Reentry is parallel and distributed. It always involves the mapping of one spatially distributed response pattern onto another one. (b) Reentry is reciprocal, unlike feedback,[48] which is unidirectional. (c) The purpose of reentry is to compare and correlate signals, but not necessarily to form a control loop. Signals transmitted by reentry do not have the predefined purpose of feedback signals. (d) Reentry can be recursive. Responses of neuronal groups originating in one brain region can be re-entered after propagating through another region and can give rise to or influence further responses in the first region. In the cerebral cortex, reentry integrates the responses of neuronal groups in spatially segregated maps representing different sensory modalities or submodalities. The constructive properties of reentry have recently been demonstrated in a detailed model of several segregated areas of the visual system.[49] A further simulation[50,51] (see next section) has shown how reentry can lead to correlated activity in neurons at distant cortical sites, consistent with recent experimental observations.[52,53]

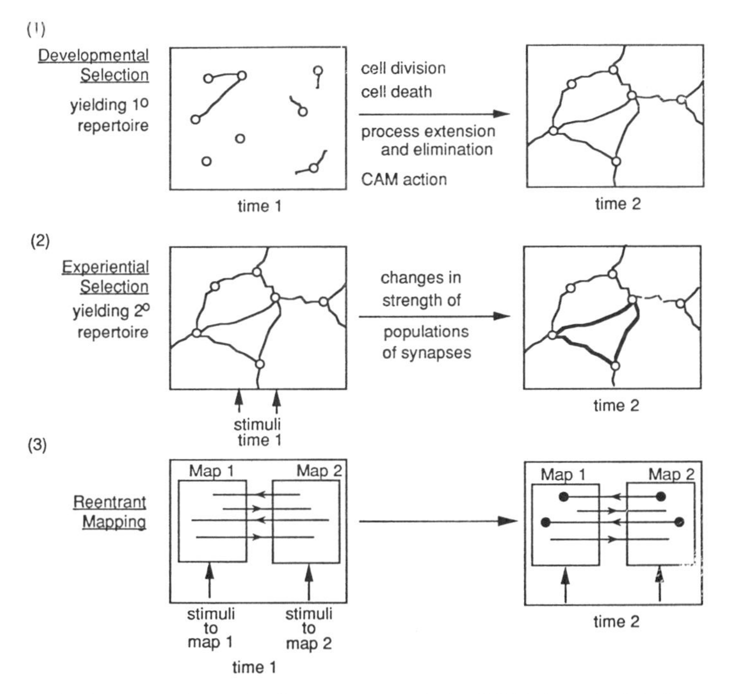

Fig. 1. Schematic diagram of the basic processes of the TNGS. (Top) Primary repertoires of neuronal groups are formed as a result of developmental selection. Cell adhesion molecules, which mediate surface interactions between developing cells, play a major role in this process. (Middle) Secondary repertoires of neuronal groups are formed as a result of experiential selection. At this stage, selection acts through differential modification of synaptic strengths. (Bottom) Correlation of the responses of neuronal groups in different maps is achieved by reentrant signalling. These maps could represent different sensory modalities or submodalities and be involved in the categorization of sensory stimuli. From ref. 2, reproduced with permission.

A MODEL FOR REENTRANT CONNECTIVITY IN CORTICAL OSCILLATIONS

The cerebral cortex of higher vertebrates is divided into a number of anatomically and physiologically distinct areas that contain neurons with distinct response properties. The primate visual cortex, for example, contains at least 17 subfields that are connected by a characteristic pattern of inter-areal fibers.[54-56] Each of these segregated areas presumably subserves a more or less distinct function in visual perception.

The distributed character of the visual system poses a problem for our understanding of its overall function: How can simultaneous and parallel, but segregated, events in the cortex be correlated such that a coherent function emerges? One obvious solution to this problem would be for a singular master area somewhere in the brain to keep track of all lower-level cortical maps and guide their function by controlling their interactions. However, candidate cortical areas receiving convergent inputs of the kind that would be required for such a master coordinating function have not been identified. Instead, most cortical areas have a variety of convergent and divergent connections with other, related areas.

A radically different point of view is that the continuity and spatial coherence of patterns of activity in the cerebral cortex is generated by reentrant neuronal connectivity.[42] By directly correlating neuronal events in distinct cortical maps, reentry obviates the need for separate detection and processing of such correlations. A system with reentrant properties can function without a central executive or master area which, if it existed, would essentially have to perform a homuncular function.

The structural substrate for this process of reentry is provided by reciprocal anatomical connections which link areas within a given region of the cortex.[57,58] In fact, these connections are one of the most striking features of cortical anatomy. Their local organization provides evidence that can guide our construction of a model for cortical reentry. The microanatomy precludes specific one-to-one linking of neurons, since axonal arbors extend over several hundred microns and appear to make thousands of synaptic junctions within their target areas. Instead, a given reentrant circuit must link groups formed from local populations of neurons. According to the TNGS, neurons within such groups tend to be more strongly connected and to share many physiological properties,[42] at least some of which are determined by cooperative interactions within the group. Consequently, long-range correlations are established by active reentrant circuits between neuronal groups rather than between single neurons (although such correlations might well be measured at the level of single cell activity).

Evidence for Coherent Oscillations in the Cortex

Recently, oscillatory neuronal activity has been described within orientation columns in the cat visual cortex.[46,53] Appropriate stimulation produced rhythmic, temporally correlated neuronal discharges in adult cats and kittens, awake or under varying conditions of anesthesia. Multi-unit activity as well as the simultaneously recorded local field potential oscillates at a frequency of 30-70 Hz (usually about 40 Hz) when a light bar of suitable orientation, velocity, and direction of movement is passed through the receptive field of one of the recorded neurons. Oscillatory activity decreases in the absence of a stimulus or during presentation of a light bar at other than the optimal orientation. There is no evidence for corresponding oscillations in the lateral geniculate nucleus (LGN), which provides the major input to the visual cortex. Thus, it is likely that the observed oscillations are generated by local excitation and recurrent inhibition within the cortex. According to Gray and Singer[46] (p. 1702), their results demonstrate "that groups of adjacent cortical neurons, when activated appropriately, engage in cooperative interactions." Even if some of these cells act as intrinsic oscillators,[59] it is likely that the synchronization and coherency of oscillatory activity in large masses of cortical neurons would require local and global network interactions as well.

Cross-correlation of multi-unit recordings from separate sites in area 17 revealed that oscillatory responses in orientation columns with similar specificity but non-overlapping receptive fields are synchronized when a single long light bar is moved across both receptive fields.[52,60] The synchronization becomes weaker if there is a gap in the stimulus contour and disappears completely if two parts of the contour are moved separately and in opposite directions.[61] An independent study[53] also revealed neuronal oscillations in cat visual cortex at similar frequencies and confirmed intra-columnar correlations as well as correlations between neighboring hypercolumns. Moreover, synchronization of oscillatory activity was observed between different cortical areas (areas 17 and 18). These oscillations were coherent with near zero phase delay.

The experimental evidence on cortical neuronal oscillations and their coherency prompts the idea that the observed temporal correlations between spatially separate cortical sites in response to linked stimulus features may be of importance in the transient association of such features. It also provides strong evidence for the existence of neuronal groups and is suggestive of the possible role of reentry in correlating the behavior of different groups. To explore further the properties of reentrant systems and the dynamics of neuronal groups we constructed a computer model based on these experimental findings. The model demonstrates how oscillatory activity can be generated within the cortex, how short-range cortico-cortical connections can give rise to phase coherency between neuronal groups within an area, and how long-range reentrant connections between different areas can give rise to global coherence and feature linking across multiple feature domains.

A Network Model of Coherent Oscillations in the Cortex

The model[50,51] consists of two separate visual areas: "OR," which contains 32 orientation- and direction-selective neuronal groups consisting of 200 excitatory and 120 inhibitory units each; and "MO," which contains 64 pattern-motion-selective neuronal groups consisting of 160 excitatory and 80 inhibitory units each (Fig. 2). The model is implemented using the Cortical Network Simulator program ("CNS")[62,63] and contains in all 96 neuronal groups, 25,600 cells, and over 3.5 million connections.

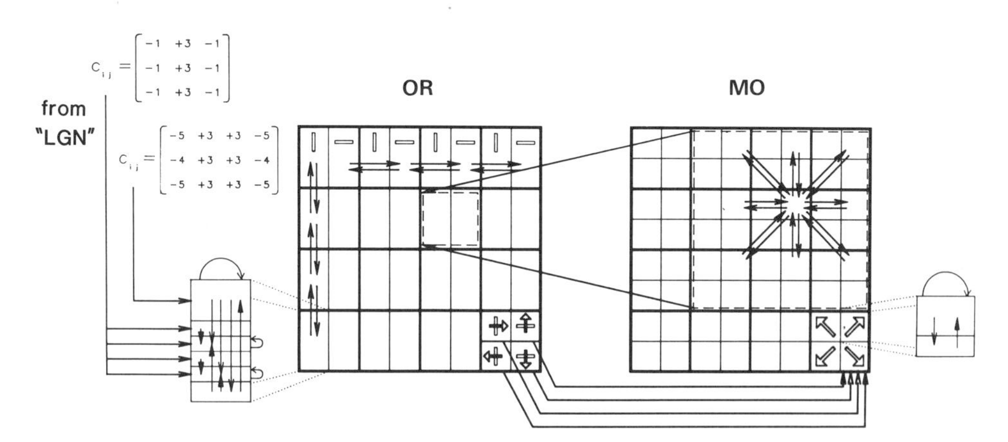

Fig. 2. Schematic diagram of neuronal connectivities in the cortical oscillation model. Boxes indicate group boundaries; small rectangles and heavy arrows within groups, preferred stimulus orientation or direction; arrows with filled heads, sets of excitatory connections; arrows with open heads, sets of inhibitory connections. Insets show connections between cell types within a group, matrices indicate typical connection strengths of afferents from "LGN." See text for detailed explanation. Figs 2-5 from ref. 50, reproduced with permission.

All units are updated synchronously and without time delay. After maximal activation, the individual units enter a refractory period during which firing is not possible. This prevents prolonged bursting of individual units. The refractory period is much shorter than the oscillatory period. Its elimination does not usually abolish the oscillatory activity of the network, which depends almost entirely on the local interactions of excitatory and inhibitory cell populations.

A simulated visual stimulus in the input array (model retina) activates a model "LGN" network which in turn projects topographically to OR. The preferred stimulus of an orientation-selective unit is a stationary or moving oriented bar. The orientation selectivity is generated based on a proposal by Hubel and Wiesel,[64] but the essential findings of this study do not depend on the details of this mechanism. The orientation-selective units in OR receive inputs from randomly chosen but overlapping positions in the model LGN. Units within a group are coupled by sparse local excitatory and inhibitory connections in a pattern that is based on observed anatomical and functional interactions among cortical neurons.[65-67] Each excitatory neuron is connected to 5 percent of its excitatory and 10 percent of its inhibitory neighbors.

Adjacent groups with the same orientation specificity are reciprocally interconnected based on anatomical and physiological evidence for the existence of such connections in primary visual cortex.[68-71] In the model, each excitatory neuron receives 32 weak connections from excitatory neurons in adjacent groups. There are no connections between groups of the same orientation specificity that have nonoverlapping receptive fields, consistent with the limited spatial extent of short-range cortico-cortical connections. Similarly, there are no connections between groups at the same position but with different orientational specificity.

Units selective for the direction of movement of an oriented bar also receive anatomically ordered inputs from the LGN. In every group of direction-selective units there is a subpopulation of inhibitory units that stay active for some time after activation. These units provide time-delayed inputs to the direction-selective units, which fire only in response to a concurrence of movement in their preferred direction and appropriate input from the orientation-selective layer. This mechanism for the generation of directional selectivity is related to other such mechanisms involving time-delayed anatomically ordered inputs.[72,73]

The direction-selective units in OR project to MO (modeled roughly on area V5 or MT in the primate visual cortex), which contains groups of units that are selective for pattern motion.[74,75] In the model, pattern-motion selectivity is generated by combining two excitatory inputs from different directions of motion and two inhibitory inputs from the opposite directions, as in other models.[49,75]

The local connectivity and physiological properties of the MO units are similar to those in OR. The excitatory neurons receive connections from other excitatory and inhibitory neurons within the same group. In addition, these neurons receive both excitatory and inhibitory connections from orientation-selective neurons in OR of appropriate selectivity. These connections are rather strong and are thresholded, such that simultaneous inputs from more than one orientation are required to fire the postsynaptic cell. Inhibitory connections can veto all excitatory inputs. Neurons in MO also receive excitatory connections from neurons in neighboring groups of the same specificity. The pattern-motion-selective groups signal back to the orientation-selective groups in OR via long-range excitatory connections. These signals are usually below firing threshold and are thus more "modulatory" in character. For simplicity and to reduce computational requirements, the model is not sensitive to line orientations other than horizontal and vertical or pattern motion other than in the four oblique directions (northeast, southeast, southwest, and northwest). None of our conclusions is dependent on this simplification.

Simulated neuronal activity is evaluated by computing auto- and cross-correlation functions of both single units and entire neuronal groups during stationary firing periods. Correlation functions are computed following published procedures.[51,76-78] Continuous activity levels, $s_i(t)$, for single units were subjected to a threshold condition (if $s_i > 0.875$ then s_i set to 1, otherwise s_i set to 0) to obtain discrete spikes. To evaluate stimulus-induced contributions to the correlograms, the stimulus-related correlogram or shift predictor was calculated.[76]

As an analogue for the local field potential, an average activity function (AAF) is computed as the average of s_i over all units within a group that are active ($s_i = 1$) at a given time. Notice that distance effects are not taken into account; all cells within a group are effectively equidistant and cells in other groups are not sampled. Auto- and cross-correlation functions for the AAF are computed similarly to those for discrete single spike data. In addition, these functions are scaled and confidence criteria are computed. The AAFs for each signal are given as pseudo-continuous functions.

Dynamics of Neuronal Groups

We describe first the behavior of the orientation-selective units of OR in response to stimulation with a moving oriented bar. Presentation of the stimulus results in non-oscillatory activity of ON-center units in the LGN network and in oscillatory activity in neuronal groups in OR. Oscillations are generated by the relatively sparse local excitatory and inhibitory connections and occur independently of inter group coupling. Oscillatory behavior of this kind is a population phenomenon and does not necessarily require inherent oscillatory properties of individual neurons.[79] Upon visual inspection of the thresholded spike trains, most units seem to discharge fairly irregularly (Fig. 3A, top), although subthreshold oscillations of unit activity are more robust. The simultaneously recorded AAF of a neuronal group exhibits clear rhythmicity (Fig. 3A, bottom). A similar statistical relationship between single unit and population discharges was found in the hippocampal slice.[80] This relationship is highly characteristic for the behavior of local populations of interconnected and cooperating neurons.

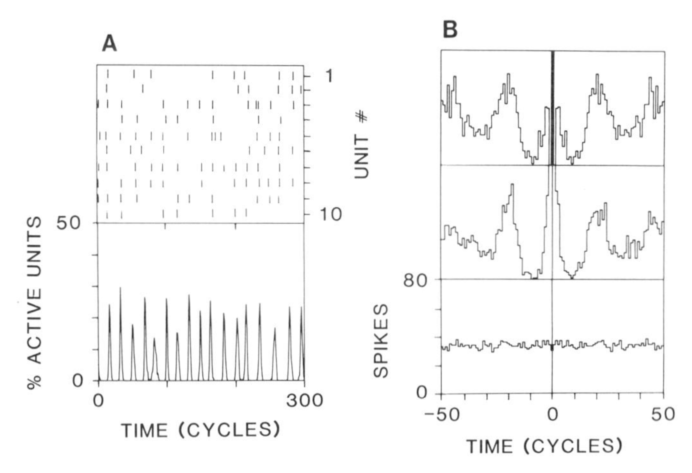

Fig. 3. (A, top) Simultaneous activity of ten different orientation-selective units within the same group in OR. (A, bottom) Average activity function (AAF) of these same ten units responding to a simulated light bar in preferred orientation moving through the unit's receptive field. (B, top) Autocorrelation, (B, middle) cross-correlation, and (B, bottom) shifted autocorrelation ("shift predictor") for single spike data of simulated neurons within a single neuronal group during stimulation with an optimally oriented light bar (responses accumulated for 10 trials).

Auto-correlation functions for single units show statistically significant oscillations (Fig. 3B, top) at levels comparable to those experimentally observed.[46] The coherency of oscillations within a group of neurons is indicated by the oscillatory cross-correlation function (Fig. 3B, middle). That oscillations are not due to some hidden periodicity in the stimulation is indicated by the flat shifted autocorrelation function (Fig. 3B, bottom).

The oscillation frequency of an isolated neuronal group depends on a variety of factors, including the intensity of the stimulus, the strength or density of intra-group connections, the decay rate of inhibitory potentials, the length of the refractory period, etc. It is a key property of the model that the parameters influencing the amplitude and frequency of neuronal oscillations vary significantly from one oscillator to the next. Such variability prevents accidental phase locking of uncorrelated groups and results in significant and largely unpredictable shifts in both the peak frequency and amplitude of the responses from trial to trial, as well as for short data segments within a single trial. This variability is also a characteristic feature of the oscillating cell populations observed in the visual cortex.[81]

Intra- and Inter-Areal Linking of Collinear Stimuli

Interactions via reentrant excitatory connections between adjacent groups of the same specificity in OR give rise to coherent oscillations with a near zero phase lag (Fig. 4). Such phase coherency can extend over distances that exceed the extent of intra-areal reentrant connectivity.[82] In the model as well as in the experiments,[52,60] correlations are found between units in groups that have nonoverlapping receptive fields if a long, continuous moving bar is presented. In the model, these distant correlations disappear if two collinear short bars are moved separately but with the same velocity; experimentally there is some weak coherency if the gap between the two bars is not too wide. In the model, only groups that have overlapping receptive fields are linked directly. If connections are added between more distant groups, some degree of coherency between groups responding to each of the collinear bars is observed. The model is consistent with the possibility that correlations might be generated via multiple and even indirect paths. In general, correlations can appear between distant groups that are linked through other intermediate groups or (as is illustrated in the following simulation) through groups in other areas (see also the "comparator model").[83]

The formation of correlations between distinct areas demonstrates the constructive properties of reentry, which include the ability to create coherency between groups of different specificity within the same area as well as between different areas. This was studied in simulations of OR and MO responding to a spatially extended moving contour (Fig. 5). The stimulus contour used in this run was shaped like a right-angle corner and thus contained line segments of orthogonal orientation. Since the model architecture does not include reentrant connectivity linking orthogonal orientations, coherency among groups of different stimulus specificity cannot develop by local interactions alone.

When this stimulus is presented, both orientation- and motion-selective groups oscillate, roughly within the same frequency range. Oscillations within MO are driven by direction-selective units in OR, whereas oscillations in OR do not depend upon those in MO. However, in the presence of inter-areal reentrant connections, coherency develops between groups responding to differently oriented parts of the contour moving together as one pattern. If the reentrant connections from MO to OR are cut (leaving connections from OR to MO intact), coherency within OR exists only between groups of the same orientation preference. Thus, the dynamic linkage of responses in OR to segments of different orientation depends on reentrant signals from another segregated visual area, MO.

The model, when functioning as shown in Fig. 5, contains reentrant connectivity both within and between areas. The interplay of intra- and inter-areal reentry links together the responses to the contour of the stimulus with those to its coherent movement, thus establishing a relationship between orientation and movement feature domains. Without reentry, the orientation and movement properties of the stimulus would remain segregated. An active reentrant circuit directly establishes a conjunction, without the need for additional hierarchical levels and specialized coincidence detectors. This mode of linking responses in different maps becomes even more advantageous as more complex situations involving collections of maps are encountered.

As an example of a possible function for the reentrant linking of perceptual maps, the establishment of coherent oscillations between different areas might provide a solution to at least some instances of the so-called "binding problem."[84-87] This problem arises whenever an entity is represented in two or more different areas, each of which is devoted to particular stimulus dimensions, and the representations must somehow be "bound" together. In the model, for example, MO is responsive to the direction of motion of the contour, but not to the orientation and precise position of the parts of which it is composed. The converse is true for OR; it is responsive to orientation and precise position of the parts of the contour, but does not respond to the overall motion of the entire stimulus. The full representation of the stimulus as a "moving corner" is only achieved through reentrant interactions and coherency between the two maps.

Comparison with Other Models

It is well known that coupled nonlinear oscillators can have complex dynamics, displaying behaviors ranging from phase-locking and symmetry breaking to chaos.[88] The recent experimental work on cortical neuronal oscillations has prompted a number of mathematical models based on the physical paradigm of coupled oscillators.[83,89-91] We have tried instead to model the oscillation phenomenon in terms of the dynamical behavior of neuronal *populations* with explicitly defined structural and physiological properties. Simulation studies of this kind have been carried out before for several oscillating systems of the brain[92-95] and are particularly useful in connecting theoretical concepts to experiments.

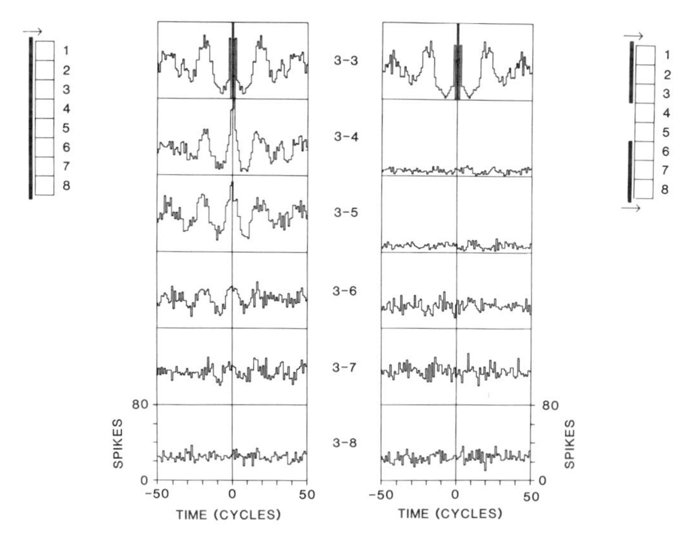

Fig. 4. Auto- and cross-correlations of single unit activity recorded from orientation-selective groups arranged in a 1 by 8 array (see insets at top) when a single long bar (left) or two short aligned bars (right) of appropriate orientation are moved through their receptive fields. Receptive field areas of adjacent groups overlap by about 60%. Correlations between units in groups with overlapping receptive field areas and direct reentrant links (3-4), and nonoverlapping receptive fields only indirectly linked (3-6), are present in the case of a single continuous stimulus (left), but are not present in the discontinuous case (right).

Our results are entirely consistent with the recent mathematical analyses, but also extend their scope in several respects. In particular, we have shown that structural diversity within a population-based oscillator can yield variable dynamical behavior of a kind that is highly characteristic of cortical neuronal groups.[81] Furthermore, the model explores the simultaneous function of reentrant projections at two distinct scales of organization: intra- and inter-areal. It thus resembles more closely the actual situation in the cortex and combines two extreme architectures which have previously only been studied in isolation.[83,96] A structurally rich model of this kind, which is open to progressive refinement and is not constrained by what is easily subject to mathematical analysis, is a useful tool for understanding the function of cortical oscillations.

REENTRY, PERCEPTUAL CATEGORIZATION, AND DARWIN III

According to the TNGS, reentry is the key to perceptual categorization. It allows the correlation of different sensory signals, sampled independently in parallel channels, and representing disjoint stimulus properties. The building block of such systems is the "classification couple" (Fig. 1, bottom), an arrangement of two distinct reciprocally interconnected neuronal maps.[1] In such a couple, for example, one pathway could map simple features (color, texture, etc.), while the other could carry responses for more global stimulus properties (contour, shape, etc.). No single feature alone is sufficient to define a category in such a system. The correlation of multiple features, however, establishes patterns that can be sampled by higher-order sensory or motor areas. If the reentrant connections between the neuronal maps are modifiable, then some correlations will have a higher probability of occurring than others, depending on experience. Thus, by its inherent associative property, categorization is intimately linked to the perceptual memory of previous encounters with similar stimulus items. The classification couple is a central component of the recognition automaton (Darwin III) which we now describe.

We have constructed several automata that display aspects of recognition processes based on principles of the TNGS. Darwin III is an example of "synthetic neural modelling,"

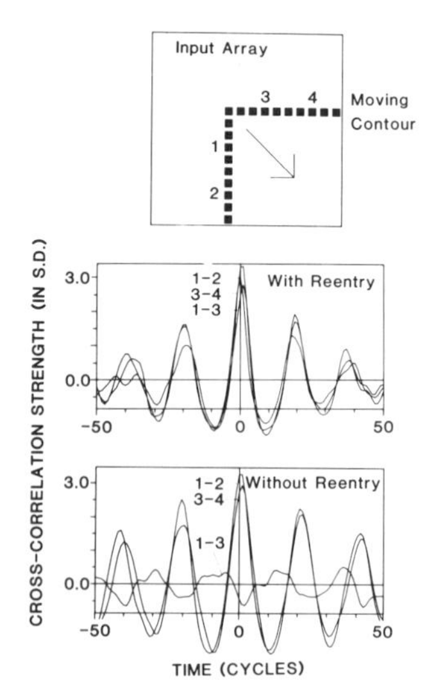

Fig. 5. The effect of long-range reentrant connections between OR and MO in response to a simulated moving contour. Cross-correlation of the AAF of neuronal groups in OR is shown in the presence (middle) and absence (bottom) of reentrant connections from MO to OR. Parts of the moving contour to which neuronal groups 1, 2, 3 and 4, located in OR, respond are indicated at the top. Cross-correlation strength is expressed in standard deviations (S.D.) above the background.

an approach to neural modelling that aims at an understanding of complex neural phenomena by taking into account processes and interactions at all relevant levels, from the synaptic to the behavioral.[62] Darwin III allows us to examine, in a single system of interconnected neuronal networks, sensory processes involving recognition and classification, as well as motor acts, such as visual saccades, reaching, and touch-exploration. In this chapter, we emphasize Darwin III's categorization system and refer the reader to other publications[29,62,97] for more detailed descriptions of the other constituent systems.

Darwin III consists of a simple but realistic nervous system that is embedded in a specific "phenotype" acting in a specific "environment." Its architecture and mode of operation are consistent with the principles of the TNGS. It has, in a crude form, some of the behavioral capabilities of an animal, and acts as an autonomous creature independent of the observer. It does not require *a priori* definitions of categories, codes, or information-processing algorithms.

Darwin III receives input from three sensory modalities: vision, touch, and kinesthesia (joint-sense). It has a movable eye and an arm with multiple joints, each controlled by motor neurons in specified repertoires. The version of Darwin III described here contains over 50,000 neurons and over 620,000 synaptic connections. Every cell in Darwin III has a single-valued activity state determined by a "response function."[62] This function has terms corresponding to neuronal properties such as synaptic inputs, Gaussian noise, decay of previous activity, depression and refractory periods, and long-term potentiation (LTP). The relative magnitudes of these terms can be varied parametrically. Rules for synaptic modification contain features reflecting some of the complex properties of real neurons: in addition to depending on local activity of the pre- and post-synaptic cells (or a short-term temporal average), changes in connection strengths also may be made to depend heterosynaptically on inputs from cells in special repertoires, the activity of which reflects the organism's evaluation of its recent behavior according to a particular "value scheme."

Darwin III consists of four neuronal subsystems (Fig. 6): a saccade and fine-tracking oculomotor system, a reaching system using a single multi-jointed arm, a touch-exploration system using a different set of "muscles" in the same arm, and a reentrant categorizing system.

By selection from a repertoire of eye motions, Darwin III's oculomotor system acquires the ability to move the eye toward objects in its visual field and to track moving objects. Visual signals are mapped topographically to a network, "SC," containing excitatory and inhibitory neurons that loosely represents the superior colliculus. Excitatory cells in SC are connected to oculomotor neurons, the activation of which causes eye motion. The initial strengths of these connections are assigned randomly, so that eye motions are initially uncorrelated to visual stimulation. Of course, improved coordination in the sensorimotor apparatus following training increases the likelihood of object classification by Darwin III's categorization module. This is one way by which categorization in Darwin III is intimately linked to exploratory motor action.

An innate value scheme imposes a global constraint on the selection of eye movements: only those motions are selected which bring visual stimuli closer to the foveal part of the retina. Adaptive value is detected by a repertoire containing cells connected densely to the fovea and more sparsely to the periphery. The responses of these cells provide a hetero-synaptic component which modulates the modification of connections from SC to the motor neurons. Those populations of connections which are active in a short time interval before foveation occurs are selected and strengthened. Thus, selection affects synaptic and cellular populations after they have contributed to a motor act.

The multi-jointed arm of Darwin III develops smooth reaching movements by selection from a prior repertoire of spontaneous gestural motions. The training scheme is essentially the same as the one used for the oculomotor system. The reaching problem is considerably more complex than the problem solved by the oculomotor system because there are multiple degenerate solutions due to the mechanical redundancy of arm motions.[98,99] The nervous system needs to master such situations routinely. One way to do so is to reduce the number of degrees of freedom of the mechanical system. Our modelling studies suggest that the mechanism by which the nervous system achieves this reduction is essentially selective and occurs by amplification of gestural motions with adaptive couplings of relevant movement parameters to form motor synergies. This process is related to pattern recognition, in the sense that classification of sensory patterns involves the mapping of variant stimuli (many degrees of freedom) to invariant responses (fewer degrees of freedom).

The arm of Darwin III, once it has established contact with the surface of an object, starts moving along the edges of objects guided by tactile signals. To facilitate the tracing of objects during such exploration it assumes a straightened posture (canonical position) by a built-in response analogous to a reflex. Tracing object contours provides Darwin III with sensory signals concerning the surface (or outline) properties of objects and these signals are used as a second sensory modality in subsequent object categorization by a classification couple. As with real animals, the sensory input reaching the categorization system depends on exploratory motor actions.

The classification couple in Darwin III is linked to a simple reflex output. Only two categories are distinguished by the version of this system presented here: "rough-striped" objects (bad for this species) and all others (good). The result of a "rough-striped" categorization is an arm motion (a "swat") aimed at the stimulus. The usual result is that the stimulus is swept from the immediate vicinity of the automaton.

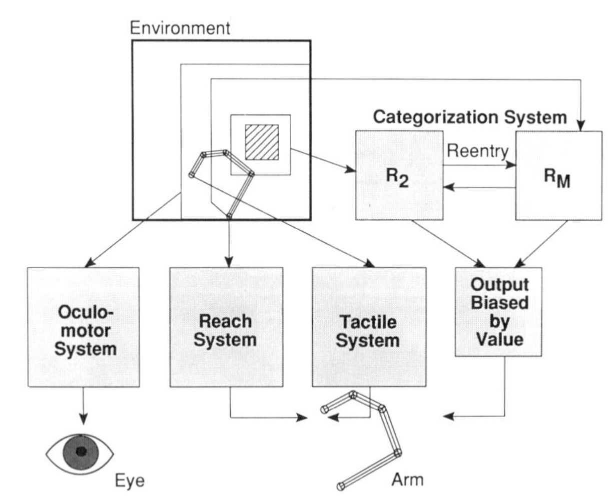

Fig. 6. Schematic diagram of Darwin III and its subsystems. The "environment" (heavy square at upper left) contains moving objects, one of which is indicated by the cross-hatched square. The two squares surrounding the hatched object represent the limits of peripheral and foveal vision. Movements of the eye and of the four-jointed arm are controlled by the oculomotor and reach systems, respectively. The distal digit of the arm contains touch sensors used by the tactile system to trace the edges of objects. The categorization system receives sensory inputs from the central part of the visual field and from joint receptors signalling arm movements. These inputs connect, via intermediate layers (not shown), to a higher-order visual center, R_2, and to an area correlating motion signals over time, R_M. R_2 and R_M are reciprocally connected to form a classification couple according to the scheme shown in Fig. 1. Figs. 6 and 8 from ref. 62, reproduced with permission.

The visual component of the classification couple consists of three repertoires. Neuronal units in a repertoire analogous to the primary visual cortex (R) respond to oriented line segments by virtue of afferent connections from LGN ON- and OFF-center units. A higher-level repertoire (R_2) contains cells that are connected to groups in the primary repertoire responding to the same or different orientations. Its responses signal the presence of stimuli of particular visual classes, for example, those with stripes or other visual textures.

The touch component of the classification couple begins with a repertoire of groups responsive to two broad classes of stimulus shapes--smooth and rough. The inputs to these groups come from kinesthetic (joint-sense) receptors in the touch-exploration motor system. Smooth-sensitive cells combine kinesthetic inputs corresponding to some angle of tracing with other excitatory inputs corresponding to the same angle but with some time delay. Other trace directions provide inhibitory inputs. These cells thus respond most strongly when tracing continues in a single direction and are inhibited when the direction of trace changes. Conversely, rough-sensitive cells combine excitatory time-delayed inputs from cells with different orientation from the prompt input, and inhibitory inputs from cells with the same orientation. These cells thus respond most strongly when the direction of tracing changes, and are inhibited when it remains the same.

The basis for associations between the visual and kinesthetic systems is a large number of reentrant connections in both directions between higher-order visual cells and rough or smooth touch cells. The firing thresholds of these connections are lowered by continued correlated stimulation from an object with an appropriate combination of visual and tactile characteristics, in a process similar to LTP. A "triggering" repertoire gives a "completion" response when visual inspection and tracing of a stimulus object reveals no more novel features.

The activity in the triggering repertoire serves as an indicator that the examination of a stimulus is completed. Coupled back to the sensory repertoires, it reexcites only those cells that were previously stimulated enough both to sensitize them by LTP and to strengthen their reentrant connections from their opposite numbers in the other modality. The category-specific combined output enters an ethological-value association area that in turn couples it to a motor center that generates the swat reflex.

As a result of all these connections, stimuli are subject to both visual and tactile examination, which is terminated as soon as novel features cease to be found. When this happens, a pattern is evoked in the reentrant categorization area that is characteristic of (but not unique for) the particular class to which the object belongs. This pattern, if it is one that is recognized by the ethological value system, evokes the swat reflex to remove the stimulus. Stimuli not recognized as noxious are left undisturbed.

The behavior and internal states of Darwin III can be observed in a variety of ways. Darwin III's environment, the configuration of its motor organs, and its complete neural state are all subject to simultaneous observation and recording. For example, Fig. 7 shows selected frames from a movie we have made that shows only the phenotype of Darwin III as it senses, explores, and categorizes an object. From data of this kind, an observer could score the behavior of the automaton as it responds to a variety of stimuli at various positions on the input array without any knowledge of its neural states, thus pursuing a kind of "machine behaviorism." On the other hand, the movements of the automaton and the changing events in the environment could also be related to the state of the automaton's networks and the activity of its neuronal units, corresponding to a kind of "machine physiology."

An example of a classification experiment with Darwin III that followed the paradigm of "machine behaviorism" is shown in Fig. 8. This figure summarizes the behavioral responses following eight presentations of each of 55 similar-sized objects of various shapes and textures. The objects have been grouped together according to the frequency with which a rejection response occurred. On average, objects that were striped and had a rough surface were rejected more frequently than other objects; these objects thus form a behaviorally defined probabilistic category. Responses and distribution of stimuli are unique to each individual version of the automaton; the partitioning of the stimuli is different in detail if the random number seeds used to initially generate the automaton's connections or the precise sequence of presentation of the stimuli are changed.

Darwin III categorizes objects differently from most other pattern recognition machines. It does not match an input pattern to a template or prototype, represent it as a vector in a multi-dimensional feature space, or compare its properties to feature lists characteristic of each exemplar. Object features are not explicitly isolated and weighed independently. Instead, they remain intimately linked although visual and tactile features enter separate channels. Categorization of an object (its recognition as belonging to one class or another) involves a complex behavioral act with an intricate temporal structure. This process is inherently variable and depends on the configuration of the creature, its history, and its neuronal state at any given time.

What indication do we have that Darwin III categorizes objects in a nonclassical way? Nonclassical categories depend on context and the state of the animal. In a series of experiments, one would expect to obtain a probabilistic gradient of a given behavioral response across a space of objects (a stimulus continuum). In fact, some of the objects in Fig. 8 appear to be regarded as "more typical" than others, and from looking at the distribution of objects it is hard to guess exactly what features have been selected by Darwin III to categorize these objects. In accordance with results on nonclassical categories discussed earlier, membership in the category of "striped and bumpy" objects is not all-or-none, but instead is a matter of degree. Thus, the results obtained with Darwin III provide an example of how "open-ended" categories are formed and show that classical models of categorization based on the analysis of separable features are not the only practical solution to the categorization problem.

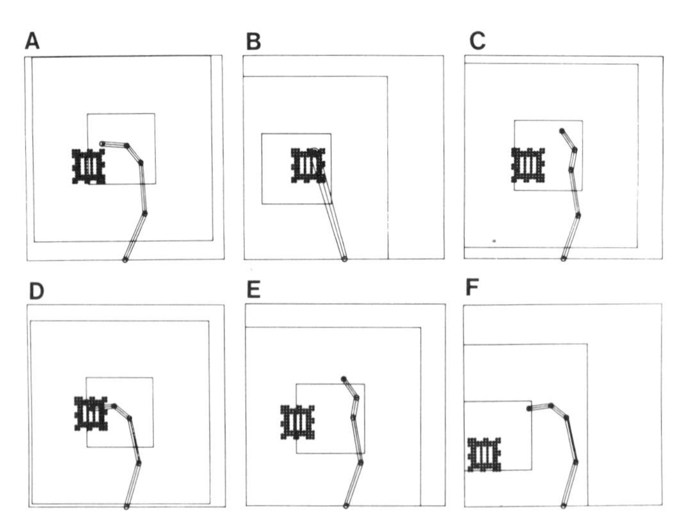

Fig. 7. Sequence of diagrams taken from a movie of Darwin III, showing a categorization and rejection reflex sequence. (A) Initial situation. (B) Eye fixates on target object and arm assumes straightened posture. (C) The automaton arrives at a categorization and a rejection response is triggered. (D-F) The arm swats the noxious object away. Eye follows motion of object in (F). From ref. 100, reproduced with permission.

CONCLUSIONS

The theory of neuronal group selection provides a consistent approach to understanding the operation of the nervous system that is applicable to both sensory and motor systems. The principle of reentry that is a key component of this theory suggests how responses in segregated sensory areas may be correlated to generate unitary sensory percepts. Recent experimental descriptions of 40-Hz cortical oscillations provide strong support for the existence of neuronal groups and reentry. A model based on the TNGS reproduces many of these experimental observations. Application of similar synthetic neural modelling principles to the construction of categorization automata has yielded a system, Darwin III, that is capable of simple perceptual categorization coupled with appropriate motor responses. These results may open up new approaches to the design of robots. Such systems could acquire knowledge about their surrounding environments without being supervised or taught by their constructors. They could also be coupled to standard von Neumann computers to provide these computers with an interface to the world. Selective recognition systems could thus enhance the capabilities of present day computing machinery and bring about significant advances necessary for the ultimate construction of perception machines.

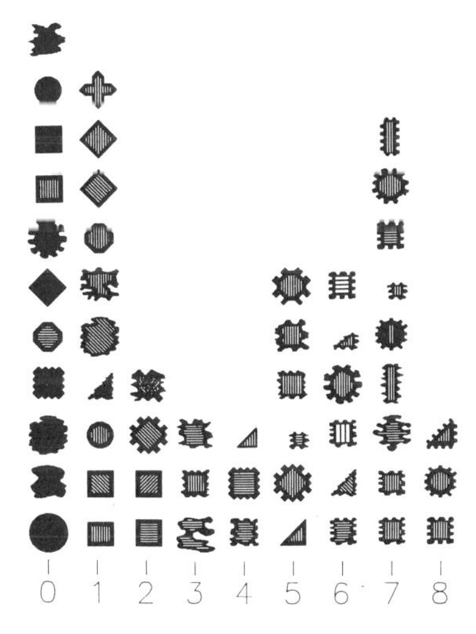

Fig. 8. A set of objects grouped according to the responses of Darwin III's categorization system. Each object was presented eight times. Activation of the rejection response at any time within 50 cycles following presentation was counted as a "rejection" response. If no response occurred within the 50 cycle time limit, the trial was ended and a new object was presented. Objects are arranged in nine columns according to the frequency with which they met with a rejection response. In this version of Darwin III, an intrinsic negative ethological value is attached to "bumpy and striped" objects.

ACKNOWLEDGMENTS

We thank G.M. Edelman for his collaboration and for useful advice on the manuscript. This research has been supported by the Neurosciences Research Foundation, the Office of Naval Research, the John D. and Catherine T. MacArthur Foundation, the Lucille P. Markey Charitable Trust, The Pew Charitable Trusts, the van Ameringen Foundation, and the Charles and Mildred Schnurmacher Foundation. Some of this research was carried out using facilities of the Cornell National Supercomputer Facility, a resource of the Center for Theory and Simulation in Science and Engineering at Cornell University, which is funded in part by the National Science Foundation, New York State, and the IBM Corporation and members of the Corporate Research Institute.

REFERENCES

1. G. M. Edelman, "Neural Darwinism: The Theory of Neuronal Group Selection," Basic Books, New York (1987).
2. G. M. Edelman, "The Remembered Present: A Biological Theory of Consciousness," Basic Books, New York (1989).
3. E. E. Smith, and D. L. Medin, "Categories and Concepts," Harvard University, Cambridge, Mass. (1981).
4. L. Wittgenstein, "Philosophical Investigations," Macmillan, New York (1953).
5. G. Lakoff, "Women, Fire, and Dangerous Things: What Categories Reveal About the Mind," University of Chicago, Chicago (1987).
6. E. Rosch, and C. B. Mervis, *Cogn. Psychol.* 7:573 (1975).
7. C. B. Mervis, and E. Rosch, *Ann. Rev. Psychol.* 32:89 (1981).
8. E. Rosch, C. Simpson, and R. S. Miller, *J. Exp. Psychol. Hum. Percept. Perf.* 2:491 (1976).
9. B. C. Malt, and E. E. Smith, *Mem. Cogn.* 10:69 (1982).
10. S. L. Armstrong, L. R. Gleitman, and H. Gleitman, *Cognition* 13:263 (1983).
11. W. S. McCulloch, and W. Pitts, *Bull. Math. Biophys.* 5:115 (1943).
12. W. H. Pitts, and W. S. McCulloch, *Bull. Math. Biophys.* 9:127 (1947).
13. F. Rosenblatt, *Psychol. Rev.* 65:386 (1958).
14. F. Rosenblatt, "Principles of Neurodynamics: Perceptrons and the Theory of Brain Mechanisms," Spartan Books, Washington, D.C. (1962).
15. M. Minsky, and S. Papert, "Perceptrons: An Introduction to Computational Geometry," MIT Press, Cambridge, Mass. (1969).
16. G. Widrow, and M. E. Hoff, *IRE Western Electronic Show and Convention, Convention Record, Part 4* 1960:96 (1960).
17. M. Bongard, "Pattern Recognition," Spartan, Washington (1970).
18. J. W. Gyr, J. S. Brown, R. Willey, and A. Zivian, *Psych. Bull.* 65:174 (1966).
19. J. J. Gibson, "The Senses Considered as Perceptual Systems," Houghton Mifflin, Boston (1966).
20. D. E. Rumelhart, J. L. McClelland, and The PDP Research Group, "Parallel Distributed Processing: Explorations in the Microstructure of Cognition. Volume 1: Foundations," MIT Press, Cambridge, Mass. (1986).
21. D. E. Rumelhart, G. E. Hinton, and R. J. Williams, *Nature* 323:533 (1986).
22. T. J. Sejnowski, P. K. Kienker, and G. E. Hinton, *Physica* 22D:260 (1986).
23. T. J. Sejnowski, and C. R. Rosenberg, *Complex Syst.* 1:145 (1987).
24. N. Qian, and T. J. Sejnowski, *J. Mol. Biol.* 202:865 (1988).
25. L. H. Holley, and M. Karplus, *Proc. Natl. Acad. Sci. USA* 86:152 (1989).
26. D. Zipser, and R. A. Andersen, *Nature* 331:679 (1988).
27. G. N. Reeke, Jr., and G. M. Edelman, *Daedalus, Proc. Am. Acad. Arts and Sciences* 117:143 (1988).
28. F. Crick, *Nature* 337:129 (1989).
29. G. N. Reeke, O. Sporns, and G. M. Edelman, *in*: "Connectionism in Perspective," R. Pfeifer, Z. Schreter, F. Fogelman-Soulie, and L. Steels, eds., Elsevier, Amsterdam (1989).
30. J. J. Hopfield, *Proc. Natl. Acad. Sci. USA* 79:2554 (1982).
31. J. S. Denker, *Physica* 22D:216 (1986).
32. J. Buhmann, and K. Schulten, *Biol. Cybern.* 54:319 (1986).
33. W. Kinzel, *Z. Phys. B* 60:205 (1985).
34. Y. S. Abu-Mostafa, and J.-M. St. Jacques, *IEEE Trans. Inf. Theory* IT-31:461 (1985).
35. R. J. McEliece, E. C. Posner, E. R. Rodemich, and S. S. Venkatesh, *IEEE Trans. Inf. Theory* IT-33:461 (1987).
36. D. W. Tank, and J. J. Hopfield, *Sci. Am.* 257(12):104 (1987).
37. G. A. Carpenter, and S. Grossberg, *Comput. Vision Graphics Image Process.* 37:54 (1987).
38. R. J. Herrnstein, and D. H. Loveland, *Science* 146:549 (1964).

39. R. J. Herrnstein, D. H. Loveland, and C. Cable, *J. Exp. Psychol. Anim. Behav. Proc.* 2:285 (1976).
40. A. M. Schrier, and P. M. Brady, *J. Exp. Psychol. Anim. Behav. Proc.* 13:136 (1987).
41. R. A. Gardner, and B. T. Gardner, *J. Comp. Psychol.* 98:381 (1984).
42. G. M. Edelman, *in*: "The Mindful Brain: Cortical Organization and the Group-Selective Theory of Higher Brain Function," G. M. Edelman, and V. B. Mountcastle, eds., MIT Press, Cambridge, Mass. (1978).
43. L. Darden, and J. A. Cain, *Philos. Sci.* 56:106 (1989).
44. R. E. Michod, *Evolution* 43:694 (1989).
45. W. Singer, *in*: "The Neural and Molecular Basis of Learning," J.-P. Changeux, and M. Konishi, eds., Wiley, Chichester (1987).
46. C. M. Gray, and W. Singer, *Proc. Natl. Acad. Sci. USA* 86:1698 (1989).
47. M. M. Merzenich, G. Recanzone, W. M. Jenkins, T. T. Allard, and R. J. Nudo, *in*: "Neurobiology of Neocortex," P. Rakic, and W. Singer, eds., Wiley, Chichester (1988).
48. N. Wiener, "Cybernetics," MIT Press, Cambridge, Mass. (1948).
49. L. H. Finkel, and G. M. Edelman, *J. Neurosci.* 9:3188 (1989).
50. O. Sporns, J. A. Gally, G. N. Reeke, Jr., and G. M. Edelman, *Proc. Natl. Acad. Sci. USA* 86:7265 (1989).
51. O. Sporns, G. Tononi, and G. M. Edelman, *in*: "Nonlinear Dynamics of Neural Networks," H. G. Schuster, ed., VCH, Weinheim (in press).
52. C. M. Gray, P. König, A. K. Engel, and W. Singer, *Nature* 338:334 (1989).
53. R. Eckhorn, R. Bauer, W. Jordan, M. Brosch, W. Kruse, M. Munk, and H. J. Reitboeck, *Biol. Cybern.* 60.121 (1988).
54. S. Zeki, *Brain Res.* 14:271 (1969).
55. D. C. Van Essen, *in*: "Cerebral Cortex, Vol. 3, Visual Cortex," A. Peters, and E. G. Jones, eds., Plenum, New York , Vol. 3 (1985).
56. S. Zeki, and S. Shipp, *Nature* 335:311 (1988).
57. E. G. Jones, and T. P. S. Powell, *Brain* 93:793 (1970).
58. L. L. Symonds, and A. C. Rosenquist, *J. Comp. Neurol.* 229:39 (1984).
59. R. Llinàs, and A. A. Grace, *Soc. Neurosci. Abstr.* 15:660 (1989).
60. A. K. Engel, P. König, C. M. Gray, and W. Singer, *Eur. J. Neurosci.* (in press).
61. P. König, C. M. Gray, A. K. Engel, and W. Singer, *Soc. Neurosci. Abstr.* 15:798 (1989).
62. G. N. Reeke, Jr., L. H. Finkel, O. Sporns, and G. M. Edelman, *in*: "Signal and Sense: Local and Global Order in Perceptual Maps," G. M. Edelman, W. E. Gall, and W. M. Cowan, eds., Wiley, New York (in press).
63. G. N. Reeke, and G. M. Edelman, *Int. J. Supercomputer. Appl.* 1:44 (1987).
64. D. H. Hubel, and T. N. Wiesel, *J. Physiol. (London)* 160:106 (1962).
65. K. Toyama, M. Kimura, and K. Tanaka, *J. Neurophysiol.* 46:191 (1981).
66. K. Toyama, M. Kimura, and K. Tanaka, *J. Neurophysiol.* 46:202 (1981).
67. A. Michalski, G. L. Gerstein, J. Czarkowska, and R. Tarnecki, *Exp. Brain Res.* 51:97 (1983).
68. K. S. Rockland, and J. S. Lund, *Science* 215:1532 (1982).
69. C. D. Gilbert, and T. N. Wiesel, *J. Neurosci.* 3:1116 (1983).
70. D. Y. Ts'o, C. D. Gilbert, and T. N. Wiesel, *J. Neurosci.* 6:1160 (1986).
71. H. J. Luhmann, J. M. Greuel, and W. Singer, *Eur. J. Neurosci.* 2:344 (1990).
72. D. Marr, and S. Ullman, *Proc. R. Soc. Lond.* B211:151 (1981).
73. C. Koch, and T. Poggio, *in*: "Models of the Visual Cortex," D. Rose, and V. G. Dobson, eds., Wiley, Chichester (1985).
74. J. A. Movshon, E. H. Adelson, M. S. Gizzi, and W. T. Newsome, *in*: "Pattern Recognition Mechanisms," (1985).
75. T. D. Albright, *J. Neurophysiol.* 52:1106 (1984).
76. D. H. Perkel, G. L. Gerstein, and G. P. Moore, *Biophys. J.* 7:391 (1967).
77. W. J. Melssen, and W. J. M. Epping, *Biol. Cybern.* 57:403 (1987).
78. G. F. Poggio, and L. J. Viernstein, *J. Neurophysiol.* 27:517 (1964).
79. R. R. Llinàs, *Science* 242:1654 (1988).

80. R. D. Traub, R. Miles, and R. K. S. Wong, *Science* 243:1319 (1989).
81. C. M. Gray, A. K. Engel, P. König, and W. Singer, *Eur. J. Neurosci.* (in press).
82. D. M. Kammen, E. Niebuhr, and C. Koch, *Soc. Neurosci. Abstr.* (1990).
83. D. M. Kammen, P. J. Holmes, and C. Koch, *in*: "Models of Brain Function," R. M. J. Cotterill, ed., Cambridge University, Cambridge UK (1989).
84. A. Treisman, and G. Gelade, *Cogn. Neuropsychol.* 12:97 (1980).
85. F. Crick, *Proc. Natl. Acad. Sci. USA* 81:4586 (1984).
86. T. J. Sejnowski, *in*: "Parallel Distributed Processing II. Applications," J. L. McClelland, and D. E. Rumelhart, eds., MIT Press, Cambridge, Mass. , Vol. 2:372 (1986).
87. A. R. Damasio, *Neural Comp.* 1:123 (1989).
88. O. Sporns, S. Roth, and F. F. Seelig, *Physica D* 26:215 (1987).
89. T. B. Schillen, and P. König, *in*: "Parallel Processing in Neural Systems and Computers," R. Eckmiller, G. Hartmann, and G. Hauske, eds., Elsevier, Amsterdam (1990).
90. H. Sompolinsky, D. Golomb, and D. Kleinfeld, *Proc. Natl. Acad. Sci. USA* (in press).
91. D. Wang, J. Buhmann, and C. von der Malsburg, *Neural Comp.* 2:94 (1990).
92. B. G. Farley, *in*: "Self-Organizing Systems 1962," M. C. Yovits, G. T. Jacobi, and G. D. Goldstein, eds., Spartan Books, Washington (1962).
93. P. Andersen, M. Gillow, and T. Rudjord, *J. Physiol. (London)* 185:418 (1966).
94. F. H. Lopes da Silva, A. Hoeks, H. Smits, and L. H. Zetterberg, *Kybernetik* 15:27 (1974).
95. R. J. MacGregor, and R. J. Palasek, *Kybernetik* 16:79 (1974).
96. D. M. Kammen, P. J. Holmes, and C. Koch, *Proc. Natl. Acad. Sci. USA* (in press).
97. G. N. Reeke, Jr., and O. Sporns, *Physica D* (in press).
98. N. A. Bernstein, "The Coordination and Regulation of Movements," Pergamon, Oxford (1967).
99. H. T. A. Whiting, ed., "Human Motor Actions: Bernstein Reassessed," North-Holland, Amsterdam (1984).
100. G. M. Edelman, and G. N. Reeke, Jr., *in*: "Parallel Computers, Neural Networks, and Intelligent Systems," J. A. Robinson, and M. Arbib, eds., MIT Press, Cambridge, Mass. (in press).

ASYMPTOTIC BEHAVIOR OF NEURAL NETWORKS AND IMAGE PROCESSING

Frédéric BERTHOMMIER, Olivier FRANCOIS, Thierry HERVE, Tomeu COLL*, Isabelle MARQUE, Philippe CINQUIN and Jacques DEMONGEOT

TIMB-TIM3-IMAG
Faculty of Medicine
University J. Fourier of Grenoble
38 700 La Tronche France

* Department of Mathematics & Informatics
University "de les Illes Balears"
Palma de Majorca Spain

INTRODUCTION

We give in this paper the definition of a formal network and after, some information about the use of its asymptotic properties for segmenting 3D images reconstructed from parallel cross sections (such as those from Computed Tomography or Magnetic Resonance Imaging). The huge size of data makes algorithmic complexity and storage requirements the key points of 3D edge detection. The classical approach consists in computing the gradient by applying an operator which enhances the grey gradient. Most of all these operators are 3D generalization of 2D edge detectors : Roberts[1], Hueckel[2], Prewitt [3], Canny[4,5,6], Marr and Hildreth[7,8] operators. A critical problem of many of these detectors concerns the size of the convolution masks used to implement the operator : small kernel are noise sensitive, but large ones need prohibitive computing times. A solution is to realize an optimal filter with recursive filters [5,6].

In Section 2., we give the definition of a formal neural network, after we propose some examples of possible inputs in this network and finally we study the asymptotic behavior in the case of a partially or massively parallel updating of the network. In Section 3., we propose a new parallelized method of obtaining an enhancement of the gradient at the boundary of "plateau" (in grey level) objects, based on a neural network procedure[16,17] and we achieve in Section 4. the 2D segmentation by searching limit cycles of a gradient-hamiltonian differential system. Smoothing the so obtained skeleton of the object of interest by using a spline surface can then perform the 3D contouring. In Section 5., we propose a methodology promitted to success consisting to use a differential diffusion or reaction-diffusion operator in order to enhance the gradient at the boundary of objects of interest. Finally, we relate this last method to the first proposed one, corresponding to a parallel neural network approach and we show that other differential operators like chemotaxis-reaction-diffusion ones can be used in image processing.

Self-Organization, Emerging Properties, and Learning
Edited by A. Babloyantz, Plenum Press, New York, 1991

ASYMPTOTIC PROPERTIES OF A NEURAL NETWORK

Definition of a formal neural network

Let us recall rapidly the definition of a formal neural network [24 - 31] :

a) deterministic case
If $x_i(t)$ denotes the state of the neuron i of a set B of neurons at time t (equal to 1 if the neuron fires at this time and to 0 if not), the discrete iterative system ruling the change of states is given by the following equations :

$$x_i(t+1) = 1, \quad \text{if } H_i(t) = \sum_{j \in V(i)} w_{ij}\, x_j(t) > 0,$$
$$= 0, \quad \text{if not,}$$

where V(i) is a neighbourhood of i in B and $H_i(t)$ plays the role of the somatic electric potential. This rule is equivalent to the following one :
$x_i(t+1) = 1$, if $\exp(H_i(t))/(1+\exp(H_i(t))) > 1/2$; $= 0$, if not.

b) stochastic case
$x_i(t+1) = 1$, with the probability $\exp(H_i(t))/(1+\exp(H_i(t)))$.

We can consider also a formal neuron as a cellular automaton[24] valued in {0,1} and a formal neural network as a graph (S,G), whose the set S of vertices is made by the formal neurons and where G is a neighbourhood system associated to the connection matrix $W=(w_{ij})_{i,j\in S}$, $w_{ij}\in \mathbf{R}$, $\forall i,j\in S$. Let us denote by $\Omega=\{0,1\}^S$ the set of all possible configurations of the neural network : we can identify Ω and P(S), the set of all subsets of S, provided with the order relation inclusion denoted by "$\leq$": $(x\leq y)\Leftrightarrow(y\supset x)\Leftrightarrow(x_i\leq y_i, \forall i\in S)$. In the stochastic case, each element i of S is submitted to the following updating (or transition) rule :

$$\forall \omega \in \Omega,\ H_i(\omega) = \langle W\omega, \{i\}\rangle,$$

where $\langle\ \rangle$ denotes the scalar product and

$$\forall a\in \{0,1\},\ \text{Prob}(\{x_i(t)=a\} \mid \{x(t-1) \text{ on } V(i)\}) = e^{a.H_i(x(t-1))}/(1+e^{H_i(x(t-1))}),$$

the choice of the future state for the neuron i being independent on the choice for the other neurons (cf. Figure 1 for displaying the activity of a neural network).

Input in a neural network

If an input $I_i(t)$ is coming in neuron i at time t, it is merged with the information coming from the neighbourhood V(i) in order to build the somatic potential $H_i(t)$:

$$H_i(t) = \sum_{j\in V(i)} w_{ij}\, x_j(t) + I_i(t)$$

A very simple way of generating such inputs is to choose, for each interval E_i (supposed to be independent on the others) between two inputs 1, the truncated geometric distribution : $\text{Prob}(\{E_i<T_i\})=0$ and $\text{Prob}(\{E_i=k\geq T_i\})=p_i(1-p_i)^{k-T_i}$, where T_i and p_i denote respectively the refractory period and the spike occurrence frequency on the afferent fiber i bringing the electric input to the neuron i[71]. The corresponding truncated geometric processes are independent or correlated between fibers (cf. Figure 2). Figure 3 shows post stimulation histograms both experimental and simulated, proving that a good qualitative fit can be obtained with the real data by playing only with the parameters T_i and p_i. On Figure 2, we see the activity of a formal neural network activated by a non-homogeneous input, the neurons firing being in white and

the intensity of the bars between neurons being proportional to the synaptic weights w_{ij}'s. Such simulation has suggested that an analogy between pixels and neurons could be made allowing the transfer of neural filtering techniques in image processing.

Asymptotic properties of neural networks

The transition rule described above can be applied in two ways :
1) in partially parallel (or asynchronous) mode, S is shared in K iteration blocks F_k, k=1,...K of N neurons (K.N=|S|) such that the transition is synchronous in each block and blocks are treated sequentially as follows :

* let us denote by F_k the k-th iteration block and suppose that :

$\forall i,j \in F_k,\ j \notin V(i) \Leftrightarrow w_{ij}=0,\ \forall i,j \in F_k$

* $\forall k$ such that $0 \leq k \leq K$, $\forall i \in S$, $\forall t \equiv k \pmod{K}$:

if $i \notin F_k$, $x_i(t)=x_i(t-1)$

if $i \in F_k$, $x_i(t)$ is chosen following the local transition rule

2) in massively parallel (or synchronous) mode, the rules are applied simultaneously for all neurons.
Then the asymptotic behavior of the network in the two cases above is ruled as follows :
1) it is well known [24,28] that the Markovian process on Ω associated to the asynchronous mode has a unique invariant measure μ defined by : $\mu(x) = \sum_{z \subset \Omega} \mu(z)W(z,x)$, $\forall x \in \Omega$.

It is the Gibbs measure associated to the quadratic energy $E(x)=\langle x,Wx \rangle$, i.e. μ verifies :

$$\forall x \in \Omega, \quad \mu(x) = e^{\langle Wx,x \rangle}$$

2) if W is symmetrical, then the Markovian process on Ω associated to the synchronous mode has also a unique invariant measure μ' verifying (see[28] for the proof) :

$$\forall x \in \Omega, \quad \mu(x) = \sum_{z \in \Omega} e^{\langle Wx,z \rangle}$$

Remarks :
1) in the asynchronous case, μ is spatially Markovian contrary to the synchronous case, where the measure μ' is in general not spatially Markovian
2) μ and μ' can be very different ; for example, it is easy to verify that :

$$\mathbf{d}_{\text{Kullback}}(\mu,\mu') \longrightarrow +\infty, \text{ when } w_{11} \longrightarrow -\infty$$

3) in both asynchronous and synchronous cases, an estimate of the speed of convergence is given by : $\mathbf{d}_{\text{Kullback}}(\mu_0 W^k,\mu) \leq \beta.k.\mathcal{H}_{KS}$,
where $\mathcal{H}_{KS}$, β, k and μ_0 denote respectively the Kolmogorov-Sinaï entropy, a positive constant, a sufficiently large integer and any initial measure on Ω.

GRADIENT ENHANCEMENT BY A NEURAL NETWORK

Image enhancement procedure

We present now in the following the essential of the method (parallelized in the natural way above) developped on the basis of results proposed in[32 - 36] :

1) reduction of a 512 x 512 NMR image in a 256 x 256 image by averaging each block of 4 neighbour pixels, in order to obtain the input image (see also[63] for a similar treatment)
2) use of this image as the mean configuration of an input geometric random field transformed

by a 256 x 256 unilayer neural network implemented on a T-node T 40 FPS ; this network has an internal evolution rule realizing a treatment of the input signal very close to a cardinal sine convolution, favouring the occurrence of a very deep gradient on the boundary of homogeneous (in grey level) objects of interest in the image we have to process
3) use of the gradient built by the neural network as the potential part of a differential system, whose hamiltonian part is given by the initial grey level (before action of the network) (see Section 4.) ; after, study of the attractors of this differential system
4) obtaining the frontiers of homogeneous objects as limit cycles of the differential system defined just above (see Section 4.), by simulating trajectories of the system in the different attraction basins of the previously studied attractors.

The step 2 consists in defining the input by a geometric random field, i.e. a collection of geometric random processes such that, if $p_i(t)$ denotes the probability to generate a spike on the afferent fiber i to the neuron i at time t, we have :

$p_i(t) = 0$, if $t-s_i \leq R$, where s_i is the time of the last 1 on the fiber i before time t and R denotes the refractory period,

$p_i(t) = \alpha_i \sin_+(\omega_i(t-s_i -R))$, if $t-s_i > R$, where $\sin_+$ denotes the positive part of the sine.
In order to incorporate an adaptation learning effect, we choose an evolution of the w_{ij}'s based on the reinforcement of equal grey activities in the same neighbourhood, we take :

$$w_{ij}(t+1) = \log(\sum_{s \leq t} p_i(s)p_j(s)/t)$$

This formula corresponds to the fact that $w_{ij}(t)$ is just the non-standardized correlation function between the $p_i(s)$'s and the $p_j(s)$'s ; if $\omega_i-\omega_j$ and R are small, $w_{ij}(t)$ tends to :

$$\log((\alpha_i\alpha_j\sin(\omega_i-\omega_j)/(\omega_i-\omega_j))/2) \qquad (1)$$

when t tends to infinity.

Image coding

After normalization of the grey level G(i) in the pixel i between 0 and 1, we take :
$\alpha_i = G(i)$ and $\omega_i = \lambda\, G(i)$
and we start the procedure by iterating the deteministic or the stochastic neural network. It is easy to prove that the probability π_i to have 1 as output of the neuron i at time t is, just before a normalization, about proportional to $\pi'_i = \sum_{j \in V(i)} \sum_{s \leq t} p_i(s)p_j(s)/t$; this last formula has been used to make the gradient enhancement visible on Figures 4,5. The behaviour of the function π'_i is similar to a convolution by a cardinal sine function (used for example in[66]), because of the approximative formula (1) above. It is easy to verify that this convolution reinforce the “plateau” or "mesa" activities in grey (or white if necessary) level. Such activities correspond in medicine to pathological objects considered as target of the treatment (like tumour in which the same clone of cells gives a homogeneous response in absorbance or resonance) or to physiological objects (like a tissue of cells having the same function) to avoid during the treatment. That proves that the procedure above can concern a great lot of possible applications. Figure 6 shows the result of a gradient enhancement by the network for a brain tumour and for the wrist. Let us finally remark that the final observed image corresponds to the asymptotics of the network and hence it is closely dependent on its implementation (partially or massively parallel)[28,30] as mentioned above.

APPLICATION TO THE IMAGE DIFFERENTIAL SEGMENTATION

The continuous modelling allows stable evaluation of differential operators such as gradient or laplacian. Our 3D edge detection consists in building a differential equation system whose stable manifold is the surface of the object we are looking for. Finding this manifold turns out to be a particular case of surface intersection problems[37] and provides an immediate analytical representation of the surface.

The other major advantages of this method are to perform segmentation and surface tracking simultaneously, to describe complex structures in which branching problems can occur if the segmentation is purely local[67], and to provide accurate and reliable results.

Methodology

Let us first consider the 2D problem. The central idea of the method is based on the Thom-Sebastiani conjecture[38,39,40] ; let us consider the differential system :

$$x'(t) = F(x,y), \qquad y'(t) = G(x,y)$$

In the neighbourhood of a singularity of the corresponding vector field supposed to be continuous, it is easy to decompose the system in 2 parts, a gradient and a hamiltonian ones, such as :

$$(x'(t), y'(t)) = - \text{grad } P(x,y) + \text{ham } H(x,y) + R(x,y),$$

where the residue R(x,y) tends to 0 when (x,y) tends to the singularity, the Thom-Sebastiani-Françoise-Saïto theorems ensuring the unicity of the decomposition.
The Thom-Sebastiani conjecture assumes that this results still holds when we replace the singularity by another attractor (or repellor) of the diffential system, like a limit-cycle. We will exploit systematically in the following this possibility to consider a contour as the limit-cycle of a gradient-hamiltonian mixed system.

In fact, we are looking for a boundary surrounding an object with an approximatively homogeneous density f, thus verifying:

$$f(x,y) = \text{constant} \tag{2}$$

This curve is represented with parametric coordinates by :

$$x = x(t) \; ; \; y = y(t)$$

The continuous modelling implies the existence of the first derivatives of f ; so a solution should verify the following equation (3) obtained by differentiation of (2):

$$x'(t) * \frac{\partial f}{\partial x} + y'(t) * \frac{\partial f}{\partial y} = 0 \tag{3}$$

A particular solution of (3) is :

$$x'(t) = \frac{\partial f}{\partial y}$$
$$y'(t) = - \frac{\partial f}{\partial x} \tag{4}$$

But this system does not provide a stable solution : a perturbation (due to noise) which would move the curve away from the contour line could not be corrected. That is why we add a component to the system (4) which brings the curve back to the contour line "f(x,y)=constant" according to the greatest slope line of the $(f\text{-constant})^2$ function. We thus obtain:

$$x'(t) = \frac{\partial f}{\partial y} - \beta * \frac{\partial f}{\partial x} * (f - \text{constant}) / G_f$$
$$y'(t) = - \frac{\partial f}{\partial x} - \beta * \frac{\partial f}{\partial y} * (f - \text{constant}) / G_f$$
$$\text{where } G_f = \frac{\partial f}{\partial x} * \frac{\partial f}{\partial x} + \frac{\partial f}{\partial y} * \frac{\partial f}{\partial y} \tag{5}$$

This system consists in two parts: the first one corresponds to "edge tracking" component and the second one is a kind of "elastic force" which allows noisy image processing. The β parameter allows to balance these two terms. The system (5) may be solved by numerical analysis methods with initial conditions. The parametric representation of the curve is directly obtained (cf. Figure 7). This differential system may be built in any plane of the 3D volume.

Particular case : looking for local extrema

This method of differential system building can be applied to look for particular points of the object surface. Let's consider the particular case of local extrema of the surface defined by :

$$f(x,y,z) = \text{constant} \quad (6)$$

Let's assume we are looking for a local minimum for z. It is caracterized by :

$$\frac{\partial z}{\partial x} = \frac{\partial z}{\partial y} = 0 \quad (7)$$

The curve which goes to this minimum can be represented by : x = x(t), y = y(t), z= z(t).
To reach this minimum from any point of the surface, we have to follow the curve defined by :

$$x'(t) = -\frac{\partial z}{\partial x}$$

$$y'(t) = -\frac{\partial z}{\partial y} \quad (8)$$

This partial derivatives are obtained by considering z as a function of x and y : if we derive (6) according to x and y, we obtain :

$$\frac{\partial f}{\partial x} + \frac{\partial z}{\partial x} * \frac{\partial f}{\partial z} = 0$$

$$\frac{\partial f}{\partial y} + \frac{\partial z}{\partial y} * \frac{\partial f}{\partial z} = 0 \quad (9)$$

From (9), we deduce :

$$x'(t) = -\frac{\partial z}{\partial x} = \frac{\partial f}{\partial x} / \frac{\partial f}{\partial z}$$

$$y'(t) = -\frac{\partial z}{\partial y} = \frac{\partial f}{\partial y} / \frac{\partial f}{\partial z} \quad (10)$$

The condition on the z(t) function is given by deriving (6) according to t (it means that the curve has to stay on the surface f(x,y,z)=constant) :

$$x'(t) * \frac{\partial f}{\partial x} + y'(t) * \frac{\partial f}{\partial y} + z'(t) * \frac{\partial f}{\partial z} = 0 \quad (11)$$

then the last equation of the differential system is :

$$z'(t) = -\left(\frac{\partial f}{\partial x} * \frac{\partial f}{\partial x} + \frac{\partial f}{\partial y} * \frac{\partial f}{\partial y}\right) / \left(\frac{\partial f}{\partial z} * \frac{\partial f}{\partial z}\right) \quad \text{for } \frac{\partial f}{\partial z} \neq 0 \quad (12)$$

For stability problems, we add an "elastic force" of the same type of previously performed one.

3D Strategy

We are looking for a surface verifying : f(x,y,z) = constant
The data geometry has induced our parametrization of the surface we are looking for :

$$x = x(t,h),\ y = y(t,h),\ z = z(h).$$

Our boundary tracking method is as follows : the algorithm starts with a point of the surface and with a characteristic density value. For each slice k, the differential system is solved in order to obtain a closed curve. From some points of this curve, we follow the object surface until the next (k+1) slice by building new 2D differential systems in slice-perpendicular planes. The ends of these curves belong to the (k+1) slice and are thus used as starting points to find one or several contours on this slice. Such a search is also performed towards previous slices : all object components can then be found. It is already an error recovery mechanism because links previously found between contours are verified in the same time. The algorithm stops when all slices have been processed or when the object surface has been entirely described.

Thus this method allows to find automatically all the components of a complex object in which branching problems may occur and to determine how they are linked together. This possibility is one of the major advantages of our method because surface reconstruction from a set of contours is a critical step for complex structures. Classically, interpolation between contours is performed by triangulation technics[41] or by creating intermediate contours by dynamic elastic interpolation[42]. But these methods need sometimes interaction. In our method suface modelling is performed in the segmentation step. This algorithm was tested on MRI (cf. Figure 9) and CT images. Homogeneity is not always a stable characteristic of an anatomical structure. So we are working now on differential systems performing H(f)=0, where H is an operator as the laplacian or Marr-Hildreth detector.

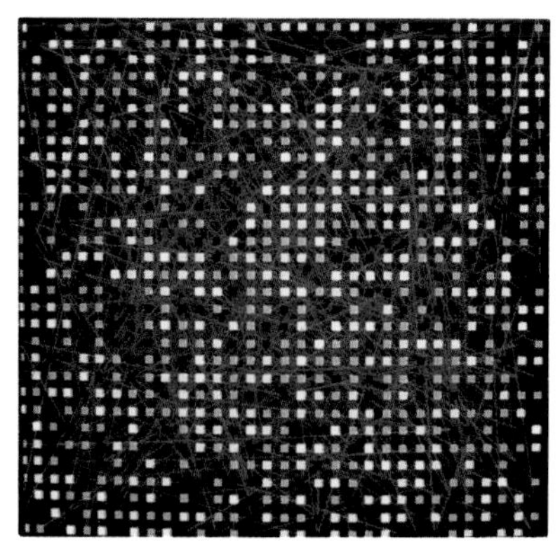

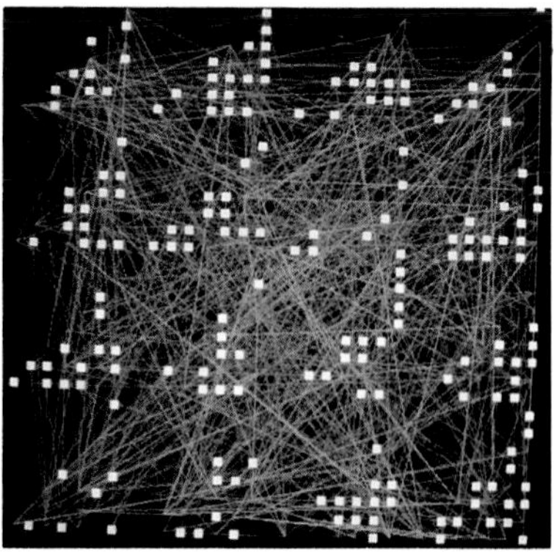

Figure 1. Neural network activity (the white neurons are firing, bars representing their correlations) corresponding to a spatial homogeneous input at the left and a spatial heterogeneous input at the right.

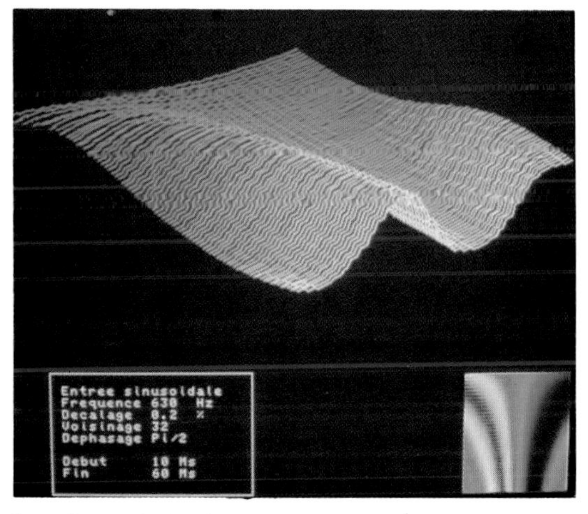

Figure 2. Correlation function of the input; the time-axis is coming to the observer and the function giving the correlation of a neuron with its 1D-neighbours is represented on the z-axis. The progressive occurrence of the "mexican hat" corresponding to the olateral inhibition is appearing progressively.

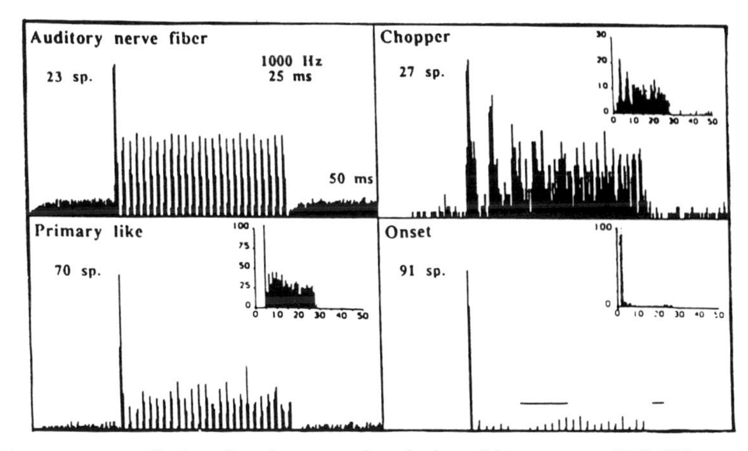

Figure 3. Four types of simulated post-stimulation histograms (PSTH) generated by a 2-parameter geometric model; the real data obtained by (W.S. Rhodes, J. Acous. Soc. Am. 78:320 (1985)) are given in the superior right part of the diagrams.

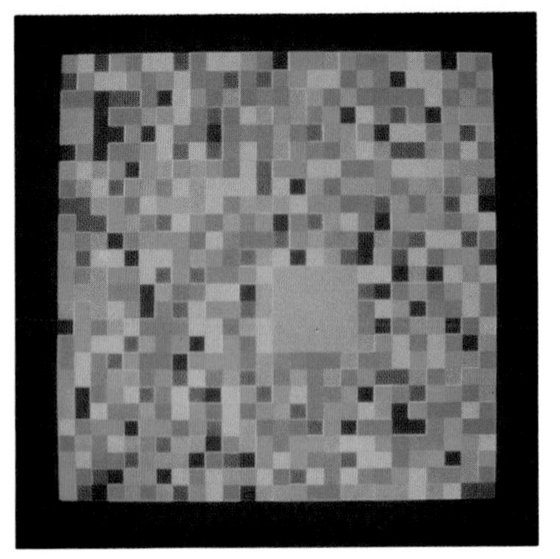
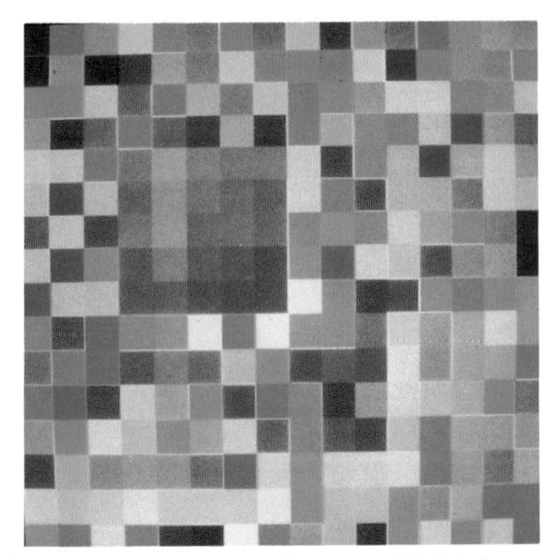

Figure 4. Enhancement of the gradient of a homogeneous square in an artificial image (initial image left, treated image right).

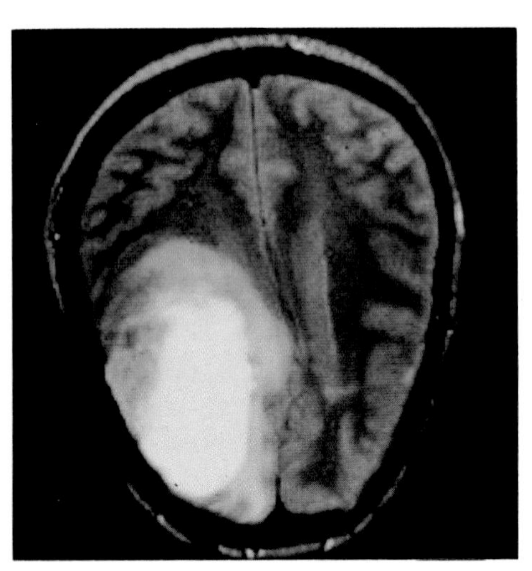
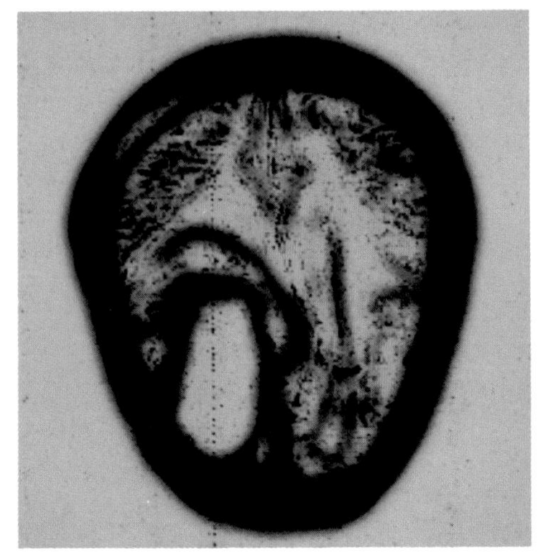

Figure 5. Initial MRI image of a tumor at the left and enhanced gradient by applying a neural network method at the right.

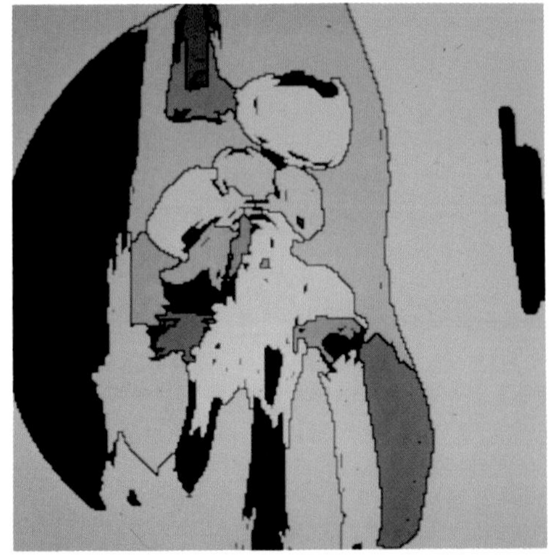

Figure 6. Segmentation after enhancement by a neural network of tumoral image at the left and of the wrist bones at the right.

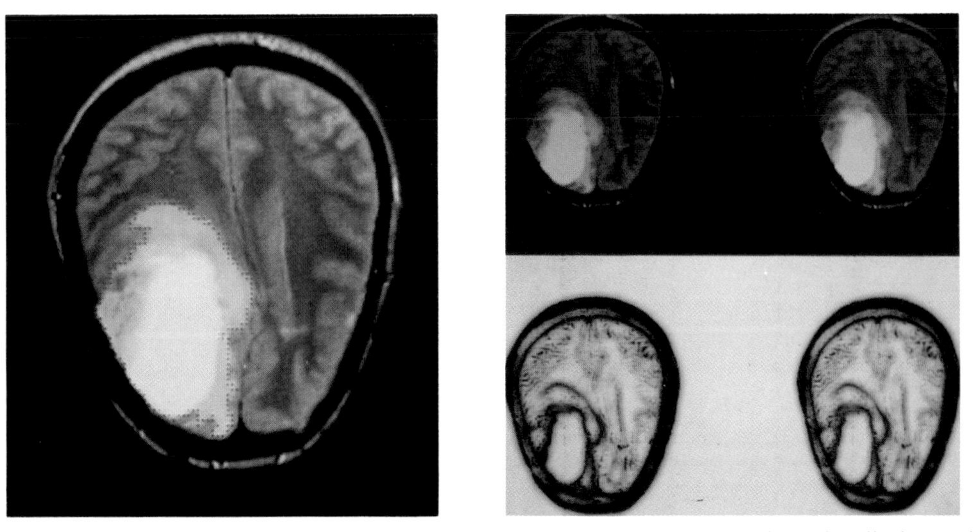

Figure 7. Obtaining the external contour of the tumour by applying the limit cycle method at the left and obtaining internal and external contours at the right.

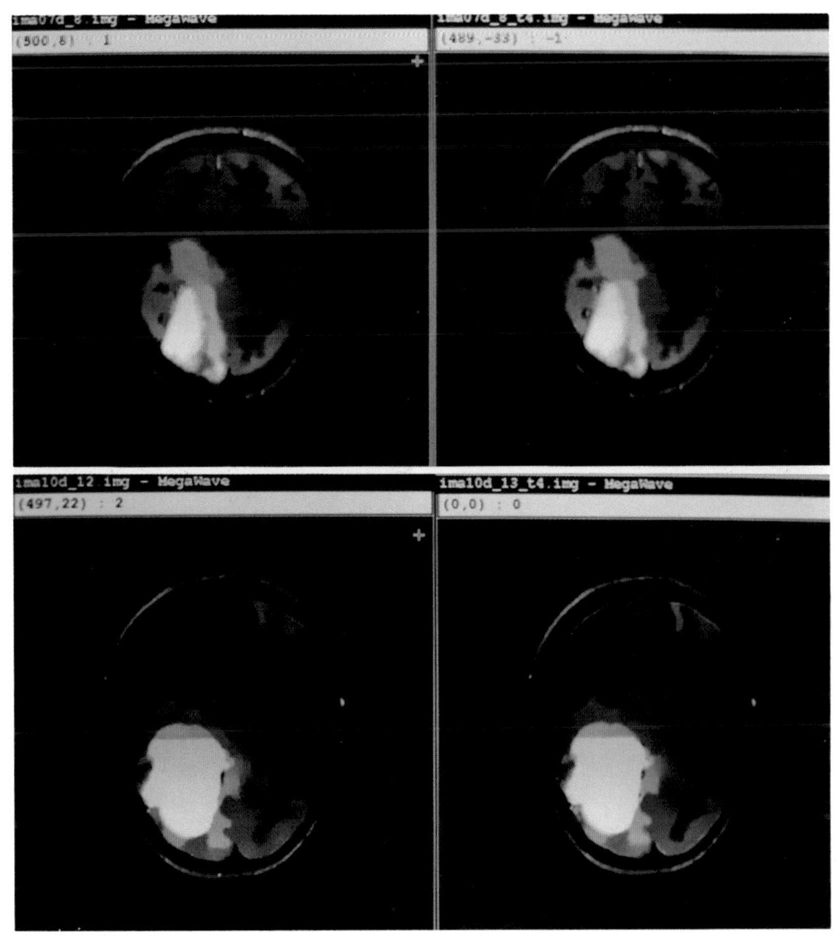

Figure 8. Obtaining the internal (above) and external (below) contours by using a non-linear diffusion operator.

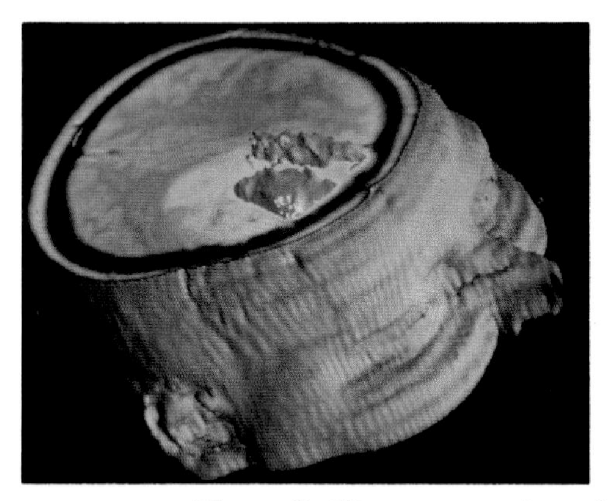
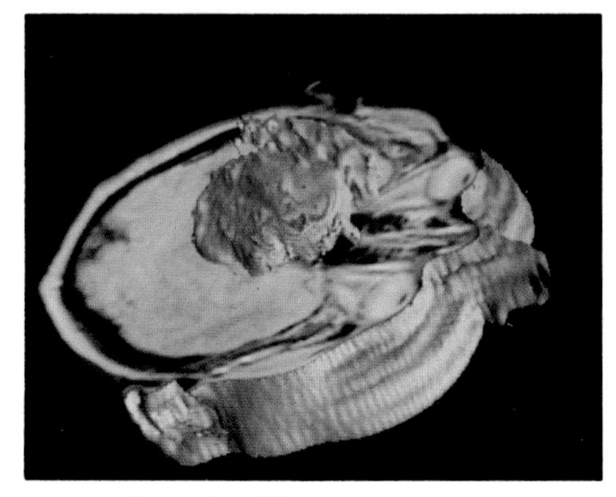

Figure 9. 3D reconstruction of the tumour from the 2D contours.

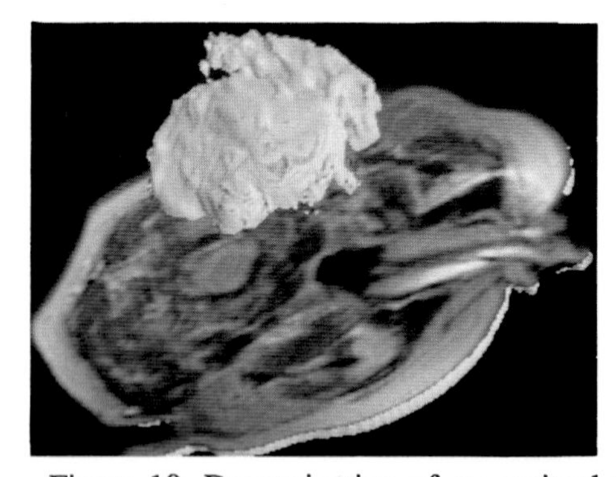
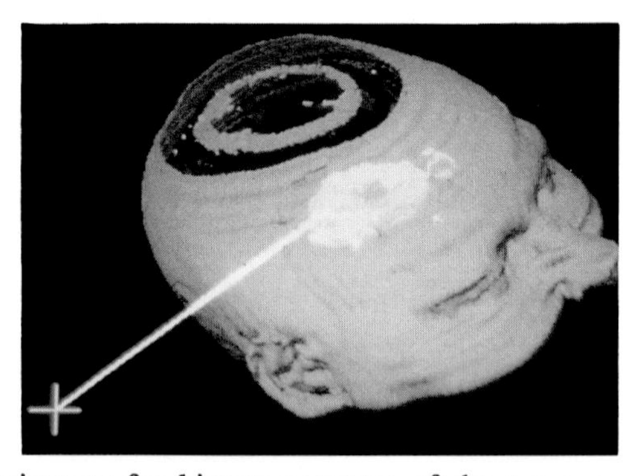

Figure 10. Determination of an optimal trajectory for biopsy-puncture of the tumour.

REACTION-DIFFUSION OPERATOR METHODS

In the same spirit as in[64], several methods of image processing by using differential linear or non linear operators have been recently proposed. These methods can be parallelized as for the neural networks presented in Section 2. and we will show in the following that there exists a deep relationship between the discrete neural network approach and the continuous differential operator approach.

The Catté-Lions-Morel-Coll non-linear diffusion operator

It has been shown[8] that successive gaussian filtering of standard deviation σ of images was completely equivalent to the application of the heat differential operator, where $\sigma = (2t)^{1/2}$:

$$\partial u/\partial t = \Delta u = \mathrm{div}(\mathrm{grad} u),$$

by choosing as initial conditions u(0) the grey level. This result then suggested the use of another differential non-linear diffusion operator[68] :

$$\partial u/\partial t = \mathrm{div}(g(|\mathrm{grad}(G * u)|).\mathrm{grad} u),$$

where G is the gaussian distribution function with standard deviation $\sigma = (2t)^{1/2}$ and g a non-negative non-increasing function on R_+ verifying : g(0)=1 and g tends to 0 at infinity ; in practice we can choose for g a set function, whose value is 1 on the interval [0,S] and 0 on]S, +∞[: there is diffusion if and only if $|\mathrm{grad}(G*u)| < S$ and, after a certain transient, it remains a gradient only on the boundary of sufficiently discriminable objects. For example, Figure 8 presents the image after some hundreds of iteration, showing the internal and the external gradient on the boundary of the tumour. The end of the procedure is then similar with the one from the neural networks method.

In order to improve the method, we must add a reaction term in order to obtain the final expected image as the asymptotic behavior of a differential reaction diffusion operator, like for the continuous operator derived from the iterative discrete neural network system[29].

The Cottet reaction-diffusion operator

By searching a continuous operator having as discrete finite elements scheme a deterministic neural network system similar to that presented in Section 2., it has been proposed in[29] the following reaction diffusion operator ; let's consider a deterministic neural network defined by :

$$x_i(t+1) = 1, \quad \text{if } H_i(t) = \sum_{j \in V(i)} w_{ij}\, x_j(t) > 0,$$
$$= 0, \quad \text{if not,}$$

where V(i) is a neighbourhood of i in B.

f we suppose the network to be 2D and infinite, lets us denote by ih the positions of the neurons, where $i \in Z^2$; if w_{ij} are symmetrical and translation invariant with finite range R (i.e. R is the radius of V(0)), there exists T defined on $[-1,1]^2$ and valued in [-1,1] such as :

$w_{ij}=T((i-j)h/R)=T((i_1-j_1)h/R,(i_2-j_2)h/R)$, the mean value of T is $m=\int T(y_1,y_2)dy_1dy_2>0$ and the variance of T is $2M-m^2$, where $2M=\int y_1^2T(y_1,y_2)dy_1dy_2=\int y_2^2T(y_1,y_2)dy_1dy_2>0$.

Let us denote by g a continuous regularized version of the Heaviside function and let us take $G=g^{-1}$, $a(u)=\lambda R^4S/(h^2G'(u))$ and $b(u)=(-G(u)+\lambda R^2/h^2mu)/G'(u)$; then the reaction diffusion operator defined by :

$$\partial u/\partial t = a(u)\Delta u + b(u)$$

has a natural discretization corresponding to the neural network above, by identifying $x_i(t)$ and u(ih,t) and by remarking that the neural network system has the same asymptotic behavior as the differential system : $dx_i(t)/dt = (\lambda\Sigma\, w_{ij}\, x_j(t) - G(x_i(t)))/G'(G(x_i(t)))$, when λ is sufficiently large. In[29], it is shown that, for adapted values of R, homogeneous in grey 1D objects can be enhanced in a heterogeneous environment, in the same way as for a neural network system.

Proposal for a new image diffusion-chemotaxis operator

In order to have, like for the last operator, the final treated image as the asymptotics of a differential operator (and not to obtain it as a transient state like for the firstly proposed operators above), we propose to consider the grey level u as a chemotactic substrate concentration consumed by animals whose concentration will be denoted by v[69].

The principle of this method consists in placing initially a uniform concentration v(0) of animals on the initial grey level u(0) : the substrate u can diffuse with a term $\varepsilon\Delta u$ and is consumed with a saturation velocity - Kuv/(u+k) ; the animal concentration v can diffuse attracted by the substrate with the term $D\Delta v$, is submitted to a drift in the direction of substrate peaks with the term $-\chi\text{div}(v\,\text{grad}u)$ and increases (because of the reproduction) with the term K'uv(u+k'). Let us remark that the two first terms ruling the animal motion can be replaced, if we do'nt want introduce a drift term, by an attraction-diffusion term like :

$$D(\partial^2 v/\partial x^2 \cdot \partial u/\partial x + \partial^2 v/\partial y^2 \cdot \partial u/\partial y).$$

The corresponding differential partial derivative operator is then given by :

$$\partial u/\partial t = \varepsilon\Delta u - Kuv/(u+k)$$
$$\partial v/\partial t = D\Delta v - \chi\text{div}(v\,\text{grad}u) + K'uv(u+k')$$

or by the following PDE :

$$\partial u/\partial t = \varepsilon\Delta u - Kuv/(u+k)$$
$$\partial v/\partial t = D(\partial^2 v/\partial x^2 \cdot \partial u/\partial x + \partial^2 v/\partial y^2 \cdot \partial u/\partial y) + K'uv(u+k')$$

In the two cases above, the asymptotics in u is 0 and the asymptotics in v gives the "treated image". The corresponding image processing primitive leads to a contrast enhancement before segmentation (for example before searching microcalcifications in breast X rays imaging). If we are adding to the second equation of these differential systems the Dupin term $Cv/\Delta u$, we will encourage animals to follow Dupin lines, i.e. inflexion curves, which is suitable before a grey anticlines segmentation (for example in vessels segmentation). More sophisticated chemotaxis-diffusion operators with non-local effects and long range diffusion can be found in[70].

The final aim of these methods is to offer a set of operators adapted to segmentation of grey singularities or grey peaks (0-dimensional objects like microcalcifications), grey anticlines (1-dimensional objects like vessels) or grey mesas (2-dimensional objects like tumors or functional regions). The problem of segmentation of more complicated objects (fractal objects like diffuse tumors affecting for example the conjunctive tissue) is open and demands that another variables like a texture based one (for example the local fractal dimension) would be taken into account instead of the grey level or beside it.

CONCLUSION

Building dynamical systems and solving them can provide new methods of segmentation which look particularly suitable for 3D image processing. For example, the splines snakes look easier to apply in 3D than the classical snakes and might provide an efficient alternative. These methods should find applications for segmenting images of complex objects such as the ones where branching problems occur (where segmentation with differential equations could be applied) or where continuous models could be used (where splines snakes could be interesting). Work is under way to apply the splines snakes approach and the differential operator approach to real 3D images and to find out in which cases splines snakes or differential equations should be used. Examples of possible fields of application in helping medical or surgical manipulations can be found in[49 - 62]. Let us recall that homogeneous in grey structures correspond in medicine either to pathological objects considered as target of the treatment (cf. Figure 10) (like tumour in which the same clone of cells gives a homogeneous response in X rays absorbency or in magnetic resonance) or to physiological objects (like a tissue of cells having the same function) to avoid during the treatment. That proves that the procedures described above can concern a great lot of possible applications.

ACKNOWLEDGEMENTS

This work has been supported by an external research project with DEC company.

REFERENCES

1. H.K. Liu, Two and three dimensional boundary detection, Computer Vision, Graphics and Image Processing 6:123 (1977).
2. S.W. Zucker and R.A. Hummel, A three dimensional edge operator, IEEE PAMI 3:20 (1981).
3. M. Morgenthaler and A. Rosenfeld, Multidimensional edge detection by hypersurface fitting, IEEE PAMI 4:482 (1981).
4. J. Canny, A computational approach to edge detection, IEEE PAMI 8:679 (1986).
5. O. Monga and R. Deriche, 3D edge detection by recursive filtering : application to scanner images, in : "Conf. on Computer Vision and Pattern Recognition," IEEE, San Diego (1989).
6. O. Monga, R. Deriche, G. Malandain and J.P. Cocquerez, Recursive filtering and edge closing : two primary tools for 3D edge detection, INRIA reports, Rocquencourt (1989).
7. K.H. Höhne and R. Bernstein, Shading 3D Images from CT Using Gray-Level Gradients, IEEE Trans. on Medical Imaging 5:45 (1986).
8. D. Marr and E.C. Hildreth, A theory of edge detection, Proc. Royal Soc. B 207:187 (1980).
9. M. Kass, A. Witkin and D. Terzopoulos, Snakes : active contour models, in : "Conf. on Computer Vision and Pattern Recognition," IEEE, London (1987).
10. A. Ayache, J.D. Boissonnat, E. Brunet, L. Cohen, J.P. Chièze, B. Geiger, O. Monga, J.M. Rocchisani and P. Sander, Building Highly Structured Volume Representations in 3D Medical Images in : "CAR 89," Springer Verlag, Berlin (1989).
11. R. Szeliski and D. Terzopoulos, From splines to fractals, Computer Graphics 23:51 (1989).
12. K.H. Höhne, M. Bomans, A. Pommert, M. Riemer and U. Tiede, 3D Segmentation and Display of Tomographic Imagery, in : "Conf. on Computer Vision and Pattern Recognition," IEEE, Rome (1988).
13. R. Gordon and H. Udupa, Fast Surface Tracking in 3D Binary Images, Computer Vision, Graphics and Image Processing 45:196 (1989).
14. J.P. Cappelletti and A. Rosenfeld, 3D boundary following, Computer Vision, Graphics and Image Processing 48:80 (1989).
15. M. Basseville, Détection de contours : méthodes et études comparatives, Ann. Télécom. 34:559 (1979).
16. G. Hégron, La technique du suivi de contour en synthèse d'image, T.S.I. 4:351 (1985).
17. D. Girard, From template matching to optimal approximation by piecewise smooth curves, in : "Curves and surfaces in computer vision and graphics," SPIE, Santa Clara (1990).
18. P. Cinquin, "Application des Fonctions-Spline au Traitement d'Images Numériques," PhD Thesis Université Joseph Fourier, Grenoble (1987).
19. I. Marque, S. Lavallée, C. Goret-Lezy and P. Cinquin, Towards a 3D medical image analysis system based on a continuous modelling, in : "SCAR 90," R.L Arenson and R.M. Friedenberg, eds., Symposia Foundation, New York (1990).
20. I. Marque and P. Cinquin, Segmentation d'images 3D par construction d'un système différentiel, in : "Conférence AFCET IA & Reconnaissance des Formes," AFCET, Paris (1989).
21. P. Cinquin, C. Goret, I. Marque and S. Lavallée, Morphoscopie et modélisation continue d'images 3D, in : "Conférence AFCET IA & Reconnaissance des Formes," AFCET, Paris (1987).

22. P. Cinquin and C. Goret, Une nouvelle technique de modélisation continue d'images 3D, in : "Proceedings Cognitiva-Image Electronique," CESTA, Paris (1987).
23. P. Cinquin, Un modèle pour la représentation d'images médicales 3D, in : "Proceedings Euromédecine 86," Sauramps Médical, Montpellier (1986).
24. J. Demongeot and M. Tchuente, Cellular Automata Theory, in : "Encyclopedy of Physical Science & Technology," Academic Press, New York (1987).
25. T. Hervé, J.M. Dolmazon and J. Demongeot, Neural network in the auditory system : influence of the temporal context on the response represented by a random field, in : "IEEE-ICASSP Proceedings," IEEE, New York (1987).
26. T. Hervé and J. Demongeot, Random field and tonotopy : simulation of an auditory neural network, Neural Networks 1S1:297 (1988).
27. T. Hervé, J.M. Dolmazon and J. Demongeot, Random field and neural information, a new representation for multi-neuronal activity, Proc. Natl. Acad. Sc. 87:806 (1990).
28. O. Francois, Ergodicité des processus neuronaux, C.R.Acad.Sc. 310:435 (1990).
29. G.H. Cottet, Modèles de réaction-diffusion pour des réseaux de neurones stochastiques et déterministes, C.R.Acad.Sc. (in print).
30. T. Hervé, O. Francois and J. Demongeot, Markovian spatial properties of a random field describing a stochastic neural network : sequential or parallel implementation ?, in : "Eurasip workshop," L.B. Almeida and C.J. Wellekens, eds., Springer Verlag, Lecture Notes in Computer Science 412, New York (1990) .
31. J. Demongeot, T. Hervé, F. Berthommier and O. François, Neural networks : from neurocomputing to neuromodeling, in : "Theoretical models for cell to cell signalling," A. Goldbeter, ed., Academic Press, NATO Series, London (1989).
32. F. Berthommier, O. Francois, D. Francillard, T. Coll, I. Marque, P. Cinquin and J. Demongeot, Asymptotic behavior of neural networks and image processing, in : "Self-Organization, Emerging properties and Learning," A. Babloyantz, ed., Plenum Press, NATO Series, New York (in print).
33. F. Berthommier, J. Demongeot and J.L. Schwartz, A neural net for processing of stationary signals in the auditory system, in : "IEE Proc. Conf. Signal Proc. London," IEE, London (1989).
34. F. Berthommier, J.L. Schwartz and P. Escudier, Auditory processing in a post-cochlear neural network : vowel spectrum processing based on spike synchrony, in : "Eurospeech 89," J.P.Tubach et al., eds., Eurospeech, Paris (1989).
35. F. Berthommier, Un modèle de la relation entre tonotopie et synchronisation du système auditif, C.R.Acad.Sc. 309:695 (1989).
36. F. Berthommier, Un nouveau principe pour comprendre la perception : la synchronisation des messages nerveux, in : "Actes Workshop Sciences Cognitives Chichilianne," E. Decamp, ed., Lasco 3, Grenoble (1989).
37. M.B. Phillips and G. Odell, An algorithm for Locating and Displaying the Intersection of 2 Arbitrary Surfaces, Computer Graphics 18:48 (1984).
38. J.P. Françoise, Systèmes maximaux d'une singularité quasi-homogène, C.R. Acad. Sc. 290:1061 (1980) .
39. R. Thom, "Modèles mathématiques de la morphogenèse," Ediscience, Paris (1978).
40. J. Demongeot, F. Estève and P. Pachot, Comportement asymptotique des systèmes : applications en biologie, Rev. Int. Syst. 2:417 (1988).
41. J.D. Boissonnat, Shape reconstruction from planar cross-sections, Computer Vision, Graphics and Image Processing 44:1 (1988).
42. W.C. Lin, C.C. Liang and C.T. Chen, Dynamic Elastic Interpolation for 3D image Reconstruction from Serial Cross Sections, IEEE Medical Imaging 7:225 (1988).
43. C. De Boor, "A Practical Guide to Splines," Springer Verlag, Berlin (1978).
44. H. Fuchs, M. Kedem and S.P. Uselton, Optimal surface reconstruction from planar contours, Commun. Ass. Comput. Mach. 20:693 (1977).
45. P.J. Laurent, "Approximation et optimisation," Hermann, Paris (1972).
46. T. Poggio, H. Voorhees and A. Yuille, A regularized solution to edge detection, MIT Reports AIM-833, Cambridge (1985).
47. D. Francillard, "La détection de contours," Master Dissertation UJF, Grenoble (1989).
48. A.A. Amini, S. Tehrani and T.E. Weymouth, Using dynamic programming for minimizing the energy of active contours in the presence of hard constraints, in : "Conf. on Computer Vision and Pattern Recognition," IEEE, Rome (1988).

49. S. Lavallée, P. Cinquin, J. Demongeot and A.L. Benabid, Neurochirurgie stéréotaxique par ordinateur et robot, in : "Actes du Ve FJC," ITBM, Paris (1990).
50. B. Mazier, S. Lavallée and P. Cinquin, Chirurgie de la colonne vertébrale assistée par ordinateur : application au vissage pédiculaire, in : "Actes du Vème Forum des Jeunes Chercheurs," ITBM, Paris (1990).
51. P. Cinquin and S. Lavallée, Gestes Médicaux et Chirurgicaux assistés par ordinateur : bases méthodologiques, in : "Informatique et Imagerie Médicale : Journées Francophones d'Informatique Médicale de Nîmes," ENSP, Rennes (1990).
52. S. Lavallée, P. Cinquin, J. Demongeot, A.L. Benabid, I. Marque and M. Djaïd, Computer assisted interventionist imaging : the instance of stereotactic brain surgery, in : "Medinfo 89 Singapore," North Holland, Amsterdam (1989).
53. S. Lavallée, P. Cinquin, J. Demongeot, A.L. Benabid, I. Marque and M. Djaid, Computer assisted driving of a needle into the brain, in : "Computer assisted radiology, Proceedings CAR 89," H.U. Lemke, M.L. Rhodes, C.C. Jaffe and R. Felix, eds., Springer Verlag, New York (1989).
54. S. Lavallée, A New System for Computer Assisted Brain Surgery, in : "Proceedings 11th IEEE Eng. in Med. and Biol. Conference Seattle," IEEE, New York (1989).
55. P. Cinquin, S. Lavallée and J. Demongeot, A new system for computer assisted neurosurgery, in : "IARP Medical and Healthcare Robotics Newcastle," DTI, London (1989).
56. S. Lavallée, P. Cinquin, C. Goret, J. Demongeot, G. Crouzet and P. Peltié, Ponction assistée par ordinateur, in : "Conférence AFCET IA & Reconnaissance des Formes," AFCET, Paris (1987).
57. P. Cinquin, S. Lavallée, C. Goret, J. Demongeot, G. Crouzet and P. Peltié, Computer assisted intervertebral puncture, in : "Proceedings MIE 87," A. Serio et al., eds., EFMI, Rome (1987).
58. P. Cinquin, J.C. Saget, G. Plasse and P. Antoine, Chirurgie plastique maxillofaciale assistée par ordinateur, in : "Proc. A. I. Biomed 86," CRIM, Montpellier (1986).
59. A.L. Benabid, P. Cinquin, S. Lavallée, J.F. Le Bas, J. Demongeot and J. de Rougemont, A computer driven robot for stereotactic surgery connected to cat-scan magnetic resonance imaging, J. Applied Neurophysiology 50:153 (1987).
60. F. Moutet, A. Chapel, P.Cinquin and L. Rose-Pitet, Imagerie du carpe en trois dimensions, Ann. Chir. Main et Membres Sup. 9:32 (1990).
61. S. Lavallée and P.Cinquin, Computer assisted medical interventions, in : "3D-imaging in medicine," K.H. Höhne et al., eds., Springer Verlag, New York (1990).
62. MURIM, "Multidimensional Reconstruction and Imaging in Medicine: State of the Art," Advanced Informatics in Medicine EEC DG XIII Reports, Brussels (1990).
63. J. Loncelle, "Détection de contours par rétro-propagation," Thomson Report, Paris (1989).
64. C.B Price, P. Wambacq and A. Ossterlinck, Computing with reaction-diffusion systems : applications in image processing, in : "Continuation and bifurcations," D. Roose et al., eds., Kluwer Academic Publishers, Amsterdam (1990).
65. L. Brunie and S. Miguet, "3D reconstruction of the wrist : a new method of segmentation," IMAG DEA Report, Grenoble (1989).
66. H.P. Chan, K. Doi, C.J. Vyborny, K.L. Lam and R.A. Schmidt, Computer-aided detection of microcalcifications in mammograms, Investigative Radiology 23:664 (1988).
67. S. Bouakaz, "Approche stochastique de la segmentation," PhD Thesis Université Joseph Fourier, Grenoble (1987).
68. F. Catté, P.L. Lions, J.M. Morel and T. Coll, Image selective smoothing and edge detection by non linear diffusion (submitted).
69. J.L. Martiel and A. Goldbeter, A model based on receptor desensitization for c-AMP signalling in *Dictyostelium* cells, Biophys. J. 52:807 (1987).
70. J. Murray, "Mathematical biology," Biomathematics Springer Verlag, New York (1989).
71. T. Hervé, T. Irino and H. Kawahara, Representing temporal information in auditory periphery based on random field theory, J. Acoust. Soc. Jap. (to appear).

ENTROPY AND LEARNING

C. Van den Broeck*

University of California at San Diego
Dept. Chem. B-040
San Diego, CA 92093

INTRODUCTION

Recently, neural networks have attracted a lot of attention from the scientific community, as is witnessed by the large number of conferences on the subject, and the introduction of a dedicated journal (neural networks). It has become a truely inter-disciplinary field, involving artificial intelligence, neurophysiology, psychology and statistical physics. One of the reasons for this success is the relative high performance of neural networks in a number of search and learning problems (e.g. the construction of associative memories[1], optimization[2], and classification[3]). These tasks are performed rather easily by the human brain, despite its relatively low processing speed (the time for a neuron to fire is in the range of msec., while it takes a fraction of a second to recognize a face). The widely accepted explanation for this fact is the high level of parallel processing in the brain, each neuron being connected to thousands of other neurons. This high connectivity is also the characteristic feature of what are called neural networks.

One of the most amazing properties of neural networks is their ability to generalize from examples. I cite three examples from the literature. In reference [4], a feedforward neural network is trained to convert English text to speech. The learning is based on the backpropagation algoritm that changes the connection strength between the neurons such as to minimize the average squared error between the value of the output units and the correct pattern. Using weights from a network with 120 hidden units trained on 1000 words, the average performance on the Miriam Webster's Pocket Dictionary of 20012 words was 77%, and up to 90% following five training passes. In reference [5], a feedforward network of neurons with continuous input and output values was trained by backpropagation to predict points in a chaotic time series. With this method, the prediction was orders of magnitudes more accurate then that of traditional extrapolation methods. Finally, in reference [6], a feedforward Boolean network (with Boolean functions of 2 variables as building blocks) was trained to learn the addition of two 8-digit (binary numbers) by a simulated annealing type of algoritm. It was found that for a sufficiently large network (160 hidden units), a correct output is achieved for all the pairs of inputs, after the network was trained to add correctly only .3% of the total numbers of additions. The authors make the very interesting suggestion that this remarkable generalization property (extracting of a rule from a reduced set of examples), may be related to the fact that a relatively large number of (microscopic) configurations give rise to the addition.

It is the purpose of this paper to investigate in more detail the structure of the corresponding "phase space", and to elaborate on the "entropy driven" nature of the learning process in random Boolean networks. I have choosen for Boolean functions as building blocks, rather than the more conventional threshold sensitive neurons, for computational

Self-Organization, Emerging Properties, and Learning
Edited by A. Babloyantz, Plenum Press, New York, 1991

facility. However, I think that the learning features will not depend very much on the details of the building blocks.

The outline of the paper is as follows. In section 2, I present analytic results for the case of Boolean functions of 1 variable. This is the simplest, and rather uninteresting case, but it serves the purpose of introducing the model. In section 3, I consider the case of random networks of 2-variable Boolean functions, and the generalization properties of such a network are discussed and analyzed in section 4. The results are discussed in a more general context in section 5.

BOOLEAN FUNCTIONS OF ONE VARIABLE

A Boolean function of n variables is a logic operator that converts n binary inputs (0's or 1's) into one binary output (0 or 1). Our purpose is to investigate the properties of random feedforward Boolean networks. I start by constructing such a network with Boolean functions of 1 variable. The 4 Boolean functions of 1 variable are represented in table I. A random feedforward network of N 1-variable Boolean functions (or neurons, as I will call them) is obtained by linking them to each other from the left to the right, but otherwise in a random way, while each neuron is choosen randomly from the four 1-variable Boolean functions (cf. figure 1a for a typical configuration). The resulting network is again a Boolean function of 1 variable, but the 4 possible 1-variable Boolean functions do not appear with equal probability. In the combined N-neuron network, the probability distribution P(i) for the Boolean functions, i=0,3, are:

$$P(1) = P(2) = \frac{1}{2}\frac{1\ 3\ 5\ \ldots\ (2N-1)}{2\ 4\ 6\ \ldots\ \ \ 2N}$$

$$P(0) = P(3) = \frac{1}{2} - \frac{1}{2}\frac{1\ 3\ 5\ \ldots\ (2N-1)}{2\ 4\ 6\ \ldots\ \ \ 2N}$$

As N increases, the "trivial" Boole fucnctions 0=00 and 3=11 gain weight at the expense of the two other ones 1=01 and 2=10 (the r.h.s. being binary code) .

Table I. The logic tables for the 4 Boolean functions of 1 variable.

#0=00		#1=01		#2=10		#3=11	
In	Out	In	Out	In	Out	In	Out
0	0	0	1	0	0	0	1
1	0	1	0	1	1	1	1

BOOLEAN FUNCTIONS OF TWO VARAIBLES

In table II, I have represented 4 of the 16 Boolean functions of 2 variables. As in the 1-variable case, one can construct a random network, choosing the Boolean functions and connections randomly. The result is again a Boolean function of 2 variables. The probability distribution P(i), i=0,15, that a randomly choosen network correspond to the Boolean function i, is represented in figure 2 for a network of 4 and 7 neurons. Again the "trivial" Boolean functions 0=0000 and 15=1111 are gaining weight with increasing network size.

Table II. The logic tables for 4 Boolean functions of 2 variables.

NONE 0000		AND 1000		XOR 0110		OR 1110	
In	Out	In	Out	In	Out	In	Out
00	0	00	0	00	0	00	0
01	0	01	0	01	1	01	1
10	0	10	0	10	1	10	1
11	0	11	1	11	0	11	1

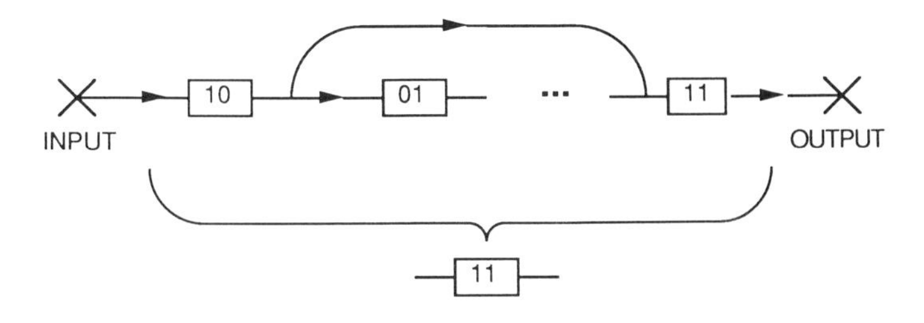

FIGURE 1a
Example of a random feedforward network of 1-variable Boolean functions.

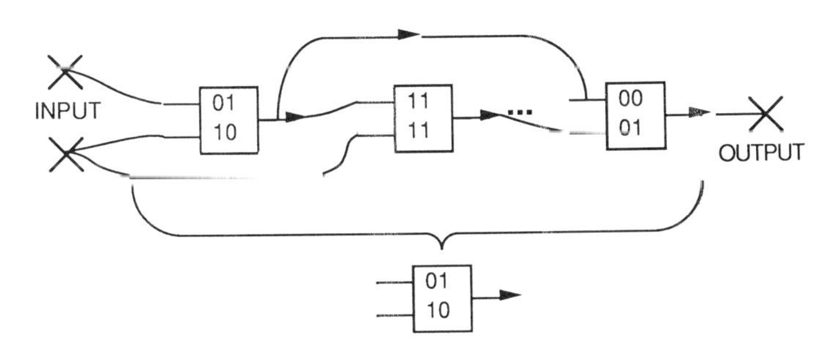

FIGURE 1b
Example of a random feedforward network of 2-variable Boolean functions.

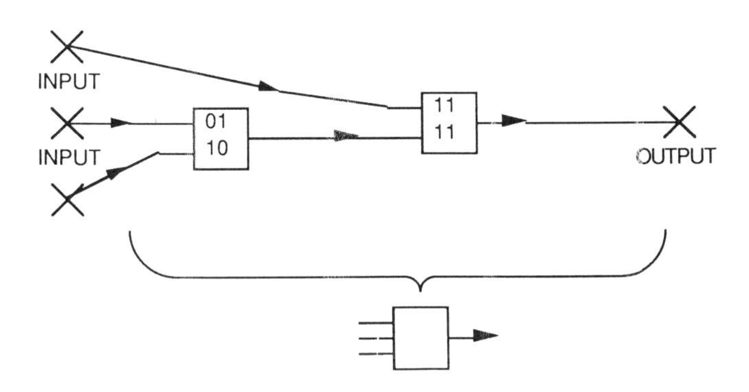

FIGURE 1c
Example of the construction of a 3-variable Boolean function from a feedforward network of 2-variable Boolean functions.

Using the same building blocks, it is however also possible to construct a more complicated object, for example a Boolean function of 3 variables, see figure 1c. There are 256 such Boolean functions, each of which can be represented by a 8-digits binary number. For example, the addition corresponds to the Boolean table represented in table III. In figure 3, I have plotted the number of configuratons that generates each of the 256 different the 3-variable Boolean function in a 4-neuron network (the total number of configurations being 11059200; we have only used those 2-variable Boolean functions that do not distinguish left from right input, i.e. the 4 Boolean functions represented in table II and their negations). Similar results are found for networks with a larger number of neurons. For example, in figure 4, I give the graph for a similar random network generating a 4-variable Boolean function (of which there are 2^16 different ones). It is apparent that, notwithstanding the random nature of the underlying network, the resulting probability distribution has a very rich structure, which I will analyze and interpret further in the next section.

Table III. The logic table for the addition, modulo2, of the 3 binary input variables. This 3-variable Boolean function is represented by the binary number 10010110=150.

In	Out
000	0
001	1
010	1
011	0
100	1
101	0
110	0
111	1

ANALYSIS OF THE RESULTS

Information entropy

A measure of the overall complexity of the probability profile P(i), i=0,255, is given by the information entropy I :

$$I = - \sum_{i=0}^{255} P(i) \log_2 P(i)$$

Note that P(i) is equal to the number of configurations (i.e. the number of choices of 2-variable Boolean functions and connections) that generates the 3-variable Boolean function i, over the total number of configurations. In other words, it is equal to the relative volume in phase space, occupied by that Boolean function. The information entropy I is represented in figure 5 as a function of the number of neurons. Note that it quickly reaches a maximum of about 5.3, and then remains almost constant. This confirms that the overall structure of the probability distribution does not depend very much on the number of neurons. On the other hand, it agrees with the observation that the learning ability of a network does not change very much once a minimal size for the network is reached.

Boolean function complexity

One of the striking features of the profile represented in figures 3 and 4 is the hierarchy in the Boolean functions. In table IV, I have listed the Boolean functions in order of decreasing phase volume (i.e. number of configurations that generate them). The Boolean functions with the largest volume (highest entropy) are 0=00000000 and 255=11111111. These Boolean functions carry no information of the inputs. Next come the ones depending on 1 input only (e.g. output = input#1), and then some Boolean functions depending on 2 inputs (e.g. output = (input#1) and (input#2)). Remarkably, the 3-input Boolean function with the highest entropy is the addition. The hierarchy between the Boolean functions is not

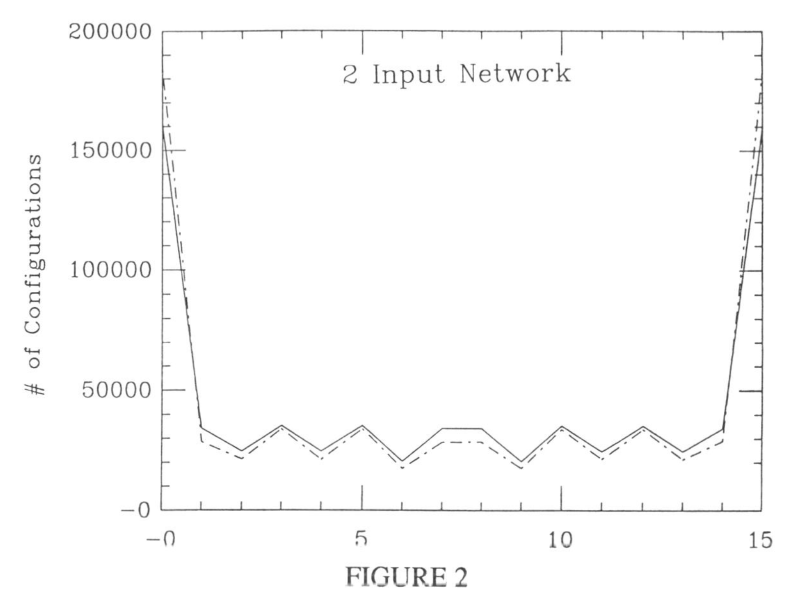

FIGURE 2
The probability P(i) that the 2-variable Boolean function i, i=0,15, is realized in a random Boolean network, for a network of 4 neurons(full line) and 7 neurons (dotted line).

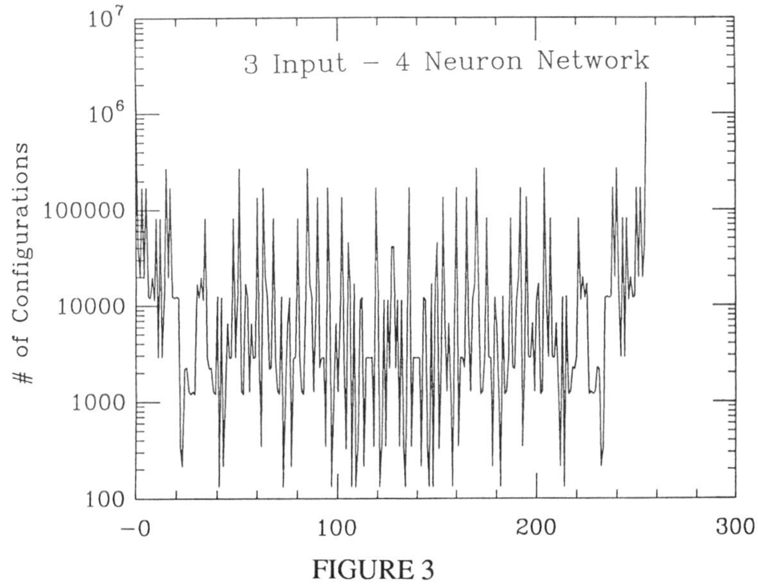

FIGURE 3
The number of configurations for which the 3-variable Boolean function i, i=0,255, is realized in a random 4-neuron Boolean network.

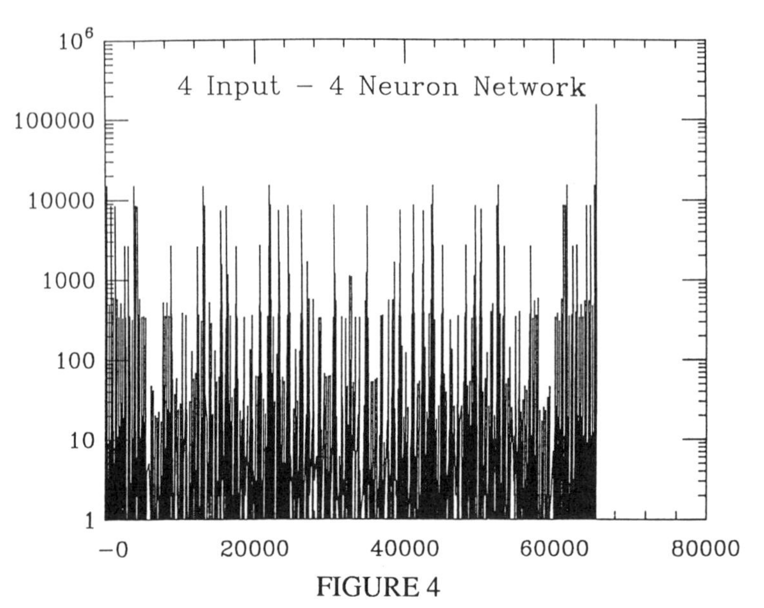

FIGURE 4
The number of configurations for which the 4-variable Boolean function i, i=0,65565, is realized in a random 4-neuron Boolean network.

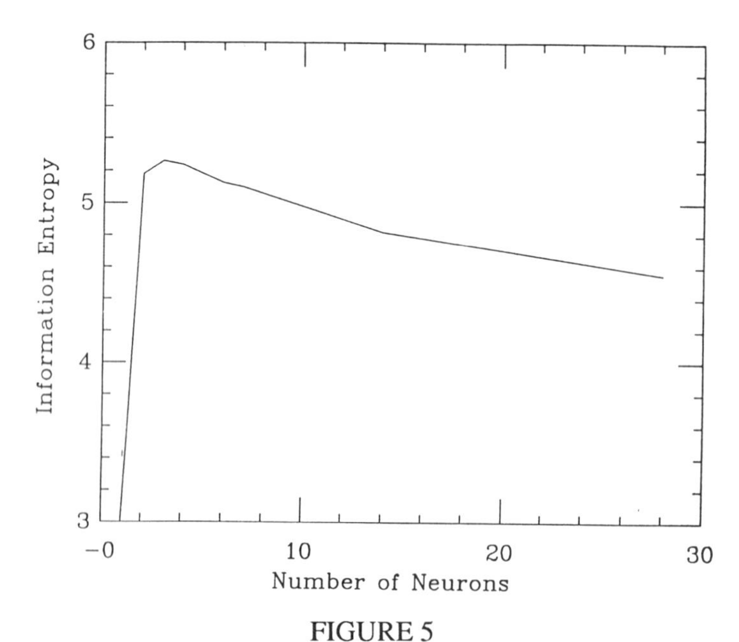

FIGURE 5
The information entropy I in function of the number of neurons for a random 4-neuron Boolean network

strict. For example, some Boolean functions can not be implemented at all in small networks, but can occupy a sizable part of the phase space as the size of the network grows. Also, the overall tendency is a slow growth of the trivial functions 0 and 255, and a decrease of the other ones.

What is the significance of the phase volume occupied by a Boolean function? As we will see, it plays an important role in the learning process. But one could also use it to give an alternative definition of its complexity (such that a large volume in phase space corresponds to a low complexity).

Table IV. The hierarchy of Boolean functions on the basis of their phase space volume for a 4-neuron network.

BOOLEAN FUNCTION	VOLUME IN PHASE SPACE =# OF CONFIGURATIONS	% OF VOLUME	TOTAL
0, 255	2 054 496	18.6%	37.2%
I1	268 176	2.4%	14.4%
I1 AND I2	169 500	1.5%	18.4%
I1 + I2	132 976	1.2%	7.2%
I1 AND NI2	80 716	.7%	8.4%
I1+I2 +I3	44 928	.4%	.8%
I1 AND I2 AND I3	40 608	.4%	1.5%
I1 AND I2 AND NI3	19 512	.2%	2.1%
....	....	......	....
	11 059 200		100%

Learning

In most of the learning procedures (simulated annealing, backpropagation, perceptron algoritm), and maybe also for learning in biological organisms, one may assume that the learning process corresponds to a random search in the phase space (being the space of the possible configurations, Boolean functions, synaptic strenghts, and coupling patterns), until a state is reached that is compatible with the examples that are being presented.

Let us investigate the consequences of this assumption, considering again our random Boolean network. If no examples are taught, the most likely outcome are the "trivial" functions (i.e. 00000000 and 11111111), since they occupy the largest fraction of the phase space. As examples are being presented, the accessible phase space shrinks. For instance, if one example has the output 0, and another one the output 1, then the trivial functions are eliminated.

To discuss this feature qualitatively, it is convenient to introduce the so-called Hamming distance between two Boolean functions. A n-variable Boolean function can be represented by a binary number with 2^n digits. The Hamming distance D between two such Boolean functions is equal to the minimum number of 0's and 1's that have to be changed in their binary representation to go from one to the other (0 < D < T=2^n). For example , the 8 nearest neighbours at a Hamming distance D=1 of the addition 10010110 (cf. table III) are: 00010110, 11010110, 10110110, 100100110, 10011110, 10010010, 10010100, 10010111. We also define the neighbouring phase space volume V(D) of a given Boolean function, as being the total volume occupied by all the Boolean functions at a Hamming distance D. In particular, V(0) is the phase space volume of the function under

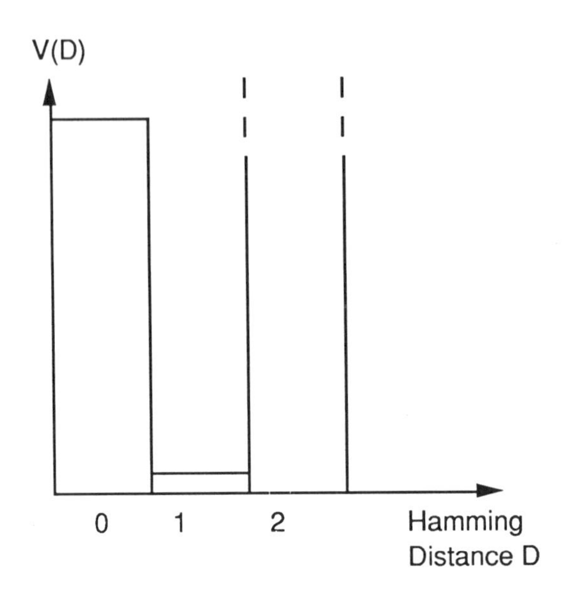

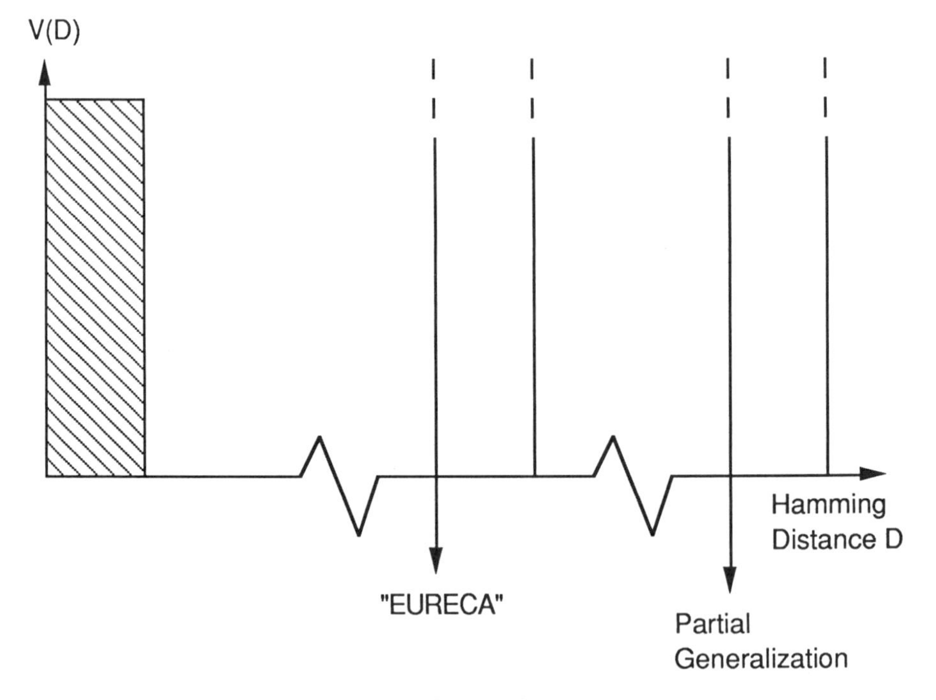

FIGURE 6
The neighbourhood function V(D) of the addition (modulo 2) in a random 4-neuron Boolean network, and a typical neigbourhood of a generalizable function in a more complex network.

consideration. In figure 6, we have represented the function V(D) for the addition (in a 4-neuron network).

Suppose now that we train a network to reproduce a Boolean function by teaching it a number of randomly choosen examples L < T=2^n. This amounts to specify L digits from the corresponding binary number and automatically eliminates all the Boolean functions at a Hamming distance larger than T-L. However, the other digits still have to be "guessed". If we assume that the learning is a random search in phase space, a combinatorial argument leads to the following probability for no errors :

$$P(\text{NO ERRORS}) = \frac{V(0)}{\sum_{D=1}^{T-L} V(D)\,\frac{(T-D)!\,(T-L)!}{(T-D-L)!\,T!}}$$

From this result, it is clear that the learning of examples reduces the possibility of a wrong guess by progressively eliminating the Boolean functions that are far away in the Hamming space. We can draw the following two important conclusions:

(1) A Boolean function can be inferred from a non-exhaustive list of examples (i.e. it is "generalizable") if it is a local maximum in the Hamming space. For example, the addition of three inputs in a 4 neuron random Boolean network is "generalizable", cf. figure 6.
(2) Complete learning (i.e. "guessing" all the digits correctly) will set in after a critical number of examples have been presented, thereby eliminating the closest peak in the Hamming space. This transition is expected to be more abrupt and may occur after presenting a relatively small number of examples in more complicated networks (more inputs and outputs, more neurons, cf. reference [6]). Partial learning may correspond to the elimination of the second closest large peak.

DISCUSSION

The present study [7] suggests that learning may be the outcome of the law of "maximum entropy", subject to the constraints, imposed by the teaching of examples. If a critical threshold of teaching is exceeded, full "understanding" occurs like a "eureca"-type of (phase?) transition. This type of intelligent organization should be distinguished from the intelligent modes of operation that result from self-organization in nonequilibrium structures or in truely unsupervised learning procedures (see e.g. the example of the ant colony, discussed by J.L. Deneubourg, and the Kohonen learning procedure applied to robotics by K. Schulten in this proceedings). The human brain clearly belongs to this much richer world. In this respect, it is even more remarkable that the "equilibrium world of entropy" produces a "crystal of intelligence" (i.e. the ability to extract a general rule from examples), that is accessible through supervised learning.

Not all the Boolean functions can be learned by generalization. It would be interesting to further characterize those functions that can be easily implemented on random networks with a minimum of supervision. A lot of research has been aimed at defining the complexity of a rule or sequence (i.e. information entropy, algoritmic complexity). We propose to relate the complexity of a logic function to how easily it is implemented in a random network.

ACKNOWLEDGMENTS

I am grateful to Dr. R. Kawai for his help with the numerical calculations. I would like to acknowledge financial support from the U.S. Dept. Energy, Grant DE-FG03-86ER13606, from the Inter-University Attraction Poles, Prime Minister's Office, Belgium, and from the N.F.W.O. Belgium.

REFERENCES

* Permanent address: L.U.C., B-3610 Diepenbeek, Belgium.
[1] T. Kohonen, **Self-Organization and Associative Memory** (Springer-Verlag, New York,1984).
[2] D. E. Rumelhart and J.L.Mc. Clelland, **Parallel Distributed Processing** (M.I.T. Press, Cambridge, MA, 1986).
[3] E. Aarts and J. Korst, **Simulated Annealing and Boltmann Machines** (J. Wiley, New York,1989).
[4] T. J. Sejnowski and Ch. Rosenberg, Complex Systems **1**, 145 (1987).
[5] A. Lapedes and R. Farber, in the Proceedings of IEEE Denver Conference on Neural Nets.
[6] S. Patarnello and P. Carnevali, Europhys. Lett. **4**, 503 (1987), Europhys. Lett. **4**, 1199 (1987).
[7] more details will be given in the following paper: C. Van den Broeck and R. Kawai, Learning in Feedforward Boolean Networks, to be published.

Molecular Evolution as a Complex Optimization Problem

Peter Schuster

Institut für Theoretische Chemie, Universität Wien,
Währingerstraße 17, A 1090 Wien, Austria

ABSTRACT : Polynucleotide folding provides an example of a physically simple but mathematically still complex phenotype-genotype relation. The notion of the phenotype is extended to molecules. It is straightforward to identify the spatial structure of a polynucleotide with the phenotype. Tertiary structures are hard to encode in compact notion and their predictions by theoretical models are notoriously bad. If, however, only secondary structures are considered as phenotypes, fairly reliable predictions of stable folded structures can be made and quantitative estimates of free energies and kinetic parameters – like rate constants for replication or degradation – are possible. Kinetic parameters cannot be derived from *first principles*. Instead empirical, qualitative models are used which are based on the current knowledge in biophysical chemistry. In order to simplify the combinatorial problem binary sequences (G,C) rather than natural four-letter sequences (G,A,C,U) are used. A simple genetic algorithm based on replication and mutation is applied to optimize some phenotypic properties. Relations between genotypes are measured in Hamming distances. The sequence space is a discrete space having the Hamming distance as metric. Individual genotypes are represented by points. Evolutionary optimization is visualized as a process in sequence space. In addition cost functions or value landscapes are explored by simulated annealing and computation of autocorrelation functions in the space of genotypes.

1. Genotypes and Molecular Phenotypes

The base sequences of the information carriers in molecular genetics – DNA and RNA in case of some classes of viruses – are commonly denoted as genotypes. They are also called the *primary* structures of polynucleotides. A genotype is understood

Self-Organization, Emerging Properties, and Learning
Edited by A. Babloyantz, Plenum Press, New York, 1991

best as a string of letters which are chosen from the four letter alphabet of naturally occuring bases (G,A,C,U in RNA or T in DNA, respectively). Phenotypes are organisms or virions. In case of molecular evolution in the test tube[1–3] *naked* RNA molecules are replicated, modified by mutation and subjected to selection. It is useful therefore to extend the notion of phenotype also to the spatial structures of these nucleic acid molecules.

Spatial structures of polynucleotides and proper phenotypes – as they are encountered with organisms – share many features. Single stranded RNA molecules exist in a great variety of different conformations – as a rule there is a single most stable one under given experimental conditions. Many organisms – in particular bacteria – can exist in different phenotypes. As the unfolding of phenotypes depends on environmental conditions, RNA and DNA molecules form different stable conformations under different experimental conditions – as there are temperature, pH, ionic strength etc.

The process of folding the string of bases of an RNA molecule into a three dimensional *tertiary* structure can be partitioned into two steps:

(I) folding of the string into a quasi-planar – two dimensional – *secondary* structure by formation of complementary base pairs, G≡C or A=U, respectively, and

(II) formation of a three dimensional spatial structure from the quasi-planar folding pattern.

Secondary structure formation is modelled much more easily than their transformation into tertiary structures. At present there are no theoretical models available which can predict tertiary structures reliably. In addition three dimensional structures are very hard to encode in compact form – commonly Cartesian coordinates of all atoms have to be stored. In contrast to the weakness in predictions and the difficulties in encoding of tertiary structures the predictions of secondary structures are simpler and reached already a level of fairly high fidelity[1]. This is mainly a consequence of the fact that the intermolecular forces stabilizing secondary structures – base pairing and base pair stacking – are much stronger than those involved in three dimensional structure formation. Algorithms are available which find not only the most stable conformation but also suboptimal foldings[2]. In addition secondary structures can be decomposed into strutural elements like free ends, loops, stems and joints which contribute to a good approximation additively to free energies or kinetic properties of the RNA molecules. The secondary structures can be readily encoded in strings which are not longer than those of the primary base sequences.

2. Evaluation of Phenotypes

The process in evolution which decides about the fitness of a given genotype may be characterized as evaluation of its phenotype. We realize an important *dichotomy*: selection acts only indirectly on the genotype by evaluation of the phenotype whereas all modification is done on the genotype. Again the modifications are subjected to selection only after unfolding of the corresponding phenotypes. Modifications, in general, come along by means of two mechanisms:

(I) mutation which is a change in the base sequence due to an error in the replication process, and

(II) recombination which does not create new parts of sequences but

instead combines anew the genetic information stored in two DNA molecules.

The fitness of a phenotype is the crucial property determining the outcome of selection in the orthodox Darwinian scenario. It is worth to make two remarks here: Darwin himself attributes some role to neutrality with respect to selection and uses the notion of *random drift* in his *Origin of Species* to characterize population dynamics in the absence of differences in fitness. From comparisons of base sequences in different organisms it is well known that neutral mutations are very frequent. Secondly, it is well established now that there are phenomena which are able to suppress competition. They lead to various forms of symbiosis in which former contestants cooperate for their benefits.

Let us return now to the Darwinian scenario which we shall study further here. Fitness is an exceedingly complex function involving many phenotypic traits. There is practically no chance to evaluate the fitness of an higher organism or of a bacterium independently from the selection process. This fact has often led to rather unjustified attacks on the theory of evolution since it seems to run inevitably into a tautology of *survival of the survivor*. Test tube evolution of molecules[3,4] opened up a new access to the fitness problem. On the basis of the known mechanism of RNA replication by $Q\beta$-replicase[5] models can be developed which allow to derive formulas relating molecular structures to the fitness determining quantities. Moreover, it is not unlikely that the physical – thermodynamic and kinetic – parameters which contribute to fitness can be measured directly in the near future.

3. Models of Realistic Fitness Landscapes

An alternative way to study relations between genotypes and phenotypes is to build a computer model. We made such an attempt with the ultimate goal to simulate evolutionary optimization in small populations[6,7]. This model was also conceived as a test of a theory of molecular evolution which describes replication with errors by means of kinetic differential equations[8,9] or as stochastic processes[10,11]. The relation between this concept of a *molecular quasi-species* and the computer simulation model has been reviewed recently[12].

Any comprehensive model of molecular evolution has to handle two formidable problems:

(I) The spatial structure of the phenotype G_k has to be expressed as a function of the genotype I_k: $G_k = \mathcal{G}(I_k)$, and

(II) the fitness of the phenotype has to be evalueted as a function of its structure: $f_k = \mathcal{F}(G_k) = \mathcal{F}(\mathcal{G}(I_k))$.

Since no analytical or other quantitative relations between genotypes and phenotypes are avialable at present, one has to go through the time consuming folding procedure in order to find out which spatial structure belongs to which genotype. As mentioned previously prediction and efficient handling of tertiary structures face enormous difficulties. The phenotype in our model is identified therefore with the secondary structure of the RNA molecule. For computation of secondary structures the algorithm developed by Zuker, Stiegler and Sankoff[13,14] was used. This algorithm determines the most stable planar conformation by means of a minimum free energy criterium.

The planar structures obtained by the folding algorithm are lacking knots and pseudoknots. Thus the phenotypes G_k are graphs in this model and can be encoded by strings of the same lengths as the primary sequences. The relations between primary structure I_k, secondary structure G_k and encoded secondary structure Γ_k are shown in figure 1. The graph G_k is easily partitioned into substructures. We distinguish four classes of substructures or structural elements: stems, loops,joints and free ends. The secondary structure is encoded by assigning a lower case letter to every base of the primary sequence. The position of a given base in the sequence is the same in I_k and in Γ_k. In particular we have:

(I) stems, encoded by $aaa\cdots$, $bbb\cdots$, $ccc\cdots$, $ddd\cdots$, etc.,

(II) loops, encoded by $xxx\cdots$,

(III) joints, encoded by $yyy\cdots$, and

(IV) free ends, encoded by $zzz\cdots$.

The use of letters is here not entirely arbitrary: the later they come in the alphabet, the more flexible are the corresponding parts of the RNA molecule. Stem regions are more rigid than loops, loops in turn are more rigid than joints and joints are less flexible than the free ends. In the coding applied here we do not distinguish between individual loops, individual joints or between the two free ends (3'- and 5'-end). Only the stems are counted individually. As shown in the figure a stem region appears twice in the encoded two dimensional structure – the two occurencies correspond to the two strands forming the double helix.

One important feature of Zuker-Stiegler-Sankoff algorithm is additivity of the contributions of substructures to the free energy of the RNA molecule. This additivity – certainly an approximation but apperently justified in the case of free energies – will be retained later on, when we evaluate the secondary structures according to other criteria. Additivity of substructure contributions is one source of selective

GGGGCGGCCGGCCCCGCGGCCGGG ⋯

Primary Structure: I_k

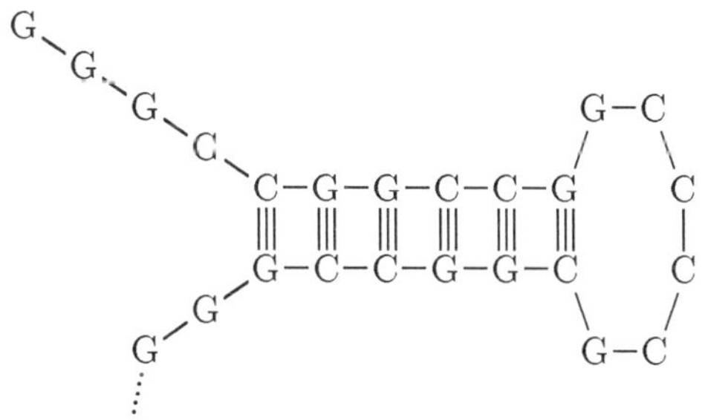

```
G
 \
  G
   \
    G                        G-C
     \                      /   \
      C                    /     \
       \                  /       \
        C-G-G-C-C-G                C
        ||| ||| ||| ||| ||| |||                |
        G-C-C-G-G-C                C
       /              \           /
      G                \         /
     /                  G-C
    G
    :
```

Secondary Structure: G_k

$zzzzaaaaaaxxxxxxaaaaaayy \cdots$

Encoded Secondary Structure: Γ_k

Fig.1. Folding and encoding of secondary structures.

neutrality on the phenotypic level. Several structures which consist of identical sets of substructures map onto the same selective values although their phenotypic appearences are different. For two selectively neutral genotypes we have in this case:

$$\mathcal{G}(I_k) = G_k \ , \ \mathcal{F}(G_k) = f_k$$

$$\mathcal{G}(I_j) = G_j \ , \ \mathcal{F}(G_j) = f_k$$

The second source of selective neutrality is even more common: several primary structures map onto the same phenotype:

$$\mathcal{G}(I_k) \ = \ G_k \quad , \quad \mathcal{G}(I_j) \ = \ G_k \ .$$

The polynucleotide landscapes considered here are characterized by an extremely high degree of neutrality. This has consequences for the evolutionary optimization process which will be discussed in the next section.

The genotypes of RNA molecules can be ordered with respect to matching sequences. The number of bases in which two sequences differ is called the Hamming distance (d). Sequences can be arranged in a discrete space such that the Hamming distance forms a metric. The object obtained thereby is known as sequence space in information theory[15]. We plot now the free energies of the most stable structures in the sequence space. The lanscape obtained thereby is highly bizarre. It represents an typical example of a *rugged landscape*[16]: closely related sequences – sequences which are mapped on nearby lying points in sequence space – may, but need not have very different free energies. Ruggedness is a consequence of the fact that small changes in the genotype may lead to large changes in the phenotype – for typical examples see Ref.12, p.115. In order to derive a quantitative measure for ruggedness we computed the autocorrelation function for a random walk of stepsize $d = 1$ on the free energy landscape[17]:

$$\rho(k) \ = \ \frac{<(F_i - F_j)^2> \ - \ <(F_i - F_{i+k})^2>}{<(F_i - F_j)^2>} \tag{1}$$

Herein the notion $< X >$ is used for the expectation value of X, F_i and F_j are the free energies of two randomly chosen genotypes I_i and I_j, and F_{i+k} is the free energy of the sequence which follows I_i after exactly k steps of the random walk. According to equation (1) we have $\rho(0) = 1$ and $\lim_{k\to\infty} \rho(k) = 0$. The average number of steps k at which the autocorrelation function takes – for the first time – a value of e^{-1} is characterized as correlation length ℓ: $\ln \rho(\ell) = -1$. In figure 2 the correlation length is plotted

as a function of the chain length ν. The two curves represent the data for two letter (G,C) and four letter (G,A,C,U) sequences: the free energy surface of two letter sequences shows much shorter correlation lengths ν and hence is more rugged.

In contrast to free energies kinetic constants of replication, A_k, and degradation,

D_k, cannot be computed straightway from known secondary structures G_k or their short-hand notations Γ_k. There is now satisfactory model available which is based purely on knowledge in biophysical chemistry. We had to use therefore a very crude estimate of these quantities. It is well known from virus specific RNA replication by Qβ replicase that only single stranded molecules are accepted as templates. The secondary structure has to *melt* in order to make replication possible. Therefore we used an estimate of the rate constant of melting as a simple model for the replication process:

$$A_k = \alpha_0 - \alpha_1 \sum_{j=1}^{s^{(k)}} \frac{n_j^{(k)}(1+n_j^{(k)})^3}{(1+n_j^{(k)})^4 + L} \; ; \qquad k = 1, 2, \ldots, 2^\nu \; . \tag{2}$$

By $n_j^{(k)}$ we denote the number of base pairs in the j-th stack of the secondary structure G_k, $s^{(k)}$ is the number of stacks in this structure, and α_0, α_1 and L are empirical constants. The function used in equation (2) takes care of the cooperativity in the melting process.

Degradation is modelled by taking into account all possible attacks of a hydrolytic agent or an enzyme with nuclease activity on the single stranded regions of the secondary structure G_k:

$$D_k = \beta_0 + \beta_1 \sum_{j=1}^{u^{(k)}} \frac{u_j^{(k)}}{u_{max}} \exp\left\{(u_j^{(k)} - u_{max})/u_{max}\right\} + \frac{\beta_2}{\nu} \sum_{j=1}^{z^{(k)}} z_j^{(k)} \; . \tag{3}$$

Herein the number of bases in the j-th loop of the secondary structure G_k is denoted by $u_j^{(k)}$. This structure has $u^{(k)}$ loops and there is a maximum loop size u_{max} above which loops are considered as completely mobile elements like free ends. The number of bases in free ends or large loops is given by $z_j^{(k)}$ and β_0, β_1 and β_2 are empirical parameters. The number $z^{(k)}$ is two – 3'- and 5'-end – plus the number of large loops.

Replication and degradation surfaces were studied by random walk statistics as well[6,7,17]. It turned out that the degradation surface is rather similar to the free energy surface whereas the replication surface appears to be much more rugged. The landscape obtained for the excess production, $E_k = A_k - D_k$ is relevant for evolutionary optimization – we shall call it the E-landscape for short. It represents the cost function of the complex optimization problem. The E-landscape was studied not only by means of random walk statistics, in addition we optimized the E_k value of a test sequence by simulated annealing. The main goal of this search was to explore the distribution of maxima of the excess production in sequence space and to compare it with a spin glass landscape. A chain length of $\nu = 70$ bases was chosen. In this case the highest possible E-value can be computed from the optimal combination of substructures: $E_{max}^{(70)} = 2245\,[t^{-1}]$ in arbitrary reciprocal time units. Such a *best* structure, however, does not occur as a minimum free

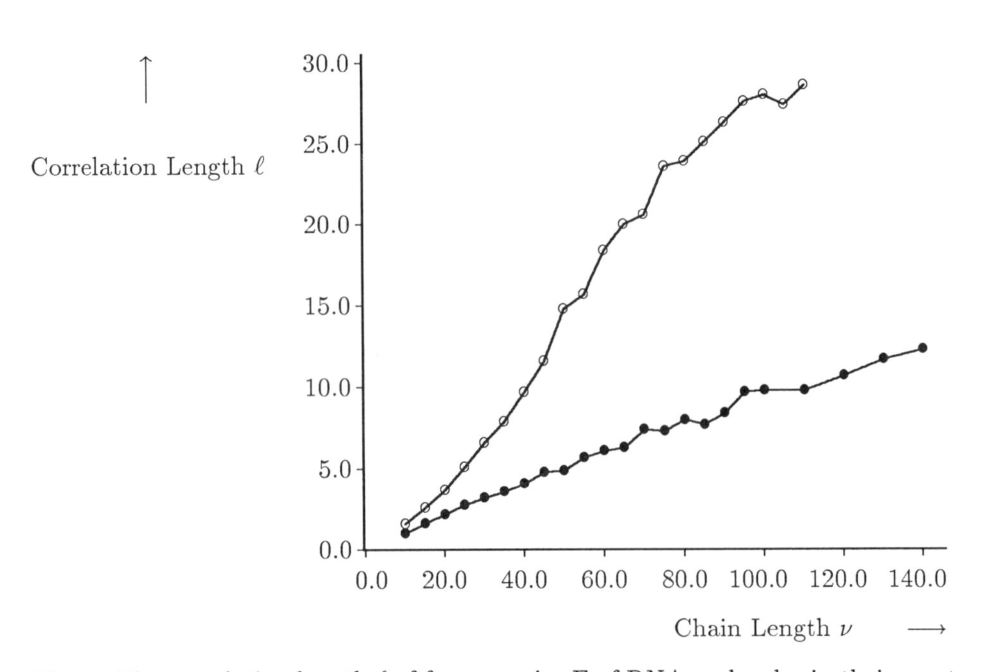

Fig.2. The correlation length ℓ of free energies F of RNA molecules in their most stable secondary structures as a function of the chain length ν. Binary sequences (G,C) and four letter sequences (G,A,C,U) are denoted by • and ○, respectively. Values are taken from[17].

energy folding pattern. These structures appear only as conformations with free energies above the corresponding conformational ground states. The highest value found by the simulated annealing technique was only $E_{opt} = 2045\,[t^{-1}]$; the genetic algorithm with variable chain length ν yielded a somewhat higher optimal value: $E_{opt} = 2059\,[t^{-1}]$.

The optimal value found with simulated annealing is degenerate: we found it with ten genotypes. They are related in pairs by symmetry – every polynucleotide sequence is transformed into one with identical structure by complementation and inversion. The remaining five sequences form two pairs of close relatives with Hamming distance $d = 2$ and one *solitary* sequence. These potential centres of quasi-species[9] are well separated in sequence spase – all Hamming distances are larger than $d \geq 30$.

In total, 879 genotypes with excess productions $E_k \geq 2011\,[t^{-1}]$ were identified. Their distribution in sequence space is characterized by clusters of peaks with high excess production which have rich internal structures. Like in mountainous areas on the surface of the earth, we find ridges as well as saddles and valleys separating zones of high excess production.

The distribution of near optimal configurations of the spin glass Hamiltonian shows characteristic features of *ultrametricity*: arbitrarily chosen triangles of near optimal configurations are either equilateral or isosceles with small basis. The distribution of clusters of high excess production in sequence space, on the other hand, shows no detectable bias towards ultrametricity. It does not deviate significantly from a random disribution. Two causes may be responsible for this apparent difference between spin glass Hamiltonians and polynucleotide folding landscapes: either binary sequences of chain length $\nu = 70$ are too short to reveal higher order structures in the value landscape, or polynucleotide folding and evaluation of the folded structures have some fundamental internal characteristics which is different from that intrinsic to the random couplings in spin glass Hamiltonians.

4. A Model of Molecular Evolution

In order to study evolution dynamics on the excess production (E) landscape a series of computer experiments was carried out[6,7]. A population of several thousand binary (G,C) sequences with chain lengths up to $\nu = 100$ nucleotides was used to optimize replication-mutation performance on the E-landscape described in the previous section. Mutations were either restricted to point mutations, or point mutations, deletions and insertions up to ten bases were allowed. In the former case the chain length remained constant. For point mutations we applied the *uniform error rate* model[9]: the single digit accuracy q – this implies an error rate of $1-q$ per digit and replication – is assumed to be the same at every position of the sequence. For the mutation probability from sequence I_j to I_k we find:

$$Q_{kj} = q^{\nu} \left(\frac{1-q}{q} \right)^{d_{kj}} . \tag{4}$$

The chain length is denoted by ν as before and d_{kj} is the Hamming distance between the two sequences I_k and I_j. The model has an important feature: all mutation rates at constant chain lenghts ν can be expressed by only two quantities, the single digit accuracy q and the Hamming distance d. In case of variable chain lenghts the frequencies of deletions and insertions are additional input parameters. In the computer experiments reported here a value of $q = 0.999$ corresponding to one error per one thousand digits was applied throughout. The stochastic dynamics of the replication-mutation-degradation reaction network was simulated under the conditions of a CSTR (Continuously Stirred Tank Reactor) by means of an efficient computer algorithm[18]. Sequences produced in excess are removed by a stochastic dilution flux. Accordingly the total population size fluctuates around a mean value $\overline{N}$ and fulfils a $\sqrt{N}$ law.

Here we report the results of computer runs[6,7] which started with an initial population of 3000 *all-G-sequences* of chain legths $\nu = 70$. The homogeneous initial population of *all-G-sequences* was chosen because this genotype cannot unfold into a phenotype which has a secondary structure. Moreover, there are no structures with reasonably good values of the excess production in its neighbourhood. In order to use a metaphor from terestrial landscapes we start our computer experiments in the planes far away from any higher peak. During a typical simulation experiment with our variant of a genetic algorithm the mean excess production of the population $\overline{E}(t)$ increases monotonously – apart from small random fluctuations – and eventually approaches a plateau value. Depending on population size, replication accuracy and structure of the value landscape two extreme scenarios were observed:

(I) The population resides for rather long time in some region of sequence space and makes little progress in optimization. It moves quickly and in jumplike manner into another area of higher mean excess production. The optimal mean excess production is approached by stepwise improvements.

(II) The population approaches the optimal value gradually by steady improvements and the population migrates rather smoothly through sequence space.

The first scenario is reminiscent of punctuated or stepwise evolution. It occurs here in small populations, at low error rates and in landscapes with distant local maxima. The second scenario dominates in large populations, at high error rates and in landscapes with close by lying local maxima. The two scenarios represent only the extreme cases. There is a steady transition from scenario I to scenario II examples of which were observed in additional computer runs: the gradual approach changes first into a sequence of small jumps when the error rate is reduced. Larger jumps appear on further decrease of the mutation frequency and eventually the population is caught in some local maximum of the fitness surface.

Three different secondary structures with almost the same excess production were obtained in three optimization runs starting from identical initial conditions: 3000 *all-G-sequences* of chain length $\nu = 70$ on the same excess production surface. The individual runs differed only with respect to the sequence of random events along the stochastic trajectory which was achived by using different initial seeds for the random number generator. The genotypes corresponding to the three structures lie far away from each other in sequence space: their Hamming distances span an almost equilateral triangle of chain length $d = 30$ which has an average distance of $d = 25$ from the initial *all-G-sequence*. An interpretation of this result can be given straightway: we start from sequences which are unable to form any secondary structure and therefore an initial random walk in the planes of the fitness surface is likely to find better sequences in almost every direction of sequence space. An initial random choice will lay down the direction into which the population migrates later on in its search for higher excess production. Apparently there is ample variability for the choice of such a direction: we have 70 independent directions in the sequence space of binary sequences with chain length $\nu = 70$.

Optimization runs with the genetic algorithm were also carried out at replication accuracies which lie below the critical accuracy for the population size and the fitness surface applied. No optimization was observed in these cases. Populations drift randomly in sequence space. This general behaviour which is in agreement with the error threshold concept in finite populations[11] was observed even if we started with homogeneous populations of near optimal genotypes obtained in previous optimization runs with smaller error rates. The initial master sequence is first surrounded by a growing cloud of mutants and then it is displaced by one of its error copies. All other sequences are lost likewise after sufficiently long time. In other words, every genotype has a finite lifetime only and localization of the population in quasi-species like manner does not occur. The predictions of the theory of molecular evolution[8,9] are valid also in populations as small as a few thousand molecules.

5. Concluding Remarks

Systematic studies on the landscapes underlying complex optimization problems were so far always restricted to the statistics of random models. In this contribution we reviewed at attempt to explore a realistic fitness surface of a biophysical optimization problem. In order to be able to study and handle such a highly complex object like a fitness surface, reduction to the most simple case was inevitable. The simplest presently known entities which show optimization behaviour in the Darwinian sense are small RNA molecules which can be replicated in the test tube by means of an enzyme. They mutate frequently since the enzyme has rather low replication fidelity. Selection and adaptation to the environmental conditions are already observed in such populations of RNA molecules replicating *in vitro*. In these molecular evolution experiments the phenotype is nothing but the three dimensional structure of the RNA molecule. It is recognized by the enzyme and thus determines the fitness of its carrier.

The feasibility of computer studies in the exploration of fitness landscapes for molecular evolution is bound to the predictability of molecular structures. Three dimensional structures of polynucleotides or proteins are very hard to predict and systematic investigations of the corresponding free energy surfaces are out of question at the present state of the art. Restriction to secondary structures of polynucleotides, in particular RNA molecules, improves the situation a lot: these structures can be predicted fairly reliably and efficient algorithms are available for computation. An additional assumption was made in order to be able to compute the free energy surface for two dimensional folding patterns of RNA molecules and to model a fitness landscape on which the optimization process can take place: we studied binary (G,C) sequences instead of the natural four letter (G,A,C,U) sequences. How significant are the results derived from binary sequences? Computations of correlation lengths for random walks on free energy surfaces for both classes of sequences showed that the binary sequences represent an interseting extreme case: the correlation length is shortest for them and this implies that their folding landscapes are most complicated. Secondary structures of binary sequences are more sensitive to mutations than those derived from sequences of the four letter alphabet.

The simulation studies of molecular evolution demonstrated the consequences of a high degree of selective neutrality. How general are these observations gained from an artificial computer model for real biology? What matters here is the structure of the fitness landscape – the degree of neutrality after all is a feature of the landscape. Selective neutrality, no doubt, does exist in nature too. Cases of neutrality on the molecular level are easy to visualize. There are the degeneracies of the genetic code and the mutations having no or little effect on protein function since they exchange amino acid residues which do not effect protein stucture very much. Surely, many additional sources of neutrality exist on the higher hierarchical levels of biology.

In molecular evolution complexity can be traced down to the properties of the carriers of genetic information. Here the simplest case of a relation between genotype and phenotype is the folding of strings into two dimensional structures by means of digit complementarity rules. This process cannot be casted into an analytical expression and creates already a highly complex free energy landscape. The ruggedness of value landscapes is the ultimate cause why all transformations of genotypes into phenotypes are exceedingly complex. Unfolding of genotypes cannot be reduced to simple algorithms therefore – not even in the most simple cases.

Acknowledgements

The research work reported here was supported financially by the Austrian *Fonds zur Förderung der wissenschaftlichen Forschung*, projects P 5286 and P 6864, by the German *Stiftung Volkswagenwerk* and by the *Jubiläumsfonds der Österreichischen Nationalbank* project 3819. Computer time on the IBM 3090/400 VF mainframe supercomputer of the *EDV Zentrum der Universität Wien* was provided within the frame of the EASI project of IBM.

REFERENCES

1. J.A.Jaeger, D.H.Turner and M.Zuker, Improved Predicitons of Secondary Structures for RNA, Proc.Natl.Acad.Sci.USA 86:7706 (1989).

2. M.Zuker, On Finding All Suboptimal Foldings of an RNA Molecule, Science 244:48 (1989).

3. S.Spiegelman, An Approach to Experimental Analysis of Precellular Evolution, Quart.Rev.Biophysics 4:36 (1971).

4. C.K.Biebricher, Darwinian Selection of Self-Replicating RNA Molecules, Evolutionary Biology 16:1 (1983).

5. C.K.Biebricher and M.Eigen, Kinetics of RNA Replication by $Q\beta$ Replicase, in: "RNA Genetics, Vol.I. RNA-directed Virus Replication", E.Domingo, J.J.Holland and P.Ahlquist, eds., CRC Press, Boca Raton (1988).

6. W.Fontana and P.Schuster, A Computer Model of Evolutionary Optimization, Biophys.Chemistry 26:123 (1987).

7. W.Fontana, W.Schnabl and P.Schuster, Physical Aspects of Evolutionary Optimization and Adaptation, Phys.Rev.A 40:3301 (1989).

8. M.Eigen, J.McCaskill and P.Schuster, Molecular Quasi-Species, J.Phys.Chem. 92:6881 (1988).

9. M.Eigen, J.McCaskill and P.Schuster, The Molecular Quasi-Species, Advances in Chemical Physics 75:149 (1989).

10. L.Demetrius, P.Schuster and K.Sigmund, Polynucleotide Evolution and Branching Processes, Bull.Math.Biol. 47:239 (1985).

11. M.Nowak and P.Schuster, Error Thresholds of Replication in Finite Populations: Mutation Frequencies and the Onset of Muller's Ratchet, J.theor.Biol. 137:375 (1989).

12. P.Schuster, Optimization and Complexity in Molecular Biology and Physics, in: "Optimal Structures in Heterogeneous Reaction Systems", P.J.Plath, ed., Springer Series in Synergetics, Vol.44, Springer-Verlag, Berlin (1989).

13. M.Zuker and P.Stiegler, Optimal Computer Folding of Large RNA Sequences Using Thermodynamics and Auxiliary Information, Nucleic Acids Research 9:133 (1981).

14. M.Zuker and D.Sankoff, RNA Secondary Structures and their Prediction, Bull.Math.Biol. 46:591 (1984).

15. R.W.Hamming, Coding and Information Theory, 2nd Ed., pp.44-47, Prentice Hall, Englewood Cliffs (1986).

16. S.Kauffman and S.Levin, Towards a General Theory of Adaptive Walks on Rugged Landscapes, J.theor.Biol. 128:11 (1987).

17. W.Fontana, P.Schuster, P.Stadler, E.Weinberger, T.Griesmacher and W.Schnabl, Characterization and Quantitative Evaluation of RNA Folding Landscapes, Preprint (1990).

18. D.T.Gillespie, A General Method for Numerically Simulating the Stochastic Time Evolution of Coupled Chemical Reactions, J.Comp.Phys. 22:403 (1976).

A NONTHERMODYNAMIC FORMALISM FOR BIOLOGICAL INFORMATION SYSTEMS: HIERARCHICAL LACUNARITY IN PARTITION SIZE OF INTERMITTENCY

Arnold J. Mandell and Karen A. Selz

Laboratory of Experimental and Constructive Mathematics
Departments of Mathematics and Psychology
Florida Atlantic University
Boca Raton, Florida 33431

1. INTRODUCTION

Generally, the thermodynamic formalism for dynamical systems, the study of nonlinear flows and/or their maps (the first transformed into the second by a Poincaré surface of section) involves modeling them using a simpler symbolic system and them applying measure theoretic tools from abstract ergodic theory and statistical mechanics to their analyses. (Kolmogorov, 1956; Sinai, 1972; Bowen, 1975; Ruelle, 1978).

The uniform hyperbolicity requirements for almost all the theorems make applications tenuous and pioneers like Ruelle (1990) express doubt about their usefulness, relevance, and appropriateness outside of the area of hydrodynamic "weak" (and temporal) turbulence. It is unfortunate that the attempts to apply this remarkable power to elucidate both uniqueness and universality in difficult–to–categorize real dynamical systems in biology (where we feel it has a natural place), (Koslow, Mandell, and Shlesinger, 1987) often take the form of e–mailed time series from mathematically intimidated biologists to mathematically and algorithmically sophisticated but lay biologically knowledgeable physicists. Even when this is not the case, researchers from both sides of this great divide feel successful when they can map real biological and behavioral events invertibly onto already established "principles" of dynamical systems without adding new possibilities to either.

It may be that the approach using *Gibbs measures on equilibrium states* of Sinai–Bowen–Ruelle as supported by the work on Axiom A, Anosov, basic sets of the ergodic theorists is too constricted for either accurate or meaningful application to systems in biology which at both small and large scales of time are *generating increasingly non–normalizable measures.* It may be that it is specifically here that the full mathematical machinery of *the thermodynamic formalism* must be restructured for biology into what we will call the *evolutionary formalism for biological dynamical systems.* Ornstein (1989) has pointed out that the exponentially relaxing, Bernoullian horseshoe (Smale, 1967), metaphorically generic in such systems, is measure zero in generalized families of dynamical systems and even on these grounds, not a candidate for the role of a model of some real system. The prominence of intermittency in the biological dynamical systems (Rinzel, 1987; Mandell and Kelso, 1990; Mandell and Selz, 1990; Selz and Mandell, 1990) also disallows the *basic set* on empirical grounds.

Self-Organization, Emerging Properties, and Learning
Edited by A. Babloyantz, Plenum Press, New York, 1991

Generally, information input to biological information processing systems such as the brain, increases the distance from a Gaussian expectation. Non–Gaussian distributions are signified by the third and fourth moments ≠ 0, (Rosenblatt, 1987). Information input leads to deviations from a strong central tendency and in this statistical sense, more ambiguity. More intuitively, in biological dynamics, more information usually leads to more questions.

We will explicate the growth of this biological ambiguity (entropy) as computable from the unfairness of a coin, p, whose expected longest run length grows like that observed in the deterministic process and transformed in the usual way, $-log p_i \, p_i$, (Selz and Mandell, 1990). A generalized and "generic" intermittency model of biological dynamics, $\varphi^t(\tau)$ with its topological mechanisms and ergodic measures will be studied. Deviation from the minimal information state of an (idealized) Gaussian distribution grows as a function of both the parameter–dependent local eigenvalue of the attractive–repelling fixed point and as a logarithmic function of n.

2. WHY NOT THE THERMODYNAMIC FORMALISM FOR BIOLOGICAL DYNAMICS?

The thermodynamic formalism for differentiable dynamical systems in the C^r topology, $r \geq 2$, involves limit theorems which require that the underlying dynamical system, $\varphi^t(\tau)$ on manifold, M, with algebra σ, supporting the many equivalent ergodic measures, μ_i, a space $\{M, \sigma, \mu, \varphi^t\}$ are uniformly expanding (Ruelle, 1976; Bowen, 1977). The Sinai–Ruelle–Bowen a "natural" measure of prominence in the ergodic theory of dynamical systems, $\mu(p)$, is in essence, a distribution function which is normalizable; $\int \varphi^t(\tau) \, p \, d\tau = 1$. The characteristic "uniform" expansion rate (Poincaré's sensitivity to initial conditions) is exponential, quantified by the "global" derivative which is averaged over the orbit, $\chi = log(D\varphi^t(\tau)/t$ (by the chain rule of differentiation) (Oseledec, 1968). This is called the characteristic exponent and serves as an invariant of these Anosov–Axiom A idealized systems (Pesin, 1976). The Hausdorff–like capacity measure, $\mu(d_H$, of set A, $d_H = lim \; sup$ ($r \to 0$) $(logN(r,A)/log(1/r)$ *where* $N(r,A)$ *is the minimum number of open balls of radius* r *necessary to cover set* A. The Hausdorff measure, d, is similar in that $N \cong r^d$. It can be viewed as a measure on the support of χ, and several theorems (Katok, 1980; Manning, 1981; Ledreppier and Young, 1985) were consistent with the general relation,

$$H_{max} = \chi \circ d_H \qquad \{1\}$$

in which, H_{max}, is the maximum, *topological*, entropy. H_{max} is the growth rate of new orbits, *the information generating rate* of $\varphi^t(\tau)$.

Constructive representations of these idealized maximal entropy, uniformly expanding dynamical systems such as Smale's horseshoe have Bernoulli characteristics (coin–flipping randomness) when the system's phase space is treated with symbolic dynamics, (alphabet–like symbol sequences mapped onto the orbit) following its (*generating*) partition. A generating partition ξ on measure space M is a maximal partition such that the usual entropy is a supremum on it (H_{max}) and practically, no more than one point is in one box. ξ is a point partition, whose atoms are the singleton sets of M.

At the heart of the ergodic theory of uniformly hyperbolic $\varphi^t(\tau){:}M \to M$, are the many measures μ on M which are invariant under $\varphi^t(\tau)$. One might emphasize the importance of four with respect to a geometric characterization of the phase space point set:

(1) The distribution of periodic (unstable fixed points) which Bowen showed to be equidistributed with respect to μ (Bowen, 1971);

(2) The maximization of the entropy, H_{max}, (Bowen, 1970);

(3) Asymptotic time averages for most initial conditions, $E(\Phi^{t}(\tau))$, (Ruelle, 1979);

(4) Smooth measure, with $\varphi^{t}(\tau)$ Anosov (uniformly hyperbolic) μ may be equivalent to Lebesgue (Anosov and Sinai, 1967).

A second signature of uniformly hyperbolic systems, one which is used to justify numerical experimentation with expanding dynamical systems, $\varphi^{t}(\tau)$ the *shadow lemma* which says that for any small $\varepsilon > 0$, there is a $\delta > 0$ such that the intersection of the stable and unstable manifold at single point, $\{\tau_i,\tau_j\}$ $W^u(\tau_i) \cap W^s(\tau_j)$, is such that there is a third point, τ_k, which follows τ_i in the future and τ_j in the past (Anosov, 1967; Bowen, 1975). This describes the process of *shadowing* which, as the lemma says, results from the actions on the orbit bundle contracting stable manifold, W^s *ironing down the orbits* onto the stretching unstable manifold, W^u such that a pseudo–orbit will ε–track the real orbit up to kinks and bends. Examples occur near homoclinic tangencies or attractive–repelling ghosts of erstwhile fixed points in inverse saddle bifurcation which both *gather and mix phase* and distort time such that the lemma breaks down.

A third fundamental assumption of the *thermodynamic formalism* implicit in the definition of $\varphi^{t}(\tau)$ as an Axiom A transformation preserving the ergodic probability measure, μ, on compact manifold, M, is that the correlation function between two observables, τ_i and τ_j, $R\tau(t) = \mu\{(\tau_i \circ \varphi^{t})\tau_j\} - \mu(\tau_i)\mu(\tau_j)$ has a Fourier transform which is holomorphic (except for the poles) on the strip and has no poles near the real axis which indicates an exponential decay of $R\tau(t)$. This was proven by transforming the time correlation function to a spatial one using a one dimensional lattice. (Pollicott, 1986; Ruelle, 1986).

The thermodynamics formalism fails in biology for robustly real and not technically fussy reasons:

Intermittent dynamical systems generate non–normalizable measure. Without characteristic scales, the Sinai–Bowen–Ruelle statistical expectation is meaningless as are (1), (2), (3), and (4);

The shadow lemma justifying the formal and numerical stability of uniformly hyperbolic systems is not applicable to intermittent dynamical systems since the characteristic topological singularities generating the multiscale structures of intermittency destroy the near hyperbolic condition on shadowing.

In biological dynamics, dynamics, $R\tau(t)$ cannot select a characteristic time due to the nonuniformity in the expansion such that its decay ranges from stretched exponential, $\cong e^{-t^{\beta}}$ to a power law, $t^{-(1+\beta)}$. The power spectra are characteristically discontinuous with multiple poles indicating singularities in the analytic continuation. These "nonharmonic" resonances are characteristic of intermittent systems (Baldi et al, 1990).

More specifically, power series expansions of intermittent processes are lacunary, their Fourier series, for example, have "gaps" (Kahane, 1985) which if they grow rapidly, prevent analytic continuation due to dense distributions of singularities on the unit circle. Noncontinuability is a characteristic property of lacunary power series. Pollicott and Ruelle's "meromorphic on the whole plane," holomorphic, theorems are conditioned upon Axiom A uniform hyperbolicity which is lost in intermittency.

Lacunarity was used by Mandelbrot (1982) as a description of the characteristic size of holes in fractal objects such as Sierpinski gasket which can be altered such that fractal dimension is preserved with changes in lacunarity. For example, if length r varies as n^2 such that the fractal ("similarity") dimension, $\log n/\log(1/r) = \log n/\log(1/n^2) = 0.5$ for, as an example, $n = 2, 3,$ and 5, we see that lacunarity (defined as the ratio of the largest gap to the a length of the entire fractal object) may change (2/4, 3/9, 5/25).

These three biologically relevant failures of the *thermodynamic formalism* are equivalent through the loss of smoothness of conditional measure, μ, along W^u. The conditional measures on unstable manifolds can fail to be absolutely continuous with respect to Lebesgue measure. Informally, since W^u is one dimensional, it and its orbital densities can be viewed as a frequency distribution (with holes).

As an example of non–normalizability, let the frequency distribution of inter–event times, $F(\tau)$, of an intermittent system be transformable into a probability distribution such that $p(\tau) = k/\sqrt{1 + \tau^2}$. $\int p(\tau)\, d\tau$ is logarithmically divergent for large $|\tau|$. As $F(\tau)$ spreads out, $p(\tau)$ as a probability distribution normalized to one tends to zero. For a normalizable $p(\tau)$, $F(\tau)$ grows like n whereas for this non–normalizable $p(\tau)$, $F(\tau)$ grows like $n/\log n$. Intuitively, as we increase our sampling length n, we see larger values of τ such that the weight of any particular value of τ is continually decreasing. *New information leads systematically to decreasing confidence in what is known, an increase in available choices, and an ever weakening (disappearing) average to use in choosing between them.*

3. NONEXPONENTIAL GROWTH IN DIVERGENCE AS AN EVOLUTIONARY CLOCK

An early study of the evolutionary divergence of homological polypeptide chains in hemoglobin, myoglobin, cytochrome c and constant regions of the immunoglobulins indicated that the number of sites subject to variation increased across evolutionary, sidereal not generational time (Jukes and Holmquist, 1972). Divergence in *minimal mutational distance* was (irregularly) increasing within the time span of the millennia (for example) from *neurospera to humans* in these common proteins. Transforming sequences to time, the polysome prints proteins at about one residue per second, Waterman et al (1987) reports that optimal matching of DNA sequences in proportion to length can vary from $\cong \log n$ *to* n.

Wright (1931) used a (continuous) diffusion approximation to the discrete–time stochastic process to try to capture the evolutionary behavior of gene frequencies. He used a multinomial sampling approach to generate a Markov chain that described the divergence of the genetic composition over time in relationship to the largest, rate–determining, noninteger eigenvalue, λ, (1945). Later work indicated that *effective population size*, N, must also be included in these models (Maruyama and Kimura, 1980).

In the simplest case without mutation or selection in which the effective number of genes, N, each with allelic types L_1 or L_2, which are randomly paired (with replacement), we ask the transition probability P_{ij} (a Markoff matrix), that if there are i L_1 genes in generation n, what is the probability that there will be j L_1 genes in generation $n+1$. This can be computed from simple probability laws, (Feller, 1968),

$$P_{ij} = \binom{N}{j} \left[\frac{i}{N}\right]^{j} \left[1 - \frac{i}{N}\right]^{N-j} \qquad i,j = 0,1,2,\ldots,N \qquad \{2\}$$

From the transition matrix, $P = P_{ij}$, with absorbing states at 0 and N, the Wright model for the effective population size, N, indicates that $N \cong 1/1-\lambda$ probability for shared parenthood $\cong 1/N$. Changes in the *effective population size* serves as a normalized measure on the rate of increases or decreases in genetic variation which can be called the eigenvalue effective population size. For example, subpopulations getting isolated in a niche which through large amplitude fluctuations disappear and turn over their site to another subpopulation reducing genetic variation. In this case, the effective, nondisappearing, population size serves as a metric on the space of genetic variation.

More generally, it appears that evolutionary divergence is a function of both the parametrically controlled eigenvalue, $\lambda(\varphi_r^t)$ and the length of time of observation, $(n = t)$ although the process is neither periodic nor exponential in rate but rather intermittent (Mandell and Selz, 1990).

We know that the rhythms of evolutionary (divergent) process are not regular. The "stickiness" in some regions that the evolutionary orbit traverses hold up the processes in bends and kinks. This generates "repeats" in the output which then demonstrate "escapes" to other phase space regions. We have referred to these nonuniformly expanding manifolds as "curdled hyperbolic." Singer (1982) reported lengths of repeated DNA sequences varying from a few nucleotides to several thousand which might follow each other in a nearly regular fashion or be scattered throughout the genome.

Recently attributed to a retroposition mechanism involving RNA information transport back to and incorporated by genomic DNA (Deininger and Daniels, 1986), as an abstract dynamical system it can be envisioned as getting stuck in a transient game of forward–backward ping–pong as orbits have a wont to do in regions of homoclinic tangencies and inverse saddle–node bifurcations (Manneville and Pomeau, 1980). Of course, it is known (Waterman et al, 1987) that the statistical probability of longest sequence matching can be modeled by two parameters: the fairness of the coin and the logarithm of the sequence length (analogous to the *effective population size*, n). The fairness of the coin, of course, is comparable to the most positive eigenvalue of the transition matrix of the Markov chain.

It is becoming clear that anomalous diffusion is a better model for the divergent evolutionary processes than diffusion (Gillespie, 1986) because without finite higher moments, there is no single time scale in the problem. In fact, as we shall show, these intermittent evolutionary processes are not only not smooth, but are *lacunary in scale such that discontinuities occur in their point (generating) partitions size.* We will propose that parameter sensitive expanding intermittent systems are generic for biological information systems. It perhaps makes intuitive sense that the expanding actions of the discrete event maps of biological systems would be absolutely continuous on the manifold of its actions and distributions, in the language of dynamical systems, uniformly or near uniformly hyperbolic with some set equivalence relations between them based on their maximal entropy (Parry, 1964). This maximal entropy state would contain no information. *Information enters with run formation beyond that expected by chance in a random or uniform hyperbolic Bernoulli system. These runs "use up" orbital points such that some values of the observables begin to disappear from particular locations in the continuous distribution of (all) scales.*

Mandelbrot's (1977) *lacunarity* injects finer distinctions between fractal structures that share the same Hausdorff dimension and involve the prefactor of scaling expressions of point set densities such as the α of $N \cong \alpha r^d$. Our *lacunarity* describes nonuniformity in the scales occupied such that *holes are made in the hierarchy.* If one imagines how it is that eliminating members of the hierarchical set of frequencies present in the range of a diatonic musical scale yields a *chord*, and how a time–dependent progression of chords constitutes the harmonic support *and suggestion* of a melody in a musical message, it is perhaps easier to see how our theory of coding with intermittency in time may take place in biological information systems.

4. CODING WITH HIERARCHICAL LACUNARITY

Neurophysiological systems from the spinal cord to the cerebral cortex tend to be regulated by inhibition (Jackson, 1931). Expression of biochemical mechanisms from nucleotide–protein transcription to feedback–regulated, rate limiting enzyme activity, can also be viewed as generating and/or transporting information by inhibition. Molecular, ontogenetic, and phylogenetic progressions can be seen to "code" in sequences of metastable states; "stopping off places" in a smoother longer range emerging dynamic. Exaggeration of this mechanism is called *paedomorphosis*, the retention of earlier stages of ancestral development in the adult form. Intermittency characterized by "runs" of similar values of various lengths is sensitive to parameters, is without the mode–locking potential of communication by limit cycle resonances, and is consistent with the generic bursting characteristics of neural and endocrine information systems. We used a generic intermittency map which is uniformly expanding to study such a system (Selz and Mandell, 1990). We restrict our studies to a short and physiologically realistic sample length on $n \lesssim 1000$.

Whether the intermittent biological system is "saying" anything at all constitutes a difficult problem since the "runs" of intermittency, of course, present in pseudorandom systems are known to grow as $c \log n$. We deal with this issue by computing both the kurtosis, σ^4, (a normalized expression of the fourth moment as the "flatness" of the distribution function and near zero for Gaussian random systems) and a comparison of the deterministic observed, $\mathbb{D}(r)$ and that which would be expected, $\mathbb{E}(r)$ in the number of runs of a random variable time series partitioned as a $\{0,1\}$ Boolean system.

Hierarchical lacunarity is seen as a pattern of gaps in the histograms of the distribution of partition sizes. An index of *global lacunarity*, $\mathscr{L}$, is computed as the ratio of the largest number of missing partitions across the range of sizes of point (generating) partitions to their number.

We compute the (local) eigenvalue, λ, of the attractive–repelling fixed point analytically and the (global) leading characteristic exponent, χ, and its variance, $var(\chi)$ using the standard Wolf algorithm across changes in the forcing parameter r of our model system.

For *differentiable* nonuniform hyperbolicity problems, the leading characteristic exponent, χ, serves as the upper bound on the entropy of the measure, $h(\mu)$ (Osledec, 1968, Ruelle, 1989). Ledrappier and Young's theorem requires that for equality, $h(\mu) = \Sigma\chi_i(dim\ i) = \mathbb{E}(\chi)$. μ must be smooth along the submanifolds corresponding to $\chi \geq o$. Lacunarity introduces problems with respect to its noncontinuous measure. In some sense, it forces another characterization than a single measure theoretic one. One perhaps better suited to the finer structure required by the task of information transport.

5. STUDIES OF A DISCRETE, NONUNIFORMLY EXPANDING MAP

We study a $C^r r \geq 2$, piece–wise, convex endomorphism of the unit interval (*mod one*) characterized by a repelling fixed point, eg. $f' > 1$ when $f = 0$ which is over all expanding, $f' \geq o$ for all f. Renyi proved the existence of a unique invariant measure, μ, for such maps (1957). It generates a sequence of inter–event times, x_i, normalized to the unit interval and in co–dimension two.

$$f(x) = x_{t+1} = (1 + r)x + (\beta - r)x^2 \quad \{mod\ one\} \qquad \{3\}$$

Table 1 lists the derivatives, *as one dimensional eigenvalues*, of the $f = 0$ critical point for a representative set of r values in {1} which can be called a mod one, "snap–back" repeller.

Table 1. Relationship between parameter r and the derivative at $x = 0$ showing the progression to a more expanding system.

r	D at f = 0
0.02	0.01
0.05	0.03
0.1	0.06
0.3	0.21
0.5	0.5
0.7	1.17
0.99	49.5

Figure 1 represents the phase portraits of map {3} in which x is plotted against $x - 4$ and embedded in four dimensions (*displayed in two*) as a sequence of stacked one dimensional manifolds; the phase points connected sequentially by lines. Notice that as r increases, the "holes" in the phase portraits fill up (within the $n = 1000$ time series). Due to the symmetry of the map, $r \leq .1$ and $r \geq 1.9$ are about equivalent with respect to "getting stuck" near 0 and 1.0 respectively for more phase points stacking and holes compared with values in the neighborhood (but not equalling) 1.0. We see the slow but continual expansion of the lowest manifold near $f = 0$ as well as the "jumps" as the phase points leave and return.

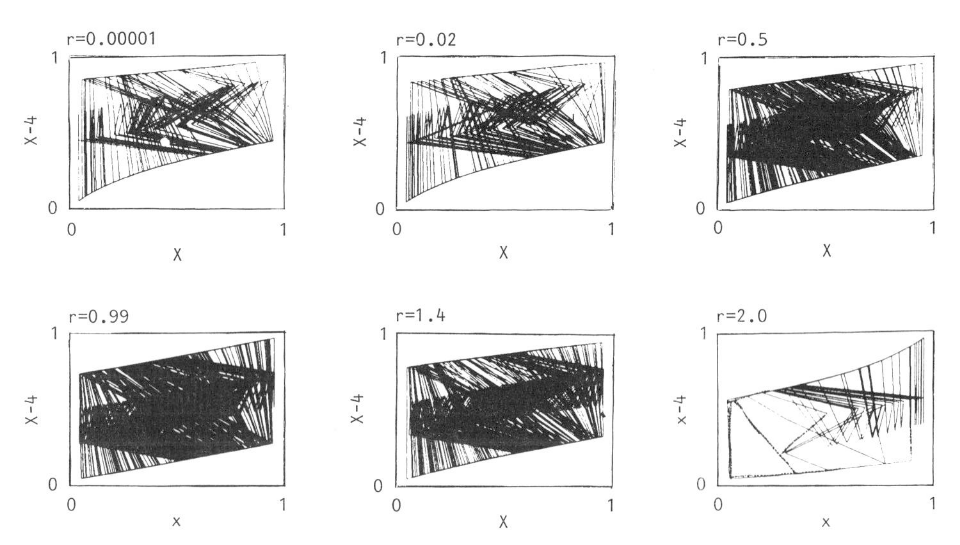

Figure 1. Phase portraits of map {3}, with x plotted versus $x - 4$ embedded in four dimensions but graphed in two demonstrating the varying densities creating lacunarity which varies with parameter r. See text.

A generating or *point partition* was formed such that no more than one point was contained in a partition. Thus studies with 300 points were partitioned into 299 "boxes." The boxes were sorted into sequences of sizes such that histograms demonstrating the distributions of box sizes demonstrated more "holes" for r–parameter values which were at the $r \leq .1$ or $r \geq 1.9$ extremes. Figure 2 portrays the graphs of the histograms of the distribution of point partition sizes at a sequence of values for r in which (for iterations of $n \leq 1000$), *lacunarity* decreased with the increase in the one dimensional eigenvalue of the map.

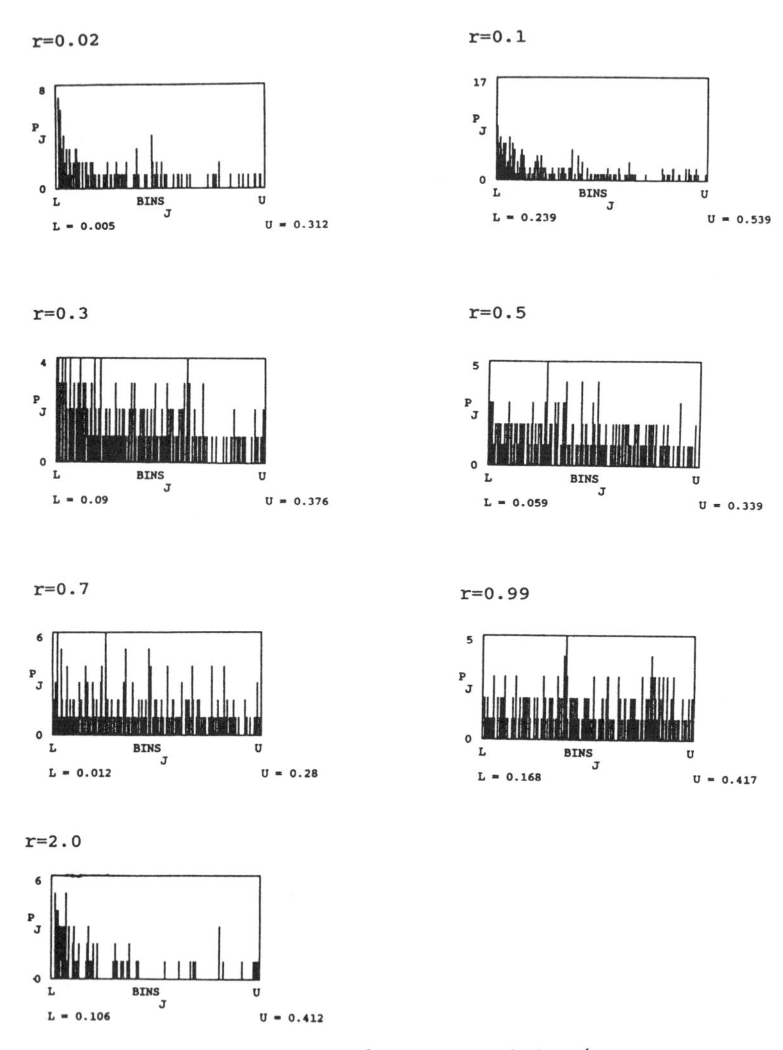

Figure 2. Histograms demonstrate holes (at $n = 1000$) which "fill up" with increases in nonlinear parameter r. See text.

The same *lacunarity* can be seen in the graphs of the Fourier transformations of the autocorrelation functions (eg. the "FFT") as in Figure 3. The forest of quasi–modes in the power spectrum is characteristic of bursting neurons, neuroendocrine release patterns, and even the intermittent representations of amino acids with certain physical properties in polypeptide chains. These "spectral gaps" in more simple power law spectra are seen as the "missing notes" in the diatonic scales that make the chords (see above).

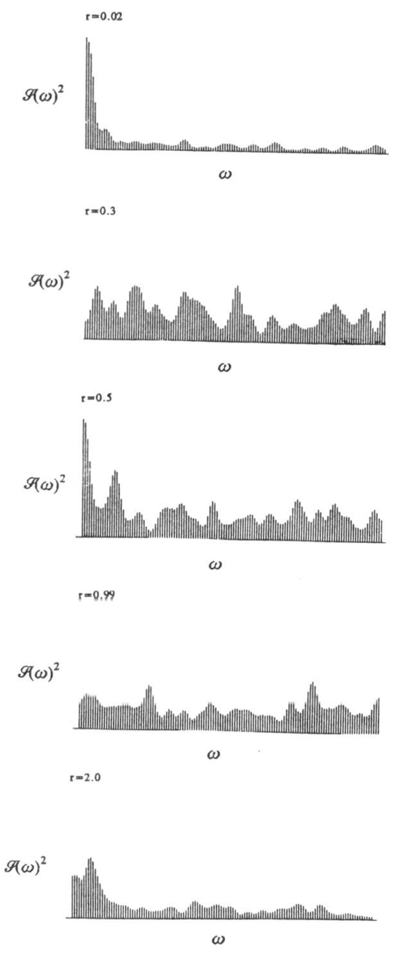

Figure 3. Power spectral transformations of the time series of Figure 2 demonstrating the characteristic "forest" of quasi–modes of intermittent dynamics. See text.

Table 2 summarizes the changes in $\{0 = 0–5;\ 1 = .5–1.0\}$ with respect to observed number of runs (a transition for 0 to 1 or the inverse) and those expected for a given number of 0's and 1's if the binary sequence was a random variable. We see that at low r (with the most gaps in the histogram and power spectra) the number of runs was below that expected. As holes disappear and the distribution becomes more continuous, the run number approximated that which might be expected in a Bernoulli process. We see also that the characteristic exponents and their variances (each point was an average of 10 consecutive computations with respect to the average derivative) were higher in the parameter regions of most holes. This is because the slow increases in expansion demonstrated these bursting regions outweighed the process of returns statistically. Lacunarity was calculated as the ratio of the biggest gap in box sizes (in the discrete metric)/ the number of gaps. We see the expected larger values in the parameter regions of most intermittency. As expected, the kurtosis was also highest in the regions of the most variable interevent times.

Table 2. A relationship between increasing nonlinear parameter r and the measures of the dynamics used in these studies. See text.

r	runs obs.	runs exp.	E(LCE)	Var(LCE)	lacunarity	kurtosis
0.02	152	316	0.0298	0.0455	0.247	11.764
0.05	202	356	0.0107	0.038	0.15	9.875
0.1	256	465	0.0137	0.014	0.146	6.925
0.3	374	471	0.0077	0.0256	0.05	5.096
0.5	422	479	0.008	0.0221	0.027	5.782
0.7	444	496	0.0094	0.0138	0.043	5.993
0.99	484	501	0.0115	0.0132	0.026	5.144
2	48	77	0.0222	0.0194	0.644	51.921

Table 3 shows that as in the case of the Erdös–Renyi growth rate of the longest run, kurtosis increases with n *with the suggestion that outside of this physiological, preasymptotic* regime of relatively short n, it may be that it converges. It is the relative rate of convergence of the longest run growth rate that we have conjectured that informational conjugacies may be found. Certainly lacunarity appears to also reflect this aspect of dynamical *mixing.*

Table 3. Kurtosis continues to grow with n, but at a decreasing rate.

r	n	kurtosis
0.02	300	10.52
0.02	500	14.115
0.02	1000	11.76
0.3	300	3.9
0.3	500	6
0.3	1000	5.096
0.5	300	2.985
0.5	500	4.079
0.5	1000	5.782
0.7	300	4.674
0.7	500	5.879
0.7	1000	5.993
0.99	300	4.759
0.99	500	6.314
0.99	1000	5.144

It may be that *unmixing* serves as the coding mechanism of intermittent systems with specific and/or global lacunarities as the basis of conjugacies.

REFERENCES

Baldi, V., Eckmann, J–P., and Ruelle, D., 1990, Resonances for intermittent systems, Nonlinearity (preprint).

Bowen, R., 1975, Equilibrium states and the ergodic theory of Anosov diffeomorphisms, Lecture Notes in Math. # 470.

Daniels, G. R., and Deininger, P. R., 1985, Repeat sequence families derived from mammalian tRNA genes, Nature 317:819–822.

Erdös, P., and Renyi, A., 1970, On a new law of large numbers.

Feller, W., 1968, "An Introduction to Probability Theory and Its Applications," Wiley, New York, pp 114–128.

Gillespie, J. H., 1986, Natural selection and the molecular clock, Mal. Biol. Eval. 3:138–155.

Jukes, T. H., and Holmquist, R., 1972, Estimation of evolutionary changes in certain homologous polypeptide chains, J. Mal. Biol. 64:163–179.

Katok, A., 1980, Lyapomov exponents, entropy, and periodic orbits for diffeomorphisms, Publ. Math. I.H.E.S. 51:137–173.

Kolmogorov, A. N., 1956, A general theory of dynamical systems and classical mechanics, in: "Foundations of Mechanics," R. Abraham and J. Marsden, eds., Benjamin, New York, 1967.

Koslow, S., Mandell, A. J., and Shlesinger, M., 1987, Perspectives in biological dynamics and theoretical medicine, Proc. N. Y. Acad. Sci. 504:1–313.

Mandell, A. J., and Kelso, J. A. S., 1990, Neurobiological coding in nonuniform times, in: "Essays on Classical and Quantum Dynamics," J. A. Ellison and H. Uberall, eds., Gordon & Beach, New York.

Mandell, A. J., and Selz, K. A., 1990, Heterochrony as a generalizable principle in biological dynamics, in: "Propagation of Correlations in Constrained Systems and Biology," E. Stanley and N. Ostrowsky, eds., Plenum, New York, in press.

Manneville, P., and Pomeau, Y., 1980, Different ways to turbulence in dissipative dynamical systems, Physica 10:219–227.

Manning, A., 1981, A relation between Lyapomov exponents, Hausdorff dimension and entropy, Ergodic Theory Dynamical Systems, 1:451–459

Maruyama, T., and Kimura, M., 1980, Genetic variability and effective population size when local extinction and recolonization of subpopulations are frequent, Proc. Nat. Acad. Sci. U.S.A. 77:6710–6714.

Ornstein, D. S., 1989, Ergodic theory, randomness and chaos, Science 243:182–187.

Osledec, V. I., 1968, A multiplicative ergodic theorem. Lyapomov characteristic numbers for dynamical systems, Trans. Moscow Math. Soc. 19:197–231.

Pesin, Ya. B., 1976, Invariant manifold families which correspond to nonvanishing characteristic exponents, Math. USSR–Isv. 10:1261–1305.

Pollicott, M., 1986, Meromorphic extensions of generalized zeta functions, Invent. Math. 85:147–164.

Renyi, A., 1957, Representation for real numbers and their ergodic properties, Acta Math. Hungar. 8:477–493.

Rinzel, J., 1987, A formal classification of bursting mechanisms in excitable systems, Lecture Notes Biomath. 71:267–281.

Rosenblatt, M., 1987, Some models exhibiting non–Gaussian intermittency, IEEE Trans. Inform. Theory IT–33:258–262.

Ruelle, D., 1989, "Chaotic Evolution and Strange Attractors," Cambridge Univ. Press, Cambridge.

Ruelle, D., 1990, Comments at Joel L. Lebowitz's 60th birthday, Rutgers, N.J.

Ruelle, D., 1978, "Thermodynamic Formalism: The Mathematical Structures of Classical Equilibrium Statistical Mechanics," Addison–Wesley, New York.

Ruelle, D., 1986, Zeta functions for expanding maps and Anosov flows, Invent. Math. 34:231–242.

Selz, K. A., and Mandell, A. J., 1990, Kindling exponents, quasi–isomorphisms and "reporting out" by neuron–like microwave popcorn discharge sequences, J. Bif. Chaos 1: in press.

Sinai, Ya. G., 1972, Gibbsian measures in ergodic theory, Russian Math. Surveys 27:21–69.

Singer, M. F., 1982, SINES and LINES: Highly repeated short and long interspersed sequences in mammalian genomes, Cell 28:433–434.

Smale, S., 1967, Differentiable dynamical systems, Bull. Amer. Math. Soc. 73:747–817.

Waterman, M. S., Gordon, L., and Arratia, R., 1987, Phase transitions in sequence matches and nucleic acid structure, Proc. Nat. Acad. Sci. U.S.A. 84:1241–1243.

Wright, S., 1945, The differential equation of the distribution of gene frequencies, Proc. Nat. Acad. Sci. U.S.A. 31:382–389.

Wright, S., 1931, Evolution in Mendelian populations, Genetics 16:97–159.

COLLECTIVELY SELF-SOLVING PROBLEMS

J.L. DENEUBOURG, S. GOSS, R. BECKERS AND G. SANDINI*

ULB, Brussels, Belgium

* University of Genova, Italy

INTRODUCTION

Observing insect societies, we are deeply impressed by their capacity to build structures and solve problems, and it is not really surprising that popular litterature is full of anthropomorphic explanations stressing the capacities of individual ants. In such blueprints, however, a central unit manages the whole system, and to achieve this it must collect all the data needed. The algorithms to treat these data are necessarily complex and therefore highly specific. They can tolerate neither internal errors, inexact or incomplete information, nor changes in the problem which is assumed to be stable. The consequences of this type of organisation are such that each solution must be constantly monitored and overhauled to cope with unforseen events, leading to a spiral of mutually increasing complexity and fragility.

The solution choosen by insect societies is in many respects diametrically opposed to the above caricature. Rather than one solitary central control unit, that is both complex and omniscient and which has its solutions programmed into it, the social insects constitute a team of simple, random units that are only locally "informed" and are not hierarchically organised. Despite this individual "ignorance", the society as a whole is able to exhibit a collective "intelligence", as illustrated by their building behaviour, their sorting, or the synchronisation of the castes' activities (See e.g. Seeley, 1985; Hölldobler and Wilson, 1990).

What are the rules governing such systems, and how can our technology benefit from understanding their mechanisms? We shall start by answering the first question, and then discuss some possible applications to illustrate the second point.

Self-Organization, Emerging Properties, and Learning
Edited by A. Babloyantz, Plenum Press, New York, 1991

THE BLUEPRINT OF AN INSECT SOCIETY

Insect societies have been frequently qualified as a superorganisms, but it is important to remark that these superorganisms have no brain and, moreover, that their members are spatially dispersed in their local environments. In order to exploit the information gathered by the different individuals, the insect societies have developed a decision making system that functions without symbolic representation by using the communication between the individuals and the physical constraints of the system.

Much of the communication or interaction between the members of insect societies has a positive feed-back character. For example if an ant recruits a nest-mate to a food source by means of a chemical trail, the recruit will in its turn reinforce the trail and recruit other nest-mates, and so on.

Such feed-back between the "units" allows not only the amplification of local information found by one or a few of them, but also the competition of different informations and the selection of one of them. To give an example that will be treated in more detail below, if one chemical trail leads to a nearby food source and another to a more distant one, the first will be amplified faster because of the physical constraints of the system. In this case the time needed to come back and recruit to a nearer source is less than that needed for a more distant one. The result of the two trails' competition for recruits at the nest will be that the nearer nearly always dominates.

Thus coordinated, the team's collective reaction to local signals is in a sense the solution to a problem which must be solved by the society, such as the selection of the closest food source. While no one individual is aware of all the alternatives possible, and no one individual contains an explicitly programmed solution, together they reach an "unconscious" decision.

We term this process of problem-solving though the interplay between environment and communication "functional self-organisation" or, following G. Beni (University of California, Santa Barbara) and A. Meystel (Drexel University) : "swarm intelligence".

The social insects offer a complete range of contrast between the individual and collective levels of intelligence and complexity. At one extreme there are species whose societies are composed of a huge number of individuals, characterised by the simplicity of their behavioural repertoire, their lack of individuality, their limited capacity for orientation and learning and their highly developed pheromone based mass communication (Beckers et al., 1990a). At the other extreme there are species whose societies are composed of small numbers of individuals, each with a high capacity for orientation and learning, and which lack mass communication, (Deneubourg et al., 1987; Theraulaz et al., 1991; Corbara et al., in prep.). Between these two extremes, there are many number of species that combine in various degrees both characteristics.

Because of this, insect societies are a particularly good material through which to approach the following problem: "Where should a group place its intelligence and complexity?". In other words to what extent should a group place its intelligence within each unit and to what extent in the interactions between units? How does this balance simultaneously influence efficiency, reliability, flexibility and fault tolerance?

Different examples of this decentralized and collective intelligence have been discussed, including building behaviour, collective choice, the formation of trail networks, sorting, collective exploration and foraging, dynamical division of labour, synchronisation and the generation of temporal and spatio-temporal oscillations (see e.g. Deneubourg, 1977; Belic et al., 1986; Pasteels et al. 1987; Wilson and Hölldobler 1988; Deneubourg et al., 1989; Goss and Deneubourg, 1989; Beckers et al., 1990b; Camazine, 1990; Camazine and Sneyd, 1990; Camazine et al., 1990; Deneubourg and Goss, 1990; Seeley et al., 1990; Skarka et al., 1990;).

Since many years, social insect specialists have emphasized the problem-solving capacity of their societies and have developed analogies with the brain. So it is not realy surprising to find analogies between insect societies and neural networks or with algorithms such as the elastic net. There are, however, some important differences. Most artificial intelligence algorithms refer to a central information processing unit, "out" the problem and calculating a priori

In insects societies, the problems apear to solve themselves, in real time, in the sense that the interactions between the environment and the actors give birth to the solution. A second difference is that our insects are mobile, and exhibit complex behaviour and an autonomy of decision which evidently doesn't exist in a nerve cell.

A SIMPLE EXAMPLE : THE EXPLOITATION OF THE CLOSEST FOOD-SOURCE

Most of the social insects use food recruitment, which can be described very generally as follows. When a scout discovers the food source i, it becomes an informed animal (X_i) and returns to the nest. There X_i invites waiting netsmates (Y) and guides them towards its discovery. These recruits became recruiters themselves after having been at the food source. The time-evolution of the population at food source i can be given approximately as:

$$dX_i/dt = a_i X_i Y - bX_i \qquad (1)$$

($i = 1,..,k$ where k is the number of food sources ; $Y = N - X_1 ...- X_k$ where N is the total number of ants).

The term $a_i X_i Y$ is the flux of recruitment towards the food source i. a_i is the rate of "reproduction" of the information and is the mean number of nest-mates that a recruiter recruits per unit of time and per nest-mate. The farther the food source is from the nest, the smaller a_i, which is more or less inversely proportional to the time needed to travel between the food source and the nest and to recruit.

$b X_i$ is the term describing the departure from the food source, b being the inverse the the average time spent at the food source. The set of i equations (1) (with $i > 1$) shows that if we have a competition between different informations, in this case the location of different sources, it is the food source charactized with the greater a_i, i.e. the closest food source, which is the source selected. The others are "forgotten". This solution is independant of both the number of food sources and the sequence of discovery.

In this case, we have no measurement of the distances and no modulation of the communication, i.e. no difference in the recruitment behaviour as a function of the distance. The "natural selection" of one food source is a direct by-product of the physical constraints, i.e. the distance between the nest and the food source, and the recruitment mechanism. The selection is a collective process and not the result of a decision made by one or more individuals. Another example in which an insect society selects the shortest path between n points is given in Goss et al. (1989) and Aron et al. (1991). A similar logic is also used by ants and bees to select the richer of two or more food sources (Beckers et al., 1990b; Camazine and Sneyd, 1990; Seeley et al., 1990).

TECHNICAL APPLICATIONS

An insect society is a strikingly efficient, robust and flexible machine, and yet is built with simple components. We have been investigating the possibility for some years now (Deneubourg et al., 1984) of "exporting" their blue-print to other fields such as robotics. This field is a little bit in a situation similar to those of the evolution when it invented insects societies with simple insects. A clever team of cheap and simple robots could well be an appropriate solution for a number of problems, and in many cases a valid alternative to making a complex and expensive robot (with on-board computer, artificial intelligence programs manipulating symbolic representations of the world, vision analysis, etc). This is especially true as cheap and simple robots are starting to appear on the market today.

Many of the problems that are self-solved in insect societies, such as building, synchronisation or sorting, are also classical engineering problems. Moreover, a number of technical systems, as for example a railway company, are like insect societies in that they consist of a large number of autonomous units interacting together and spatially distributed. However the analogy goes no further as most of these systems are governed by a central unit that is more or less perfectly informed. The recent development of automation and of microelectronics nevertheless encourages us to ask if it is not possible to rethink the organisation of such systems along the blueprint of swarm intelligence.

What should then be the rules governing the units' behaviour and communication, and how should they be tuned to generate different and efficient collective solutions? To start with, the units should only perceive and act on information from their immediate environment or from other nearby units, and their internal decisions should not be based on any calculation or representation. In this way our swarm intelligence systems are spared the problems of data accumulation and processing which can form bottle-necks, particularly when the actors are numerous as in a transportation system.

We are actually studying different particular cases which represent a large class of situations, and we shall now present two of them. The "satisfactory" rules are emprically selected from tests by computer simulation and/or by the development of prototypes.

THE A.N.T. PROJECT (AUTONOMOUS NAVIGATION AND TRANSPORTATION)

Collective Exploration

A number of technical activities can be defined as the synchronized movement of vehicles in a heterogeneous and impredictable environment. These vehicles collect information (cartography, scientific survey, border control, security patrols, ...) or objects (mining, cleaning, fishing, ...), or fight against something such as fire or enemy vehicules. Let us consider a simple case in which our swarm of robot vehicules must find their way in an unpredictable environment in which it is not easy to move.

The vehicules are distributed in space, searching an easy path. At the individual level, if a robot progresses easily, it doesn't change of direction, whereas if blocked, it turns left or right searching for a zone of easy progression. We now add communication between the vehicles, such that each vehicle emits a signal attracting that attracts others (see Sandini and Dario, 1989; Deneubourg et al., 1990, for the description of such a design). The signal strength is correlated to the vehicle's speed, the higher its speed, the higher the strength. The quality of the local environment thus modulates the emission.

What spatial distribution does such a group adopt? Firstly, and this is not surprising, the vehicules cluster in a number of chains in zones of easy progression. However if the speed of a chain decreases its atractivity declines and its members begin to disperse and explore the immediate environment. When one discovers a new path of easy progression, it increases its speed and attracts other vehicules, and a new chains is formed.

This case, for which a prototype is in preparation, is easy to simulate, and shows how a simple link between "perception" and communication can modulate the development of a collective and synchronized mouvement which can shift towards individual exploration when needed.

Transportation : Shift Between Different Modes

The interplay between collective behaviour (e.g. synchronized mouvement) and individual movement (e.g. exploration) as we have just discussed, is found in transport. Is it possible for a fleet of vehicles - without a central coordinating unit - to transport objects efficiently between a number of different points, avoiding collisions and adapting to a varyiable "supply and demand" of objects to be transported?

In such a system, a number of short and long range interactions between these units are required to organise the traffic. This differentiating it from the more classical system one of urban traffic, in which the interactions are few and limited. Urban traffic is typically a mixture of a large number of players with weak interactions, obeying the decisions of a central controller (e.g. traffic lights). For the last thirty years or so, most traffic-controllers dream of a large and perfectly informed computer directing urban traffic. Very few of the envisage increasing the communication between vehicules.

Take the case of the reduction of collisions and the optimal use of a transport network. A solution frequently proposed in transportation and robotics (see e.g. Fukuda and Kawauchi, 1989) - some prototypes having been tested - envisage vehicules tagging on after others going in the same direction so as to form "trains". This reduces both crowding and the overall probability of collision. Note that joining and leaving a train of vehicules is linked to the problem of a vehicule perceiving who is going where, and thus with who to link and when.

This problem of train formation (which is only one of the different forms of cooperation between vehicles that can be imagined) leads us to consider another classical problem, that of the mode choice. Transport can be achieved by different modes which are locally in competition (e.g. car versus bus). Each mode attracts users, and their choice modifies the pattern of the transportation network. With the introduction of automated systems and different interactions between vehicles, such trains formation, the competition and separation between modes loses its meaning. In future transportation systems, vehicles will be simultaneously a "module" of a large collective vehicle and an independant vehicle, with the supply and demand governing the user density and thus the pattern of the transportation systems.

As supply and demand for transport inevitably vary, other related problems appears such as how to generate an emergent time-table. Each point-source of objets to be transported needs to be serviced by an appropriate number of transporters, according to the varying number of objects (or people) it "produces". These demands typically fluctuate and the system must react appropriately, for example changing from an occasional to a regular service of a point. Such transport problems in fact imply scheduling, planning and synchronising the activities of the different units.

We present here a simple system of equations decribing a large fleet of AGV (automated guided vehicles), and shall see how with vehicles possessing only local perception, the transport system can shift from an individual to a collective mode to assure traffic fluidity.

These mobile robot's are capable of avoiding obstacles and each other and of performing a few "cooperative behaviours". For example, each vehicule is able to form a train with other vehicules, and to wait at a crossroad to join in a train. Without giving a too detailed description (e.g. we neglect in the model the explicit destination), we shall discuss the behavioural rules which allow the AGVs to change behaviour and their influence on the pattern of the transport system. The model derives from a set of models developed by D. Kahn of the D.O.T and co-workers to describe competition between different transportaion modes (Deneubourg et al., 1979; Kahn et al., 1981, 1983, 1985).

An AGV can be in two-states : X corresponds to an individualist behaviour (like a car), the second Y corresponds to a cooperative behaviour (e.g. a member of a train). The equations (2) give us the mean number of AGVs at any moment in the state X and Y. Each AGV emits a long range signal, giving its position and speed, and from these signals each AGV is able to estimate the vehicles' spatial distribution and speed distribution. As a function of these estimates and its behavioural rules, the vehicle changes from state X to Y or vice-versa. This is expressed by the terms G(X,Y) and F(X,Y).

$$dX/dt = - G(X,Y)X + F(X,Y)Y + INX - OUTX \qquad (2,1)$$

$$dY/dt = G(X,Y)X - F(X,Y)Y + INY - OUTY \qquad (2,2)$$

The terms INX, INY, OUTX and OUTY reflect the arrivals and the departure of new vehicles in the network, both of which vary in time. We suppose here that these terms are nul, then total number of vehicles remains constant. If we take for example that $G(X,Y) = (a + Y)^n$ and $F(X,Y) = V(K-X)^m$. The first term (G) expresses a " coperative-logic" : the higher the Y density, the more efficient the cooperative strategy (more ready to join a train, the more important the train with respect to an individual vehicule, ...). The term F reflects the individual logic: the higher the vehicule density, the lower the efficiency of this strategy and the lower the attraction of this solution.

With such behavioural rules, the model shows that without any central controller, the system is able to modify its configuration as a function of the density of vehicles, to assume an efficent fluidity of transport. At low density, more or less all the vehicules exhibit an individualist behaviour. If the density is high, the vehicles adopt a cooperative strategy. Between these extremes, different situations are observed such as multiple stationary states, i.e. for the same value of parameters, different configurations are possible and the network adopts one as a function of the system's history (a detailed description of such a system is in prep.).

DISCUSSION

Our goal is to explore a blue-print which could be both complementary and an alternative to "knowledge-based" blue-prints. We have described how insects societies solve problems and how some technical systems could be organized in a similar manner. What are the benefits for such solutions?

The solutions our swarms generate are necessarily less efficient than optimally tailored ones. However this short term sacrifice can be outweighed by other long term benefits, the first being reliability. The more decentralised and less omniscient the unit, the simpler it can be. The simpler the units, the easier it is to program them and the less likely they are to "break down". Working as a team of generalists implies that even if many the units fail, the rest continue. Note that it is not always possible for all the members of a swarm to be, and in these cases the swarm can include sub-populations of physically specialized units (e.g. soldiers/minors, lorries/cars, builders/carriers). This does not change the basic principles described earlier.

The second benefit is flexibility. As no solution is explicitly imposed, the decentralised units can react to environmental heterogeneities, either temporal or spatial. Indeeed as the units are intimately mixed with the problem and its environment, these heterogeneities contribute to the solution, allowing the team to cope with unforseen and unforseeable circumstances.

The third benefit is fault tolerance. Individual simplicity implies a high degree of randomness and error. The positive feed back interactions allow this and easily coordinate random individuals into an efficient team. Far from being undesirable, individual randomness offers an escape route to individuals caught in a maze, and can help the team to reach collective solutions that would otherwise be out of reach.

The trade off between these somewhat overlapping and interwoven benefits and efficiency is especially revealing in situations where positive feed-back can block a team in a sub-optimal solution. To escape from the sub-optimal solution to a better one it can be necessary to increase the individual complexity or randomness, at the risk of reducing the benefits of a collective organisation (Deneubourg et al., 1983; Deneubourg et al., 1982; Pasteels et al., 1987). In societies, randomness in decision and communication contributes to their "imagination".

We have very briefly discussed the role of "individual complexity", such as described by Seeley (1985) for bee scouts searching for a new nest-site, or Lumsden and Hölldobler (1983) for the estimation of enemy number in ritualized combats in ants. What are the benefits of increasing the complexity of units? This is part of the general question we have beeen discussing.

A very wide spectrum of applications are actually envisaged. The management of a potentially large number of spatially distributed units is an ideal application. Beni and co-workers, with the concept of cellular robotics, have treated the problem extensively and proposed a number of applications (Beni, 1988; Wang and Beni, 1988; Hackwood and Wang, 1988).

Apart from the problems discussed, other applications are actually under study, such as the management of a very large set of floating solar captors (Brenig et al., in prep.), sorting (Deneubourg et al., 1990), digging, building, and so on. Working in extreme conditions and in unpredictible environments seems a second group of applications. Some authors envisage spatial exploration (see Brooks et al., 1990 who have imagined a wide range of applications, or Steels, 1990).

Hackwood and Wang (1988) have proposed a system of fire-security and rapid evacuation (see also Wang and Beni, 1988). These last examples lead us to envisage rebuilding some machines in which currently the physical components are completely separated from a "central unit" controlling the behaviour. A new design, could be closer to a bee swarm. The bees locally modify their behaviour as a function of what they perceive, and so for example the swarm can change form and thereby regulate their temperature.

Most of these projects can appear as dreams, but they are testable dreams.

Acknowledgements

This work was supported in part by the Belgian Program for Inter-University Attraction Poles and by the Belgian Fonds National de la Recherche Scientifique.

REFERENCES

Aron, S., Deneubourg, J.L., Goss, S. and Pasteels, J.M. (1991). Functional self-organisation illustrated by inter-nest traffic in the argentine ant *Iridomyrmex humilis* In Biological Motion, eds. W. Alt and G. Hoffman. Lecture Notes in Biomathematics, Springer Verlag (in press).

Beckers, R., Deneubourg, J.L., Goss, S. and Pasteels, J.M. (1990a). Colony size, communication and ant foraging strategy. Psyche, 96: 239-256.

Beckers, R., Deneubourg, J.L., Goss, S. and Pasteels, J.M. (1990b). Collective decision making through food recruitment. Social Insects, 37: 258-267.

Belic, M.R., Skarka, V., Deneubourg, J.L. and Lax, M. (1986). Mathematical model of a honeycomb construction. Journal of Mathematical Biology, 24: 437-449.

Beni, G. (1988). The Concept of cellular robotic systems. In IEEE International Symposium on Intelligent Control, Arlington, VA.

Brooks, R.A., Maes, P., Mataric, M.J. and More, G. (1990). Lunar base construction robots. In Proc. Japan-USA symposium on Flexible Automation (in press).

Camazine, S. (1990). Pattern formation on the combs of honey bee colonies : self-organization through simple behavioural rules. Behavioral Ecology and Sociobiology (in press).

Camazine, S. and Sneyd, J. (1990). A mathematical model of colony-level nectar source selection by honey bees : self-organization through simple individual rules. Journal of Theoretical Biology (in press).

Camazine, S, Sneyd, J, Jenkins, M.J. and Murray, J.D. (1990). A mathematical model of self-organized pattern formation on the combs of honey bee colonies. Journal of Theoretical Biology (in press).

Deneubourg, J.L. (1977). Application de l'ordre par fluctuations à la description de certaines étapes de la construction du nid chez les termites. Social Insects, 24 : 117-130.

Deneubourg, J.L. and Goss, S. (1990). Collective patterns and decision-making. Ecology, Ethology & Evolution, 1: 295-311.

Deneubourg, J.L. de Palma, A. and Kahn, D. (1979). Dynamic models of competition between transportation modes. Environment and Planning, 11A: 665-673.

Deneubourg, J.L., Pasteels, J.M. and Verhaeghe, J.C. (1983). Probabilistic behaviour in ants : a strategy of errors. Journal of Theoretical Biology, 105: 259-271.

Deneubourg, J.L., Pasteels, J.M. and Verhaeghe, J.C. (1984). Quand l'erreur alimente l'imagination d'une société: le cas des fourmis. Nouvelles de la Science et des Technologies, 2: 47-52.

Deneubourg, J.L., Goss, S., Franks, N. and Pasteels, J.M. (1989). The blind leading the blind: modelling chemically mediated army ant raid patterns. Journal of Insect Behaviour, 2: 719-725.

Deneubourg, J.L., Parro, M., Pasteels, J.M. Verhaeghe, J.C. and Champagne, Ph. (1982). L'exploitation des ressources chez les fourmis: un jeu d'erreurs et d'amplification. In La communication chez les sociétés d'insectes, Eds. A. de Haro & X. Espadaler, 97-106. Universidad Autonoma de Barcelona, Barcelona.

Deneubourg, J.L., Fresneau, D., Goss, S, Lachaud, J.P. and Pasteels, J.M. (1987). Self-organisation mechanisms in ant societies (II): learning during foraging and division of labor. In From individual characteristics to collective organisation in social insects, Eds J.M. Pasteels & J.L. Deneubourg, 177-196. Experientia Supplementum 54, Birkhaüser, Bâle.

Deneubourg, J.L., Goss, S., Sandini, G., Ferrari, F. and Dario, P. (1990). Self-organizing collection and transport of objects in unpredictable environments. In Proc. Japan-USA symposium on Flexible Automation (in press).

Deneubourg, J.L., Goss, S., Franks, N., Sandova-Franks, A., Detrain, C. and Chretien, L. (1991). The Dynamics of collective sorting : Robots-like ants and ant-like robots. In Simulation of animal behaviour: From Animal to Animats, Eds J.A. Meyer and S. Wilson, MIT Press, Cambridge, Mass (in press).

Fukuda, T. and Kawauchi, Y. (1989). Cellular robotics, construction of complicated systems from simple functions. In Robots and Biological Systems, Eds P. Dario, G. Sandini & P. Aebisher. Proc. NATO ARW, Il Ciocco, Italy.

Goss, S. & J.L. Deneubourg, J.L. (1988). Autocatalysis as a source of synchronised rhythmical activity. Insectes Sociaux, 35: 310-315.

Goss, S., Aron, S., Deneubourg, J.L. and Pasteels, J.M. (1989). Self-organised short cuts in the argentine ant. Naturwissenchaften, 76: 579-581.

Hackwood S. and Wang, J. (1988). The engineering of cellular robotic systems. In IEEE International Symposium on Intelligent Control, Arlington, VA.

Hölldobler, B. and Wilson, E.O. (1990). The Ants, Springer-Verlag, Berlin.

Kahn, D., Deneubourg, J.L. and de Palma, A. (1981). Transportation mode choice. Environment and Planning, 13A: 1163-1174.

Kahn, D., Deneubourg, J.L. and de Palma, A. (1983). Transportation mode choice and city-suburban public transportation service. Transportation Research, 17B: 25-43.

Kahn, D., de Palma, A. and Deneubourg, J.L. (1985). Noisy demand and mode choice. Transportation Research, 19B: 143-153.

Lumsden, C.J. & B. Hölldobler, B. (1983). Ritualized combat and intercolony communication in ants. Journal of Theoretical Biology, 100: 81-98.

Pasteels, J.M., Deneubourg, J.L. and Goss, S. (1987). Self-organisation mechanisms in ant societies (I) : the example of food recruitment. In From individual characteristics to collective organisation in social insects, Eds J.M. Pasteels and J.L. Deneubourg, 155-175, Experientia Supplementum 54, Birkhaüser, Bâle.

Sandini, G. and Dario, P. (1989). Sensing strategies in cellular robotic systems. In Proc. IAS , Eds T. Kanade, F. Groen and L. Hertzberg, 2: 937-942.

Seeley, T.D. (1985). Honeybees Ecology, Princeton University Press, Princeton, NJ.

Seeley, T.D., Camazine, S. and Sneyd, J. (1990). Collective decision-making in honey bees : how colonies choose among nectar sources. Behavioural Ecology and Sociobiology (in press).

Skarka, V., Deneubourg, J.L. and Belic, M.R. (1990). Mathematical model of building behavior of Apis mekifera. Journal of Theoretical Biology (in press).

Steels, L. (1990). Cooperation between distributed agents through self-organisation. In Decentralized A.I., Eds Y. Demazeau and J.P. Müller 175-196. Elsevier (North Holland) Amsterdam.

Theraulaz, G., Goss, S., Gervet, J. and Deneubourg, J.L. (1991). Task differentiation in Polistes wasp colonies : a model for self-organizing groups of robots. In Simulation of animal behaviour : from animals to animats, Eds J.A. Meyer, S. Wilson. MIT Press, Cambridge, Mass. (in press).

Wang, J. and Beni, G. (1988). Patterns generation in cellular robotic Systems. In IEEE International Symposium on Intelligent Control, Arlington, VA.

Wilson, E.O. and Hölldobler, B. (1988). Dense heterarchies and mass communication as the basis of the organization in ant colonies TREE, 3 : 65-68.

IMITATION, LEARNING, AND COMMUNICATION:

CENTRAL OR POLARIZED PATTERNS IN COLLECTIVE ACTIONS

Ping Chen

I. Prigogine Center for Studies in
Statistical Mechanics and Complex Systems
and IC[2] Institute
University of Texas at Austin
Austin, Texas, USA

INTRODUCTION

Neoclassical models in microeconomics often describe an atomized society, in which every individual makes his or her own decision based on individual independent preference without the communication and interaction with the fellow members within the same community.[1] Therefore, static economic theory cannot explain collective behavior and changes of social trends, such as fashion, popular name brand, political middling or polarization.

Physicists have been interested in collective phenomena caused by the success of ferromagnetic theory in explaining collective phenomena under thermodynamic equilibrium. The Ising fish model[2] and the public opinion model[3] represented the early efforts in developing complex dynamic theory of collective behavior. However, the human society is an open system, there is no ground to apply the technique of equilibrium statistical mechanics, especially, the Maxwell-type transition probability, to social behavior. Therefore, these pioneer models have been discussed in physics community, but received little attention among social scientists.

However, the development of nonequilibrium thermodynamics and theory of dissipative structure open new way to deal with complex dynamics in chemical systems and biological systems,[4,5,6] that have many similarities with the behavior of social systems. In this short article, we will introduce new transition probability into master equation from the consideration of socio-psychological mechanism. The model may shed light on social behavior such as fashion, public choice and political campaign.

STOCHASTIC MODELS FOR IMITATION AND COLLECTIVE BEHAVIOR

Stochastic models are very useful in describing social

Self-Organization, Emerging Properties, and Learning
Edited by A. Babloyantz, Plenum Press, New York, 1991

processes.[2,3,4,5,7] To define basic concepts, we will start from the master equation and follow Haken's formulation of the Public Opinion Model.[5]

Let us consider the polarized situation in a society that only two kinds of opinions are subject to public choice. The two opinions are denoted by the symbols of plus and minus. The formation of an individual's opinion is influenced by the presence of the fellow community members with the same or the opposite opinion. We may use a transition probability to describe the changes of opinions.

We denote the transition probabilities by

$$p_{+-}(n_+,n_-) \text{ and } p_{-+}(n_+,n_-)$$

where n+ and n- are the numbers of individual holding the corresponding opinions + and -, respectively; p+- denotes the opinion changes from + to -, and p-+ denotes the opposite changes from - to +

We also denote the probability distribution function by f(n+, n-; t).

The master equation can be derived as following:

$$\begin{aligned} d\, f(n+,\ n;\ t)\ /dt = \\ (n_+ + 1)\ p_{+-}[n_+ + 1,\ n_- - 1]\ f[n_+ + 1,\ n_- - 1;\ t] + \\ (n_- + 1)\ p_{-+}[n_+ - 1,\ n_- + 1]\ f[n_+ - 1,\ n_- + 1:\ t] - \\ \{\ (n_+)\ p_{+-}[n_+,\ n_-] + (n_-)\ p_{-+}[n_+,\ n_-]\ \}\ f[n_+,\ n_-;\ t] \end{aligned} \tag{2.1}$$

We may simplify the equation by introducing new variables and parameters:

$$\begin{aligned} &\text{Total population :} \quad n = n_+ + n_-, \\ &\text{Order Parameter:} \quad q = (n+ - n-)\ /\ 2n \\ &w_{+-}(q) = n_+\ p_{+-}\ [n_+,\ n_-] = n(1/2 + q)\ P_{+-}(q) \\ &w_{-+}(q) = n_-\ p_{-+}\ [n_+,\ n_-] = n(1/2 - q)\ P_{-+}(q) \end{aligned}$$

where, n is the total number of the community members, q measures the difference ratio and can be regarded as an order parameter, $w_{+-}(q)$ and $w_{-+}(q)$ are the new function describing the opinion change rate which is a function of order parameter q.

This equation can describe collective behavior such as formation of public opinion, imitation, fashion, and mass movement. The problem remains unsolved here is how to know the transition probabilities $p_{+-}(n+,\ n-)$ and $p_{-+}(n+,\ n-)$. The choice of the form of transition probability is closely associated with the assumption of communication patterns in human behavior.

ISING MODEL OF COLLECTIVE BEHAVIOR

The early attempt to solve the problem of human collective behavior was directly borrowed the formulation of transition probability of Ising Model under equilibrium phase transition,[2,3,5] that was developed to explain the phase transition from paramagetic state to a ferromagnetic state on a magnetic lattice in equilibrium statistic mechanics.[7] Haken define the transition probability in analog of Ising model during equilibrium phase transition

$$
\begin{aligned}
p_{+-}[q] &= p_{+-}[n_+, n_-] = \\
&\quad v \exp\{-(Iq+H)/Q\} = v \exp\{-(kq+h)\} \\
p_{-+}[q] &= p_{+-}[n_+, n_-] = \\
&\quad v \exp\{+(Iq+H)/Q\} = v \exp\{+(kq+h)\}
\end{aligned} \tag{3.1}
$$

Where,

I is a measure of the strength of adaptation to neighbors or interaction constant. One could imagine that the interaction constant is low in an individual culture but high in an collective society.

H is a preference parameter or social field ($H > 0$ means that opinion + is preferred to -). We may label liberal trend as +, and conservative trend as -. The corresponding variable of H in physics is the magnetic field. We may explain the social field as the results of propaganda, advertising, or political campaign.

Q or $1/k$ is collective climate parameter or social temperature corresponding to kT in thermodynamics where k is the Boltzmann constant and T is the absolute temperature in physics. The social climate is warm (with high **Q** or low k) when social stress or tension is low and tolerance allow diversified public opinion.

v is the frequency of flipping process.

For mathematical simplicity, here only show the cases of $h=0$ when there are no centralized propaganda, and other means to form a dominating social opinion.

Master Equation (2.1) can be transformed into a partial differential equation. Under this simplification, the master equation has two kind of steady-state solutions [see Fig.1]. The analytic solution of steady state is:

$$
F_{st}(q) = c \exp\{2 \int [K1(y)/K2(y)]\, dy\} / K2(q) \tag{3.2}
$$

(1). The first kind of steady solution is the centered distribution when frequent changes of opinion occur. It happens under high social temperature and weak interpersonal interaction. For example, the case (a) (with $k = 0$) may imply an individualistic atomized society;

(2). The second kind of steady solution is the polarized distribution that describes polarization in public opinions by strong neighbor interaction under low social temperature and strong group interaction. Say, the case (c) (with $k > 2.5$ and $h = 0$) represents the polarization in dense population society with collective culture

(3) If we consider the influence of mass media or advertising, then we have a social magnetic field H. When social field $H > 0$ or $H < 0$, the peak distribution may shift to right or left, by strong mass media or social trend.

Although Ising model of collective behavior gives some vivid qualitative pictures of social collective phenomena, its mathematical formulation is only an intellectual game that has not been taken seriously by social scientists.

There are several problems of using the Ising model in describing social phenomena : The first objection is that society is not an closed system. Therefore, the equilibrium thermodynamics and statistical mechanics cannot be applied to social systems. Then, the transition probability of Ising model becomes groundless in human systems. The second

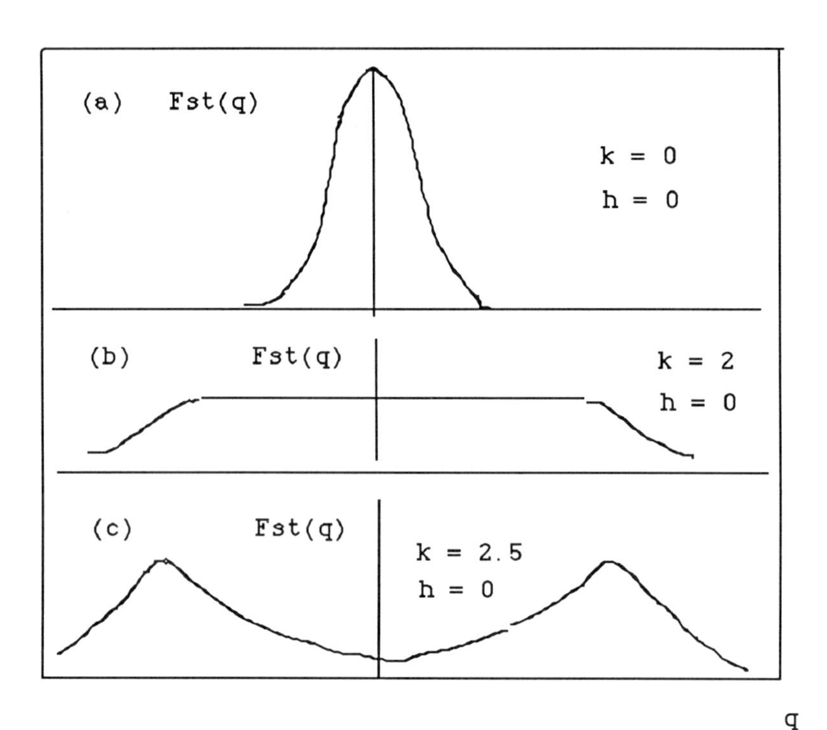

Fig.1. The steady state of probability distribution function in the Ising Model of Collective Behavior with h=0. (a) Centered distribution with k=0. middling behavior under high social temperature and weak interaction in societies. (b) Marginal distribution at the phase transition with k=2. Behavioral phase transition occurs between middling and polarization in collective behavior. (c) Polarized distribution with k=2.5. Strongly dependent behavior under low social temperature and strong group interactions in societies.

objection is that the social temperature Q and social field H have no measurable definition, therefore they are not observational indicators.

To overcome the difficulties of Ising model, we have to give up thermodynamic transition probability in equilibrium systems. We may find some plausible arguments to formulate the transition probability functions.

CHEN's SOCIO-PSYCOLOGICAL MODEL OF COLLECTIVE CHOICE

Consider a socio-psychological process with simple interaction relation.[8]

$$w[n_+, n_-] = a_1 n_+ + b_1 n_+ n_-$$
$$w[n_-, n_+] = a_2 n_- + b_2 n_+ n_- \tag{4.1}$$

This transition probability has simple explanation: the rates of changes in personal opinion depend both on the population size holding the same opinion (as shown in the first tern in the right hand side of equation) and the communication and interaction between the people holding opposite opinions. The outcome of collective choice is the result of the balance between individualistic orientation observed from independent choice and social pressure determined by the strength of communication and interaction. In other words, it depends on the relative magnitude of a and b.

Using new transition probability (4.1), we may solve the master equation (2.1) and have two kinds of steady state solutions [see Fig.2].[9,10]

(1) When $b < a$, we have central distribution without polarization. The middling implies that independent decision overcomes the influence of social pressure in social systems. This phenomena is quite familiar in the moderate social atmosphere of developed countries where the middle class plays a dominating role and middling attitude prevails in individualistic and pluralistic societies.

(2) When $b > a$, we have polarized distribution. It means that personal decision is much influenced by the social opinion. This situation is often observed in religious war and polarized revolution or mass movements. It may often caused by extremely poverty and social injustice in developing countries, when societies are polarized by racial, cultural, economic, and political polarization and confrontation in non-democratic social systems.

From above simple model and observation, we may concluded that individualistic culture is more likely to avoid polarized extremeness or protracted confrontation. The progress of individualism may increase the opportunity to achieve a compromising attitude in human societies.

POTENTIAL APPLICATIONS OF COLLECTIVE CHOICE

In this short paper, we only present the stochastic approach in addressing the collective behavior in human society. It is a modified version of Ising model by considering the interplay between individualistic orientation

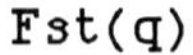

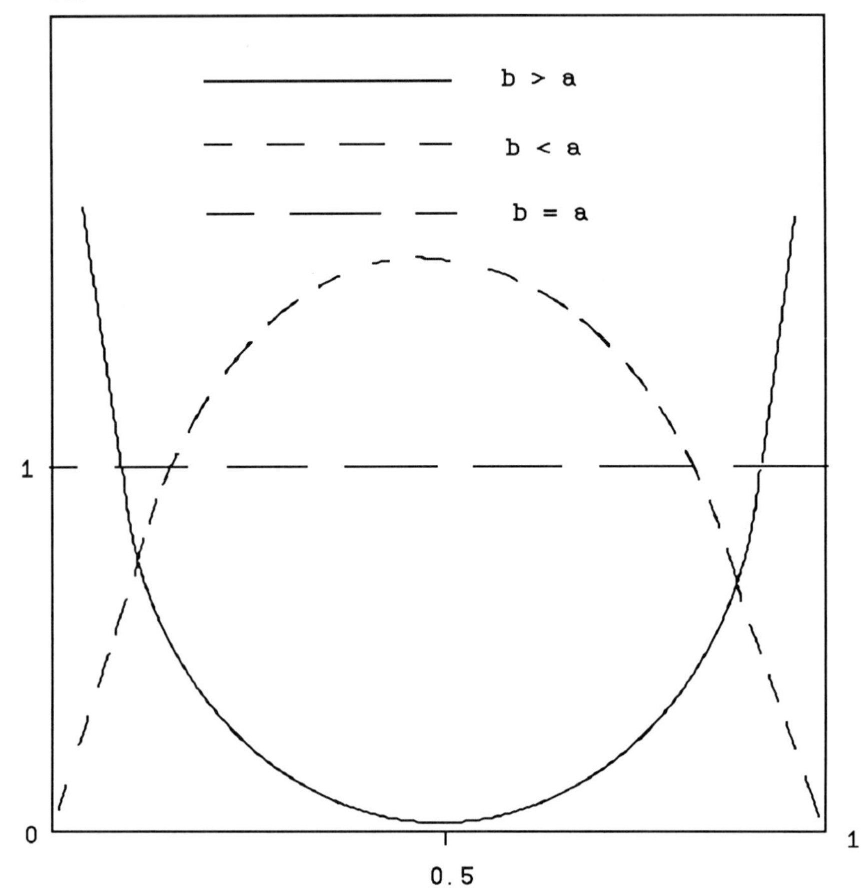

Fig. 2 The steady state of probability distribution function in socio-psychological model of collective choice. (a) Centered distribution with $b < a$ (denoted by short dashed curve). It happens when independent decision rooted in individualistic orientation overcomes social pressure through mutual communication. (b) Horizontal flat distribution with $b = a$ (denoted by long dashed line.) Marginal case when individualistic orientation balances the social pressure. (c) Polarized distribution with $b > a$ (denoted by solid line). It occurs when social pressure through mutual communication is stronger than independent judgement.

and social pressure through communication and interaction. We may also have the deterministic approach which modifies the conventional competition model in theoretical biology by introducing learning behavior and cultural pattern.[11] Another interesting phenomena in nonlinear feedback control in human behavior. The time-delay and over-reaction may cause chaotic behavior because of bounded human rationality. This situation may happen when individual behavior is governing by collective trend and reacted to the deviations from the trend.[12] To explore the role of learning and communication in collective action and control behavior in complex human systems is an exciting new field. We have tremendous job to do. A few interesting problems are listed here for further development:

Theoretical biology: modeling learning behavior in prey-predator and competition models.

Sociology: modeling information diffusion, fashion and immigration.

Political Science: modeling class struggle, arm race, voting, political campaign, social trend, and differentiation.

Economy: modeling staged economic growth, fashion switching, risk loving and risk aversion competition, marketing and advertising strategy.

CONCLUSION

Nonequilibrium physics opens new way to understand social evolution. Stochastic model with nonlinear transition probability is capable to describe the collective choice and social changes. Deterministic model of competition, learning, and nonlinear feedback control are also applicable in explaining social behavior.

REFERENCES

1. H. R. Varian, "Microeconomic Analysis," Norton, New York (1984).
2. E. Callen and D. Shapero, Ising Model of Social Imitation, _Physics Today_, July:23 (1974).
3. W. Weidlich, Ising Model of Public Opinion, _Collective Phenomena_, 1:51 (1972).
4. G, Nicolis and I. Prigogine, "Self-Organization in Nonequilibrium Systems," Wiley, New York, (1977).
5. H. Haken, "Synergetics, An Introduction," Springer-Verlag, Berlin, (1977).
6. G, Nicolis and I. Prigogine, "Exploring Complexity," Freeman, New York (1989)
7. L. Reichl, "A Modern Course in Statistical Physics," University of Texas Press, Austin (1980).
8. Ping Chen, "Nonlinear Dynamics and Business Cycles," Ph.D.Dissertation, University of Texas at Austin, (1987).
9. D. J. Bartholomew, "Stochastic Models for Social Processes," 3rd. Edition, Wiley, New York (1982).
10. M. Malek Mansour and A. de Palma, On the stochastic modelling of systems with non-local interactions, _Physica_ A, 128: 377 (1984).

11. Ping Chen, Origin of the Division of Labour and a Stochastic Mechanism of Differentiation, European Journal of Operational Research, 30:246 (1987).
12. Ping Chen, Empirical and Theoretical Evidence of Economic Chaos, System Dynamics Review, 4(1-2):81 (1988).

SPECTRAL ENTROPY AS A MEASURE OF SELF-ORGANIZATION IN TRANSITION FLOWS

John C. Crepeau, L. King Isaacson

Department of Mechanical Engineering
University of Utah
Salt Lake City, UT 84112 U.S.A.

ABSTRACT

We employ the spectral entropy of Powell and Percival to analyze and reduce numerical and experimental transition flow data. As dissipative structures form, the spectral entropy decreases from its equilibrium value. The convection cells in the Rayleigh-Bénard problem as modeled by the Lorenz equations possess a lower spectral entropy than both the equilibrium and turbulent states. Vortex structures produced along steep velocity gradients in a free shear layer in an internal cavity have a lower spectral entropy than the surrounding laminar flow. The spectral entropy algorithms can be applied to either time-series or power spectra data.

INTRODUCTION

Much has been written in these proceedings about systems which display patterns of self-organization. These include neural networks, pattern recognition and learning. A natural parameter for studying the self-organization of a system is the entropy. However, many forms of entropy have been identified: Information entropy, Kolmogorov entropy etc. How practical are these parameters in determining whether or not a system is spontaneously organizing? This depends on the type of data that is taken. Should the entropy decrease while the system becomes organized? According to Prigogine[1] and Ebeling and Klimontovich[2], the answer must be yes. By the maximum entropy condition, when a system is in thermal equilibrium, the entropy is at a maximum. Perturbing the system away from equilibrium forces it from its highest entropy state. By permitting the perturbations to die down, the system again reaches equilibrium and therefore its maximum entropy. Any system far from equilibrium is in a lower entropy state.

In this work, we propose to use the spectral entropy technique described by Powell and Percival[3] to quantify the degree of self-organization in transition flows. We will also show similarities between the spectral entropy and the statistical thermodynamic definition of the entropy.

The spectral entropy is computed from the power spectra of experimentally determined or numerically computed time-series data. Although the following focuses on results from transition fluid flow, we feel that the spectral entropy can be used to quantify self-organization in other applications where time-series or power spectra data are used.

Self-Organization, Emerging Properties, and Learning
Edited by A. Babloyantz, Plenum Press, New York, 1991

The spectral entropy is defined as,

$$S = -\sum_{k} P_k \ln P_k \tag{1}$$

where,

$$P_k = \frac{|f_k|^2}{\sum_{k'} |f_{k'}|^2}. \tag{2}$$

Here, $|f_k|^2$ denotes the value of the power spectra at wavenumber *k*. Essentially, the term P_k gives a measure of the distribution of the energy over all of the wavenumbers. If the energy is concentrated in a few wavenumbers, the spectral entropy will be relatively low. When the power spectra is broadband, the spectral entropy becomes larger.

In transition flows, where only a few dominant frequencies exist, the spectral entropy is lower than in fully developed turbulence. This is what we might expect. Ordered vortex structures occur in transition flows and the entropy of those structures should be less than the entropy of the surrounding flow[2]. When the structures break down and the flow becomes fully developed, we expect the entropy to increase. In Section I, we will show that the spectral entropy of a particular dissipative structure(convection cells) is less than the spectral entropy of both the equilibrium and turbulent states, and in Section II that the spectral entropy of vortex bursts is less than the surrounding laminar flow.

A natural question arises when we compare the spectral entropy behavior with the appearance of ordered structures: How does the spectral entropy relate to the statistical thermodynamic definition of entropy? We would like to point out some similarities.

In thermodynamics, the entropy is defined as,

$$S = -k \sum_{i} P_i \ln P_i \tag{3}$$

where,

k - Boltzmann's constant

and

$$P_i = \frac{e^{-E_i/kT}}{\sum_{i'} e^{-E_{i'}/kT}}. \tag{4}$$

In Equation 4, P_i denotes the probability that the system is in the ith energy state E_i.

In a sense, P_i represents a distribution of energy over all possible states of the system. The probabilities as defined by the spectral entropy(Eqn. 2) give a distribution of the energy over all of the relevant wavenumbers. If the power spectra is dependent on the initial conditions, as it will be in Section I, then many power spectras can be averaged together so as to obtain a probable energy at which any wavenumber resides. By averaging the power spectras to find a most probable energy state, the spectral entropy becomes qualitatively similar to the Boltzmann entropy.

One can easily begin to see striking similarities between the two. The remarkable property of the spectral entropy is that it can provide an easily computable method for describing and quantifying dissipative structures in time-series data. In addition, the characteristics of the spectral entropy correspond very well to intuitive insights concerning the entropy. Ordered structures have a low spectral entropy, while turbulent signals possess higher spectral entropies.

In the following section, we show a lowering of the spectral entropy across the initial bifurcation of the Lorenz equations. This critical bifurcation represents the initiation of convection cells in the Rayleigh-Bénard instability. Increasing the control parameter(the Rayleigh number) past the transition point forces the system into a turbulent state. The spectral entropy at this point increases, representing the dissipation of convection cells into fully developed turbulence.

Klimontovich[4] in his S-Theorem explains how entropy decreases in a fluid flow as the control parameter(in this case, the Reynolds number) increases. He argues that while a system is in equilibrium, it is in the maximum entropy configuration, and any deviations from equilibrium, especially large ones, serve to decrease the overall entropy of the system. Figure 2 shows self-organization when the Rayleigh number is increased. There are interesting comparisons between the S-Theorem of Klimontovich and the behavior of the spectral entropy, namely the reduction of the entropy in non-equilibrium states.

In Section II, one observes the behavior of the spectral entropy in experimental data of the transition regime for a free shear layer in an internal cavity. We present evidence of ordered vortex structures, with low spectral entropies, existing within laminar regions of high spectral entropy.

I. SELF-ORGANIZATION IN THE LORENZ EQUATIONS

As an example of the application of spectral entropy in a simple system, we would like to demonstrate the spectral entropy behavior during self-organizing processes in the Lorenz[5] model. We have shown previous results[6] using this model, but would like to further refine some ideas.

The Lorenz equations are a truncated model of Rayleigh-Bénard convection in the atmosphere. One wishes to find some entropy-like parameter which decreases when self-organization occurs(creation of dissipative structures), and increases as those structures collapse. The Lorenz equations provide an excellent model which exhibits both self-organization and destruction of structure. The equations are given as,

$$\frac{dx}{dt} = \sigma(y - x)$$

$$\frac{dy}{dt} = - xy + rx - y \tag{5}$$

$$\frac{dz}{dt} = xy - bz$$

where,

$$\sigma = 10$$
$$b = 8/3.$$

Here, r represents the Rayleigh number, and also serves as the bifurcation parameter for the set of Equations 5. The variable σ is the Prandtl number and b can be interpreted as a non-dimensional space parameter.

By increasing r from 0 to 30, the system undergoes a series of bifurcations. At $r = 1$, the single steady-state solution splits into two steady-state branches. Physically, this corresponds to the formation of convection cells(Figures 1a-b). Clearly, the formation of these dissipative structures is a self-organizing process. In fact, we can now extract mechanical work out of the system. Imagine placing a small vane in the middle of one of the convection cells. As the convecting fluid moves in an organized fashion, it will cause the vane to rotate, creating shaft work. In this case, thermal energy is converted into mechanical energy. Feistel and Ebeling[7] assert that dissipative structures are an energy transforming device which changes a lower valued energy into a higher valued energy. The high valued energy contained in a dissipative structure implies a lower entropy within the structure.

As the control parameter increases past $r \approx 24$, the two steady states bifurcate into a strange attractor. This models the dissipation of the convection cells into a fully developed turbulent state (Figures 1b-c).

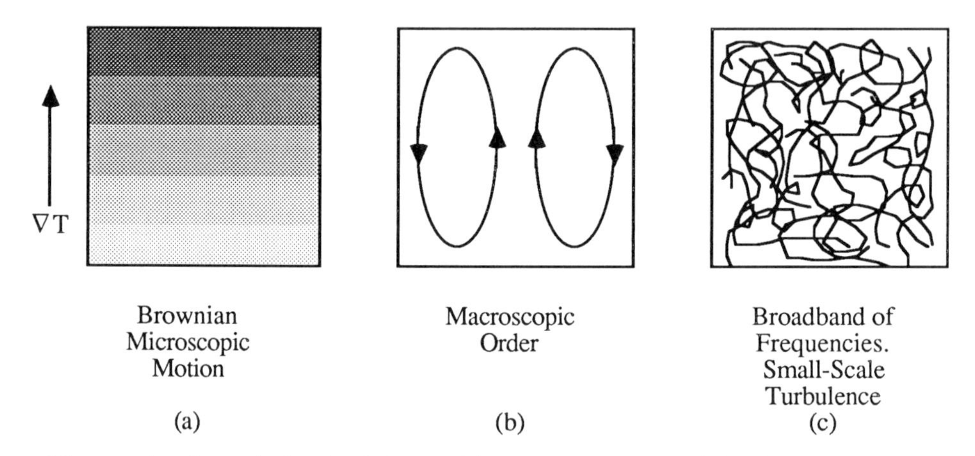

Figure 1 A rough sketch of the transition sequence in Rayleigh-Bénard convection (a fluid heated from below). Figure (a) shows a fluid with an inverted temperature gradient and no macroscopic motion(striations exist only in the artist's reproduction, and not in the actual fluid). After the initial bifurcation, dissipative structures form (Figure (b)). Upon heating the fluid beyond the transition point, the dissipative structures collapse (Figure (c)).

What happens now to the vane in our thought experiment? Since it is now immersed in a turbulent fluid, the blades bounce back and forth randomly, producing no net shaft work. The energy of the system loses its high value, so that the entropy should increase. This is exactly the sequence which occurs with the spectral entropy.

Figure 2 shows how the spectral entropy varies with r in comparison to the bifurcation diagram for the Lorenz equations. One can see how the spectral entropy decreases across the initial bifurcation, and increases when the system bifurcates into a strange attractor. The response of the spectral entropy closely mimics the entropy behavior as predicted by Feistel and Ebeling.

The spectral entropy in Figure 2 has been computed in the following manner. For any given value of the control parameter r, a fourth-order Runge-Kutta scheme integrates the set of Equations 5 through 2^{12} steps, using random initial conditions. After integrating, we compute the power spectra. This process is completed 40 times so that an average power spectra can be computed for every wavenumber in the spectra and at each value of the control parameter(which represents the Rayleigh number) r. By computing the power spectra in this manner, we can better simulate the energy distribution given in Equation 4 so that the spectral entropy behaves similarly to the entropy. The spectral entropy for each r is then computed using Equations 1 and 2.

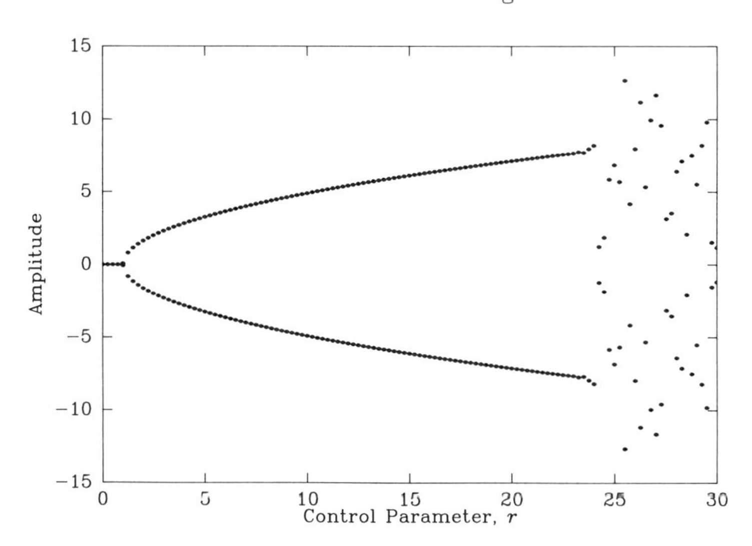

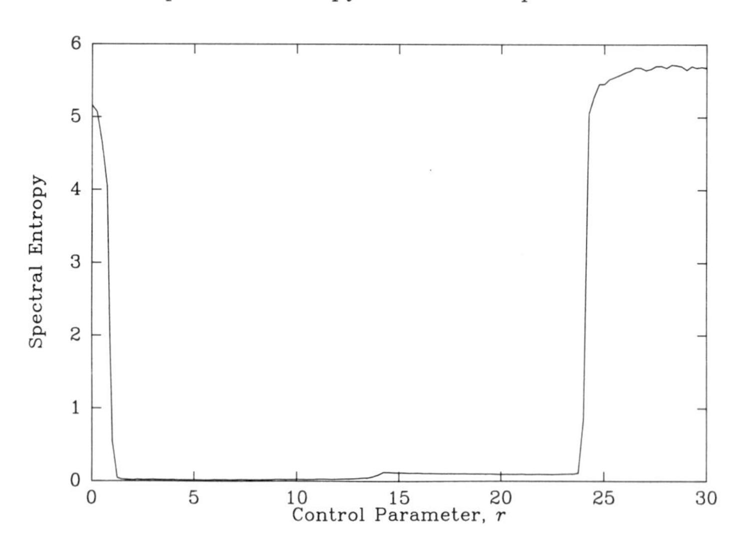

Figure 2 Top: Lorenz Bifurcation Diagram
Bottom: Spectral Entropy versus Control Parameter, r, representing the Rayleigh number

II. SPECTRAL ENTROPY CALCULATIONS OF DISSIPATIVE STRUCTURES IN TRANSITION FLOW

Now, suppose we have experimental time-series data which exhibit self-organizing behavior. Can we use the spectral entropy to characterize the degree of order? Let us observe the behavior of an experimental time-series. These data were taken by a single hot-film probe, located two millimeters downstream from the the forward restrictor(Fig. 3). This experimental set-up has been described in detail previously[8]. Figure 4 gives the time evolution of the normalized velocity fluctuations for curved air flow in a steep velocity gradient. The data clearly show four vortex structures, which exist in a laminar flow regime. We call these structures "ordered," or "dissipative."

In order to quantify the degree of order within these structures, we can invoke our entropy algorithm and compute the spectral entropy in both the burst and the laminar regions. The bottom portion in Figure 4 reveals the result of this analysis. The spectral entropies from the four burst regions are significantly lower than the two laminar regions. A lower spectral entropy correlates with the existence of a dissipative structure.

The relative degree of order between the burst and laminar regimes is not dependent on the width of the sample. For example, the width of the third burst is about 40 units. If we take 40 unit intervals in either of the two laminar regions and compute the spectral entropy, we still find that laminar regions exhibit less order.

The power spectra from each of the four bursts revealed the dominance of the same wavenumber. This implies a large amount of structural similarities in each of the bursts. The power spectra of the laminar regions possess a spread of wavenumbers, lacking a single dominant peak.

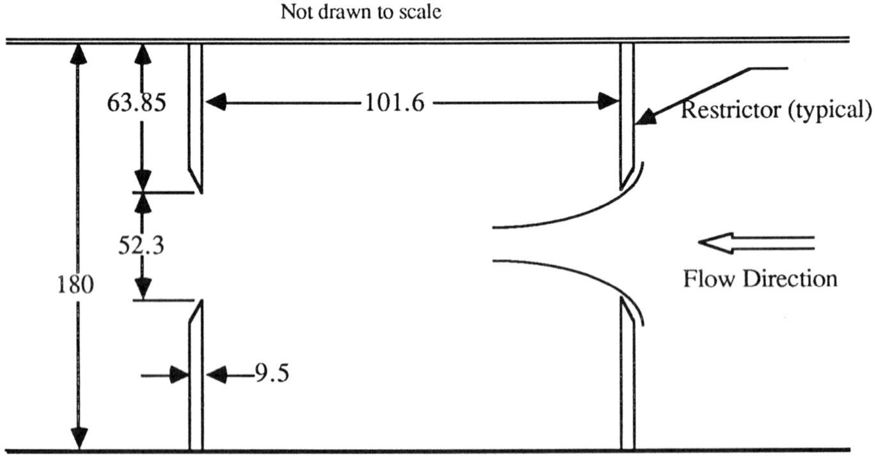

Figure 3 Experimental Flow Cavity.
The curved lines at the entrance of the restrictors denote the approximate shape of the critical layers within the flow field. A detailed description of this flow cavity along with other results are presented in Isaacson, Denison and Crepeau[8].

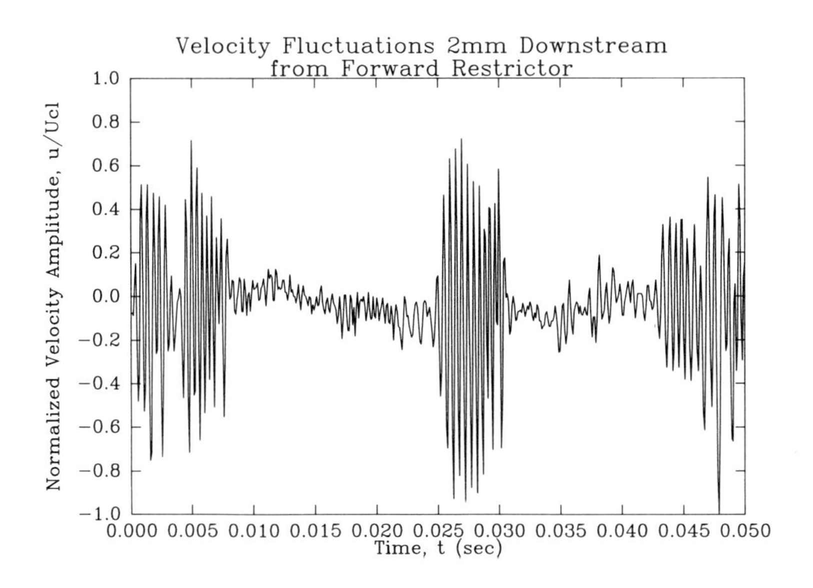

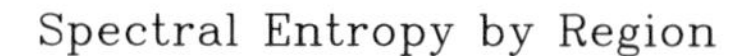

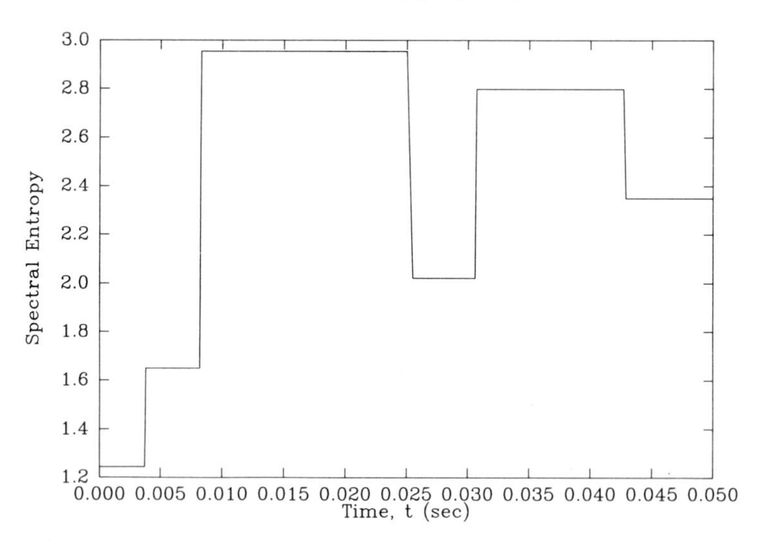

Figure 4 Hot-film anemometer data taken 2mm downstream of the forward restrictor shown in Figure 4. Notice the four structures that exist in the flow. The plot below gives the spectral entropy of each of the six regions in the time-series data.

CONCLUSIONS

Based on our observations, we conclude that the spectral entropy follows our expected behavior of the Boltzmann entropy. In regions of visual order, the spectral entropy is low, and in regions of disorder, the spectral entropy is higher. If one employs the spectral entropy as a measure of order, then dissipative structures should possess only a few dominant frequencies. The spectral entropy qualitatively emulates the entropy behavior as postulated by Ebeling and Klimontovich.

Although the applications of the spectral entropy in this paper are directed towards transition fluid flows, we feel it can be easily applied to either numerical or experimental time-series data, or power spectra results in other fields. Noting the characteristics of the spectral entropy, we may infer that dissipative structures (with low spectral entropy) contain only a few dominant frequencies. Systems with broadbanded power spectras have a higher spectral entropy.

ACKNOWLEDGMENTS

We would like to thank the Utah Supercomputing Institute for donating their time and resources in computing the Lorenz spectral entropies. One of us (J.C.) gratefully acknowledges Professor William K. Van Moorhem and the Thiokol Incorporated, University IR&D program for their financial support.

REFERENCES

[1] Prigogine, I., From Being to Becoming, 1st edition, W.H.Freeman, 1980.

[2] Ebeling, W., Klimontovich, Y.I., Selforganization and Turbulence in Liquids, 1st edition, Teubner, 1984.

[3] Powell, G.E., Percival, I.C., A spectral entropy method for distinguishing regular and irregular motion of Hamiltonian systems, J. Phys. A: Math. Gen., 12 (1979), 2053.

[4] Klimontovich, Y.L., S-Theorem, Z. Phys. B, 66 (1987), 125.

[5] Lorenz, E., Deterministic Nonperiodic Flow, J.Atm.Sci., 20 (1963), 130.

[6] Crepeau, J.C., Isaacson, L.K., On the Spectral Entropy Behavior of Self-Organizing Processes, J. Non-Equilib. Thermo., 15 (1990).

[7] Feistel, R., Ebeling, W., Exp. Technik der Physik, (1984)

[8] Isaacson, L.K., Denison, M.K., Crepeau, J.C., Unstable Vortices in the Near Region of an Internal Cavity Flow, AIAA J., 27 (1989), 1667.

Contributors

Babloyantz, A.	Université Libre de Bruxelles, Chimie Physique CP 231 Bvd du Triomphe, 1050 Bruxelles, BELGIUM
Beckers, R.	Université Libre de Bruxelles, Chimie Physique CP 231 Bvd du Triomphe, 1050 Bruxelles, BELGIUM
Berthommier, F.	TIM3-IMAG, Université J. Fourier, Faculté de Médecine F-38700 La Tronche, FRANCE
Chen, P.	Center for statistical mechanics, University of Texas at Austin, Austin, Texas 78712, U.S.A.
Cinquin, Ph.	TIM3-IMAG, Université J. Fourier, Faculté de Médecine F-38700 La Tronche, FRANCE
Coll, T.	Departement of Mathematics & Informatics University "de les Illes Balears", Palma de Majorca, SPAIN
Crepeau, J.C.	Departement of Mechanical Engineering, University of Utah Salt Lake City, UT 84112, U.S.A.
DeGuzman, G.C.	Center for Complex Systems, Florida Atlantic University Boca Raton, FL-33431, U.S.A.
Demongeot, J.	TIM3-IMAG, Université J. Fourier, Faculté de Médecine F-38700 La Tronche, FRANCE
Deneubourg, J.L.	Université Libre de Bruxelles, Chimie Physique CP 231 Bvd du Triomphe, 1050 Bruxelles, BELGIUM
Destexhe, A.	Université Libre de Bruxelles, Chimie Physique CP 231 Bvd du Triomphe, 1050 Bruxelles, BELGIUM
Eckhorn, R.	Institute für Angewandte , Physik und Biophysik, Universität Marburg, D-3550 Marburg, F.R.G.
Francillard, D.	TIM3-IMAG, Université J. Fourier, Faculté de Médecine F-38700 La Tronche, FRANCE
Francois, O.	TIM3-IMAG, Université J. Fourier, Faculté de Médecine F-38700 La Tronche, FRANCE
Gaudiano, P.	Center for Adaptive Systems, Dept. of Mathematics Boston University, Boston MA-02215, U.S.A.
Goss, S.	Université Libre de Bruxelles, Chimie Physique CP 231 Bvd du Triomphe, 1050 Bruxelles, BELGIUM
Grossberg, S.	Center for Adaptive Systems, Dept. of Mathematics Boston University, Boston MA-02215, U.S.A.
Haken, H.	Inst. Theor. Phys., Universitat Stuttgart D-7000 Stuttgart 80, F.R.G.

Holroyd, T.	Center for Complex Systems, Florida Atlantic University Boca Raton, FL-33431, U.S.A.
Isaacson, L.K.	Departement of Mechanical Engineering, University of Utah Salt Lake City, UT 84112, U.S.A.
Kapral, R.	Department of Chemistry, University of Toronto Toronto, Ontario,M5S 1A1, CANADA
Kelso, J.A.S.	Center for Complex Systems, Florida Atlantic University Boca Raton, FL-33431, U.S.A.
Kurrer, C.	Dept. of Physics, University of Illinois 405 N. Mathews Ave., Urbana, Il 61801, U.S.A.
Mandell, A.J.	Center for Complex Systems, Florida Atlantic University Boca Raton, FL-33431-0991, U.S.A.
Marque, I.	TIM3-IMAG, Université J. Fourier, Faculté de Médecine F-38700 La Tronche, FRANCE
Miles, R.	Laboratoire de Neurobiologie Cellulaire,Institut Pasteur 28 rue du Dr. Roux, 75724 Paris Cedex 15, FRANCE
Nicolis, J.S.	Dept. of Electrical Engineering, School of Engineering University of Patras, 26110 Patras , GREECE
Nieswand, B.	Dept. of Physics, University of Illinois 405 N. Mathews Ave., Urbana, Il 61801, U.S.A.
Reeke, G.	The Rockfeller University, 1230 York Ave. New York 10021-66399, U.S.A.
Sandini, G.	University of Genova ITALY
Schanze, T.	Institute für Angewandte, Physik und Biophysik, Universität Marburg, D-3550 Marburg, F.R.G.
Schulten, K.	Dept. of Physics, University of Illinois 405 N. Mathews Ave., Urbana, Il 61801, U.S.A.
Schuster, P.	Institut für Theor. Chemie - StrahlenchemieUniversität Wien,Wahringer Strasse 17, A 1090 Wien, AUSTRIA
Selz, K.A.	Center for Complex Systems, Florida Atlantic University Boca Raton, FL-33431-0991, U.S.A.
Sepulchre, J.A.	Université Libre de Bruxelles, Chimie Physique CP 231 Bvd du Triomphe, 1050 Bruxelles, BELGIUM
Sporns, O.	The Rockfeller University, 1230 York Ave. New York 10021-66399, U.S.A.
Tang, D.S.	Microelectronics and Computer Technology Corporation 3500 West Balcones Center Drive, Austin, U.S.A.
Torras, C.	Institute of Cybernetics, Planta Diagonal 647.2 Barcelona 08028, SPAIN
Traub, R.D.	IBM Watson Research Center Yorktown Heights, New York 10528, U.S.A.
Van den Broeck, C.	Univerity of California at San Diego, Dept. Chem. B040, San Diego, CA 92093, U.S.A.
Wang, X.J.	National Institute of Health Bethesda, Maryland 20892, U.S.A.
Wunderlin, A.	Inst. Theor. Phys., Universitat Stuttgart D-7000 Stuttgart 80, F.R.G.

INDEX